de Gruyter Expositions in Mathematics 18

de Gruyter Expositions in Mathematics

1 The Analytical and Topological Theory of Semigroups, *K. H. Hofmann, J. D. Lawson, J. S. Pym* (Eds.)

2 Combinatorial Homotopy and 4-Dimensional Complexes, *H. J. Baues*

3 The Stefan Problem, *A. M. Meirmanov*

4 Finite Soluble Groups, *K. Doerk, T. O. Hawkes*

5 The Riemann Zeta-Function, *A. A. Karatsuba, S. M. Voronin*

6 Contact Geometry and Linear Differential Equations, *V. R. Nazaikinskii, V. E. Shatalov, B. Yu. Sternin*

7 Infinite Dimensional Lie Superalgebras, *Yu. A. Bahturin, A. A. Mikhalev, V. M. Petrogradsky, M. V. Zaicev*

8 Nilpotent Groups and their Automorphisms, *E. I. Khukhro*

9 Invariant Distances and Metrics in Complex Analysis, *M. Jarnicki, P. Pflug*

10 The Link Invariants of the Chern-Simons Field Theory, *E. Guadagnini*

11 Global Affine Differential Geometry of Hypersurfaces, *A.-M. Li, U. Simon, G. Zhao*

12 Moduli Spaces of Abelian Surfaces: Compactification, Degenerations, and Theta Functions, *K. Hulek, C. Kahn, S. H. Weintraub*

13 Elliptic Problems in Domains with Piecewise Smooth Boundaries, *S. A. Nazarov, B. A. Plamenevsky*

14 Subgroup Lattices of Groups, *R. Schmidt*

15 Orthogonal Decompositions and Integral Lattices, *A. I. Kostrikin, P. H. Tiep*

16 The Adjunction Theory of Complex Projective Varieties, *M. Beltrametti, A. Sommese*

17 The Restricted 3-Body Problem: Plane Periodic Orbits, *A. D. Bruno*

Unitary Representation Theory of Exponential Lie Groups

by

Horst Leptin
Jean Ludwig

Walter de Gruyter · Berlin · New York 1994

Authors

Horst Leptin
Fakultät für Mathematik
Universität Bielefeld
Universitätsstraße 25
D-33615 Bielefeld
Germany

Jean Ludwig
Département de Mathématiques et d'Informatique
U. F. R. de Mathématiques, Informatique et Mécanique
Ile du Saulay
F-57045 Metz Cedex 01
France

1991 Mathematics Subject Classification: 22-02; 22Exx

Keywords: Lie groups, solvable groups, exponential groups, unitary representations, orbit method, harmonic analysis

♾ Printed on acid-free paper which falls within the guidelines of the ANSI to ensure permanence and durability.

Library of Congress Cataloging-in-Publication Data

Leptin, Horst, 1927–
Unitary representation theory of exponential Lie groups / Horst Leptin, Jean Ludwig.
p. cm. – (De Gruyter expositions in mathematics ; 18)
Includes bibliographical references.
ISBN 3-11-013938-3
1. Lie groups. 2. Representations of groups. I. Ludwig, Jean, 1947– . II. Title. III. Series.
QA387.L46 1994
512'.55–dc20 94-27983
CIP

Die Deutsche Bibliothek – Cataloging-in-Publication Data

Leptin, Horst:
Unitary representation theory of exponential Lie groups / by Horst Leptin ; Jean Ludwig. – Berlin ; New York : de Gruyter, 1994
(De Gruyter expositions in mathematics ; 18)
ISBN 3-11-013938-3
NE: Ludwig, Jean:; GT

Printed in Germany.
Disk Conversion: Lewis & Leins, Berlin. Printing: Gerike GmbH, Berlin.
Binding: Lüderitz & Bauer GmbH, Berlin. Cover design: Thomas Bonnie, Hamburg.

Table of Contents

Preface

The reason for and the purpose of this book is to present a full proof of Theorem 1 in chapter III: The Kirillov map K from the space $\mathfrak{g}^*/G$ of coadjoint G-orbits onto the unitary dual $\widehat{G}$ of the connected, simply connected exponential Lie group G with Lie algebra $\mathfrak{g}$ is a homeomorphism.

This theorem solves one of the basic problems of the unitary representation theory of locally compact groups for the class of exponential Lie groups, namely the concrete and explicit determination of the unitary dual $\widehat{G}$ of the group G as a topological space. It is, in a sense, the widest possible extension of the Kirillov–I. Brown theorem for the nilpotent case. Let us briefly sketch its content:

Let G be a connected, simply connected solvable Lie group, $\mathfrak{g}$ its Lie algebra and E the exponential mapping from $\mathfrak{g}$ into G. We say that G is *exponential*, if E is a diffeomorphism from $\mathfrak{g}$ onto G. Let l be a real linear form on $\mathfrak{g}$, i.e. an element of the linear dual $\mathfrak{g}^*$ of $\mathfrak{g}$.

A subalgebra $\mathfrak{h}$ of $\mathfrak{g}$ is called *subordinate* to l, if $l(\mathfrak{h}') = 0$, $\mathfrak{h}' = [\mathfrak{h}, \mathfrak{h}]$ the commutator algebra of $\mathfrak{h}$. In this case the formula

$$\chi(E(x)) = e^{il(x)}, \qquad x \in \mathfrak{h},$$

defines a unitary abelian character of the subgroup $H = E(\mathfrak{h})$ corresponding to $\mathfrak{h}$. We set

$$\pi(l, \mathfrak{h}) = \operatorname{ind}_H^G \chi$$

for the unitary representation of G, induced from the character χ of H.

After some basic work of Dixmier, A. A. Kirillov published in 1962 his fundamental paper [17], in which he proved:

If G, resp. $\mathfrak{g}$ is nilpotent, then

- $\pi(l, \mathfrak{h})$ is irreducible if and only if $\dim \mathfrak{h}$ is maximal among the dimensions of subalgebras of $\mathfrak{g}$ subordinate to l.
- If $\pi(l, \mathfrak{h}) \in \widehat{G}$, then $\pi(l, \mathfrak{h}) = \pi(l)$ is independent of $\mathfrak{h}$.
- G acts on $\mathfrak{g}^*$ via the coadjoint representation. The elements $l, l' \in \mathfrak{g}^*$ are in the same G-orbit if and only if $\pi(l)$ and $\pi(l')$ are equivalent.
- Every $\pi \in \widehat{G}$ is of the form $\pi(l)$ for some $l \in \mathfrak{g}^*$.

These results can be expressed in the following more concise form:

For $l \in \mathfrak{g}^*$ let $\Omega_l \subset \mathfrak{g}^*$ be the orbit of l and set

$$K(\Omega_l) = \pi(l) \in \widehat{G}\,.$$

Then $K : \Omega_l \to K(\Omega_l)$ is a bijection from the orbit space $\mathfrak{g}^*/G$ onto the unitary dual $\widehat{G}$ of G.

The mapping K (or sometimes its inverse) is called the *Kirillov map*. Kirillov's result parametrizes set–theoretically in a most satisfactory way the dual $\widehat{G}$. However, $\mathfrak{g}^*/G$ and $\widehat{G}$ both bear natural topologies and already Kirillov himself had proven that K is continuous. His *conjecture*, namely that K is a homeomorphism, was proved only in 1973 by I. D. Brown [6]. It is easy to see that, in the nilpotent case, it suffices to prove the conjecture only for q–step nilpotent algebras $\mathfrak{g}$ with k free generators (see [6]). The core of Brown's proof consists in the fact, that these algebras, resp. groups have an abundance of symmetries, more precisely: Every linear automorphism of $\mathfrak{g}/\mathfrak{g}'$ extends to an algebra automorphism of $\mathfrak{g}$.

Already in 1957 Takenouchi had shown that also for an exponential group G every $\pi \in \widehat{G}$ is monomial, i.e. of the form $\pi(l, \mathfrak{h})$ for some $l \in \mathfrak{g}^*$ and certain l–subordinate algebras $\mathfrak{h} \subset \mathfrak{g}$ [34]. Later, in 1965, Bernat was able to extend Kirillov's theorem to general exponential groups: There exists a canonical bijection $K : \mathfrak{g}^*/G \to \widehat{G}$, see [1].

While for nilpotent groups the representation $\pi(l, \mathfrak{h})$ is irreducible if and only if $\dim \mathfrak{h}$ is maximal among the l–subordinate subalgebras $\mathfrak{h}$, already the "$ax + b$"–algebra $\mathbb{R}a \oplus \mathbb{R}b$ with $[a, b] = b$ shows, that this is no longer true for exponential groups. The answer to this irreducibility question was given in 1967 by L. Pukanszky: $\pi(l, \mathfrak{h})$ is irreducible if and only if the linear submanifold $l + \mathfrak{h}^\perp$ of $\mathfrak{g}^*$ is contained in the orbit Ω_l [28]. One year later Pukanszky proved the continuity of the Kirillov map K for general exponential groups [29]. Left open for a long time was the question whether K was also open, i.e. a homeomorphism from $\mathfrak{g}^*/G$ onto $\widehat{G}$. The first substantial result in this direction came 1984 from H. Fujiwara [10]. He proved that $\mathfrak{g}^*/G$ contains a dense open set, which K maps homeomorphically onto a dense open set of $\widehat{G}$. This book, finally, contains in its chapters II and III the complete proof of the Kirillov conjecture: For all exponential groups, K is a homeomorphism from $\mathfrak{g}^*/G$ onto $\widehat{G}$.

It is worth mentioning that the analogous problem for the primitive ideal–space of the enveloping algebras of exponential Lie algebras recently has been solved by O. Mathieu [22].

Contrary to the nilpotent case there are no "free models" for general exponential groups and in general the inner structure of these groups is on the one hand various, on the other rather rigid, which implies that there is no way to extend Brown's method directly to the exponential case. This led Jean Ludwig to the idea to force the necessary amount of flexibility of the objects by extending the category of groups and algebras to the category of *variable* objects. This idea and the basic steps of the actual proof in chapter III are due to him. Ludwig reported on it first at the conference on "Harmonische Analyse und Darstellungstheorie topologischer Gruppen" in Oberwolfach, summer 1987.

The textbook literature on solvable Lie groups and their representations is very limited, still the 1972–volume [2] by Bernat et al. is the main source in this field and apparently no monograph exists, which exposes the theory beyond the type I–groups, in particular the fundamental work of Pukanszky [30], [31]. In this context we point the interested reader to the excellent survey article of C. C. Moore in the Proceedings of the 1972 Williamstown conference [23]. In any case, this situation caused us to include a relatively long first chapter in this book, the content of which will be sketched below.

Chapter I starts in § 1 with a general discussion of exponential Lie groups. § 2 contains fundamental facts on homogeneous spaces, quasiinvariant measures and group algebras. In § 3 we define and study induced representations. The central result is Mackey's imprimitivity theorem. We give a complete proof of the theorem in its general form for arbitrary locally compact groups, with a minor, unessential restriction: For the sake of transparency we suppose that on the homogeneous space G/H there exists a relatively invariant measure. Also included in § 3 are results on intertwining operators and an irreducibility criterion for induced representations. In § 4 we come back to exponential Lie groups. After studying polarizations we prove Bernat's results, i.e. the bijectivity of the Kirillov map. § 5 contains two theorems with the precise description of the kernels of the restriction $\pi|_H$ of an irreducible representation of G onto a closed subgroup H, resp. of the induction $\mathrm{ind}_H^G\pi$ of $\pi \in \widehat{H}$. This problem, resp. the problem of decomposing $\pi|_H$ and $\mathrm{ind}_H^G\pi$ of course has been studied extensively in the literature; from the many papers by Corwin, Greenleaf, Grélaud, Lipsman and others we cite only Fujiwara [10], who proves a precise formula for the decomposition of $\pi|_H$ into irreducible representations of H.

§ 6 is more or less independent of the rest of this book. Based on Ludwig's paper [21] it treats the following problem: Let $\pi = \mathrm{ind}_P^G\chi \in \widehat{G}$ be given. The quotient G/P is diffeomorphic with some $\mathbb{R}^m$ and the representation space of π can naturally be identified with $L^2(\mathbb{R}^m)$. Then for any $f \in L^1(G)$ the operator $\pi(f)$ is a kernel operator, i.e. there exists a function K_f on $\mathbb{R}^m \times \mathbb{R}^m$ such that $(\pi(f)\xi)(x) = \int_{\mathbb{R}^m} K_f(x, y)\xi(y)dy$. Problem: Which kernels occur in this fashion? For nilpotent G it is known that for every Schwartz function $F \in \mathcal{S}(\mathbb{R}^m \times \mathbb{R}^m)$ there exists $f \in L^1(G)$ with $F = K_f$, see [14]. For exponential groups such a result cannot be expected, for the following reason: The parametrization of G/H via $\mathbb{R}^m$ depends substantially upon the choice of "coexponential bases" in $\mathfrak{g}$ for the polarization $\mathfrak{p}$. In the nilpotent case a change of the basis induces a bipolynomial diffeomorphism of the parameter space $\mathbb{R}^m$, which is compatible with polynomial decay and Fourier transform. In the general exponential case however exponential functions enter the picture and these are incompatible in particular with Fourier transform. Whether our theorem 12 is optimal we do not know, in any case it is the best we could prove for the moment. Certainly it guarantees the existence of finite rank operators in the image $\pi(L^1(G))$ for every $\pi \in \widehat{G}$. For more applications of this theorem see [21]. It is in order to mention here also the remarkable result of D. Poguntke [27]: Let G be a connected locally compact group and $\pi \in \widehat{G}$.

If $\pi(C^*(G))$ contains the compact operators, then $\pi(L^1(G))$ contains rank–one projections. This is false for non connected groups [26].

Chapter II contains the foundation of the theory of variable structures, in particular of variable groups and algebras and their representation theory. Obviously this theory has intimate connections with other existing theories, e.g. with the modern theory of groupoids, with bundles and above all with contractions. Here we do not discuss these connections and don't even cite the papers of Renault, Fell, Saletan etc. In § 3 we show that there exists a functor **l** from the category of variable Lie groups into (but not onto!) the category of variable Lie algebras. For every variable Lie group **G** with algebra **l(G)** there exists the exponential mapping $E : \mathbf{l}(\mathbf{G}) \to \mathbf{G}$ which also in the variable case is a local homeomorphism. For exponential **G** the map E is a global homeomorphism.

In chapter III finally we prove the Kirillov conjecture for general variable exponential groups. § 1 contains the continuity part of the proof. Already here the use of variable structures simplifies matters substantially. In order to show that also the inverse K^{-1} of the Kirillov map is continuous we of course apply induction on $\dim G = \dim \mathfrak{g}$, where G, resp. $\mathfrak{g}$ is the underlying manifold, resp. vector space of the variable structure. With relatively simple arguments the problem is reduced to the case where all algebras ${}_p\mathfrak{g}$, constituting the variable Lie algebra $\boldsymbol{\mathfrak{g}} = P \times \mathfrak{g} = \bigcup_{p \in P} p \times {}_p\mathfrak{g}$, have the same one dimensional center $\mathfrak{z}$. Then, as usually for real solvable Lie groups, one has to distinguish two cases: Case 1: Infinitely many quotients ${}_p\mathfrak{g}/\mathfrak{z}$ contain a one dimensional minimal ideal; or Case 2: Infinitely often there exist minimal two dimensional ideals in ${}_p\mathfrak{g}/\mathfrak{z}$. Although the basic ideas of the proof appear already in Case 1, treated in § 2, considerable not only technical complications occur in Case 2, which is treated in § 3.

The process of writing this book was for more than one reason complicated and non linear. Certainly this will have left traces, although we have tried to do our best in detecting, correcting and removing misprints, inconsistent notations etc; in this respect we just hope for the lenience of the reader. The same wish we have with respect to colleagues who feel that their work is insufficiently or even not at all regarded. As one can see we have quoted only such facts and results which are indispensable for our purposes, in particular for our proofs. Moreover, to keep the book selfcontained, in particular in chapter I, we gave our own proofs whenever possible. This means under no circumstances that e.g. we claim priority. We hope that this attitude is justified by the fact that this book is not a monograph on an established, well defined mathematical field, but the representation of new methods and results in such a field.

Bielefeld and Metz, December 1993

Horst Leptin
Jean Ludwig

Chapter 1

Solvable Lie Groups, Representations

§ 1 Bases in solvable Lie algebras, exponential groups

We assume the reader to be familiar with the basic facts of the theory of Lie groups and their algebras: If G is a real n dimensional Lie group then its Lie algebra $\mathfrak{g}$ is the n dimensional vector space of the left translation invariant vector fields on G. For every $x \in \mathfrak{g}$ there exists a unique one parameter subgroup in G, which we denote by $e^{\mathbb{R}x}$ or $\{e^{tx}; t \in \mathbb{R}\}$. Thus $t \to e^{tx} \in G$ is the smooth homomorphism from $\mathbb{R}$ into G, for which

$$(xf)(g) = \tfrac{d}{dt} f(ge^{tx})|_{t=0} \ .$$

The *exponential mapping*

$$E : \mathfrak{g} \to G$$

is then defined by $E(x) = e^x$. We will also use the notation exp. It is known that E is an analytic mapping. If $\{b_i\}_1^n$ is a linear basis for $\mathfrak{g}$, then

$$\mathbb{R}^n \ni \xi = (\xi_1, \xi_2, \cdots, \xi_n) \to E\left(\sum_1^n \xi_i b_i\right) \in G$$

defines a local coordinate system in a neighborhood of the unit element e of G, the so called *canonical coordinates of the first kind.*

More important for many purposes are the canonical coordinates of the second kind. These are defined by the mapping

$$(\xi_1, \xi_2, \cdots, \xi_n) \to E(\xi_1 b_1)E(\xi_2 b_2) \cdot E(\xi_n b_n) \ .$$

They are particularly usefull, when the basis $\mathcal{B} = \{b_1, b_2, \cdots, b_n\}$ of $\mathfrak{g}$ is a *Malcev basis* or a *Jordan–Hölder* basis: Let

$$\mathfrak{g}_k = \mathfrak{g}_k(\mathcal{B}) := \sum_{i=k}^{n}{}^{\oplus} \mathbb{R}\, b_i$$

be the subspaces generated linearly by $b_k, b_{k+1}, \cdots, b_n$.

Definition. The basis $\mathcal{B}$ of $\mathfrak{g}$ is a *Malcev basis* or briefly an M basis, if all $\mathfrak{g}_k$ are subalgebras of $\mathfrak{g}$ and $\mathfrak{g}_{k+1}$ is an ideal in $\mathfrak{g}_k$ for $k = 1, 2, \ldots, n$, with $\mathfrak{g}_{n+1} = 0$. $\mathcal{B}$ is a *Jordan–Hölder basis* or JH basis, if $\mathcal{B}$ is a Malcev basis and the sequence $\{\mathfrak{g}_k\}_k$ is a refinement of a composition series.

So an M basis $\mathcal{B}$ is a JH basis, if for $k = 1, 2, \ldots, k$ either $\mathfrak{g}_k$ is an ideal in $\mathfrak{g}$ or $\mathfrak{g}_{k+1}$ is an ideal and $\mathfrak{g}_{k-1}/\mathfrak{g}_{k+1}$ is an irreducible $\mathfrak{g}$ module. Evidently M bases, resp. JH bases exist if and only if $\mathfrak{g}$ is solvable.

It is clear that for every normal series in $\mathfrak{g}$, i.e. every series $\mathfrak{g} = \mathfrak{h}_1 \supset \mathfrak{h}_2 \supset \cdots \supset \mathfrak{h}_m \supset 0$ of ideals $\mathfrak{h}_j$ in $\mathfrak{g}$, there exists a JH basis $\mathcal{B}$ such that $\mathfrak{h}_j = \mathfrak{g}_{k_j}(\mathcal{B})$ for suitable k_j, $j = 1, 2, \ldots, m$. To see this we may assume that $\{\mathfrak{h}_j\}$ is a composition series. This implies that all quotients $\mathfrak{h}_j/\mathfrak{h}_{j+1}$ are $\mathfrak{g}$ irreducible, hence dim $(\mathfrak{h}_j/\mathfrak{h}_{j+1}) \leq 2$. If dim $(\mathfrak{h}_j/\mathfrak{h}_{j+1}) = 2$, then $[\mathfrak{h}_j, \mathfrak{h}_j] \subset \mathfrak{h}_{j+1}$ and any subspace $\mathfrak{a}$ with dim $\mathfrak{a} =$ dim $\mathfrak{h}_{j+1} + 1$ and $\mathfrak{h}_j \supset \mathfrak{a} \supset \mathfrak{h}_{j+1}$ is an ideal in $\mathfrak{h}_j$. Adding such subspaces to the chain of the $\mathfrak{h}_j$ we obtain a refinement $\mathfrak{g} = \mathfrak{g}_1 \supset \mathfrak{g}_2 \supset \cdots \supset \mathfrak{g}_n$ of the series $\{\mathfrak{h}_j\}$. Then any set $\mathcal{B} = \{b_k\} \subset \mathfrak{g}$ with $\mathfrak{g}_k = \mathbb{R}\, b_k \oplus \mathfrak{g}_{k+1}$ is an JH–basis with $\mathfrak{h}_j = \mathfrak{g}_{k_j}(\mathcal{B})$ for suitable k_j.

Now let $\mathcal{B} = \{b_j\}$ be a fixed M–basis of the solvable Lie algebra $\mathfrak{g}$. Then the following proposition holds:

The mapping

$$E_{\mathcal{B}} : \xi = (\xi_1, \xi_2, \ldots, \xi_n) \to E_{\mathcal{B}}(\xi) = E(\xi_1 b_1)E(\xi_2 b_2) \ldots E(\xi_n b_n)$$

is bijective and bianalytic from $\mathbb{R}^n$ onto the connected, simply connected solvable Lie group G, defined by $\mathfrak{g}$.

The well known proof is an immediate consequence of the following, equally well known fact: Let $\mathfrak{h}$ be an ideal of codimension 1 in $\mathfrak{g}$ and H the corresponding analytic closed normal subgroup of G. If $\mathfrak{g} = \mathbb{R}\, a \oplus \mathfrak{h}$ and $R = E(\mathbb{R}\, a)$ is the one parameter subgroup defining a, then $G = R \ltimes H$ is the semidirect product of R and H and $(\xi, x) \to E(\xi a)x$ is a bianalytic map from $\mathbb{R} \times H$ onto G.

If a Malcev basis $\mathcal{B}$ is fixed, we will frequently identify $\xi \in \mathbb{R}^n$ with its image $E_{\mathcal{B}}(\xi) \in G$, so G, as a manifold, "is" $\mathbb{R}^n$ and the one dimensional subspace $\{x \in \mathbb{R}^n\, ,\ x_i = 0 \text{ for } i \neq j\}$ is the one parameter group $e^{\mathbb{R}\, b_j}$. In this sense we can write

$$E\left(\sum_1^n \xi_j b_j\right) = (y_1(\xi), y_2(\xi), \ldots, y_n(\xi)) \tag{1}$$

where the y_j are real analytic functions in ξ. Assume that $\mathfrak{h} = \mathfrak{g}_k$ for some k is an ideal, hence $H = \{x \in \mathbb{R}^n\, ;\ x_i = 0, i < k\}$ is the corresponding normal subgroup in G. Then the functorial properties of the exponential mapping imply, that the coset $E(x)H$ for $x \in \mathfrak{g}$ depends only on the coset $x + \mathfrak{h}$ in $\mathfrak{g}$. Consequently the functions $y_j(\xi)$ in (1) depend for $j < k$ only on the ξ_i, $i < k$. This implies

(2)**Lemma.** *Let $\mathfrak{B}$ be a JH basis and $j_1 = 1, j_2, \dots, j_l$ the sequence of indices, for which the $\mathfrak{g}_{j_k}$ are ideals, thus $j_{k+1} \le j_k + 2$. Then E is a local diffeomorphism in $x^0 = \sum \xi_j^0 b_j \in \mathfrak{g}$ if and only if all determinants $D_k = \frac{\partial y_{j_k}}{\partial \xi_{j_k}}(x^0)$ $(j_{k+1} = j_k + 1)$, resp. $D_k = \left(\frac{\partial y_{j_k}}{\partial \xi_{j_k}}\frac{\partial y_{j_k+1}}{\partial \xi_{j_k+1}} - \frac{\partial y_{j_k}}{\partial \xi_{j_k+1}}\frac{\partial y_{j_k+1}}{\partial \xi_{j_k}}\right)(x^0)$ are nonzero.*

This is clear, because the product $\prod_{k=1}^{l} D_k$ is the determinant of the Jacobian $\left(\frac{\partial y}{\partial \xi}\right)$ of E in x^0.

We will now study the conditions, under which the exponential mapping is a global diffeomorphism, i.e. G is an exponential group in the sense of the

Definition. An *exponential group* is a connected, simply connected solvable Lie group G, for which the exponential map E is a diffeomorphism from the Lie algebra $\mathfrak{g}$ of G onto G.

We start with the indecomposable groups of dimension 2 and 3.

$n = 2$: There exists exactly one algebra

$$\mathfrak{s}_2 = \mathbb{R}\, a \oplus \mathbb{R}\, b \text{ with } [a, b] = b$$

with the group

$$S_2 = \mathbb{R} \ltimes \mathbb{R} \text{ with } (x_1, x_2)(y_1, y_2) = (x_1 + y_1, e^{-y_1}x_2 + y_2) ,$$

thus $a = \partial_1 - x_2\partial_2$, $b = \partial_2$. Here (and in the sequel) we use the notation $\partial_j = \frac{\partial}{\partial x_j}$.

So clearly $e^{\mathbb{R}a} = \mathbb{R} \times \{0\}$, $e^{\mathbb{R}b} = \{0\} \times \mathbb{R}$ and

$$E(\xi_1 a)E(\xi_2 b) = (\xi_1, \xi_2) .$$

Let

$$\tilde{e}(z) = \sum_{k=0}^{\infty} \frac{z^k}{(k+1)!} = \frac{e^z - 1}{z} .$$

Then one verifies easily that in S_2 the coordinates of the first kind are given by

$$E(\xi_1 a + \xi_2 b) = (\xi_1, \tilde{e}(-\xi_1)\xi_2) ,$$

which shows that S_2 is exponential.

$n = 3$: There are two different types: The three dimensional Heisenberg algebra

$$\mathfrak{h}_1 = \mathbb{R}\, a \oplus \mathbb{R}\, b \oplus \mathbb{R}\, c \quad \text{with} \quad [a, b] = c$$

and group

$$\mathbb{H}_1 = \mathbb{R}^3 \text{ with } (x_1, x_2, x_3)(y_1, y_2, y_3) = (x_1 + y_1, x_2 + y_2, x_3 + y_3 + x_1 y_2) ,$$

thus $a = \partial_1$, $b = \partial_2 + x_1\partial_3$, $c = \partial_3$,

$$E(\xi_1 a + \xi_2 b + \xi_3 c) = (\xi_1, \xi_2, \xi_3 + \tfrac{1}{2}\xi_1\xi_2) ,$$

and $\mathbb{H}_1$ is exponential.

Now let $\lambda = \lambda_1 + i\lambda_2 \in \mathbb{C}^\times$. We set

$$\mathfrak{s}(\lambda) = \mathbb{R}\, b_1 \oplus \mathbb{R}\, b_2 \oplus \mathbb{R}\, b_3$$

with

$$[b_1, b_2] = \lambda_1 b_2 + \lambda_2 b_3$$
$$[b_1, b_3] = -\lambda_2 b_2 + \lambda_1 b_3 .$$

and group

$$S(\lambda) = \mathbb{R} \times \mathbb{C} \text{ with } (x, u)(y, v) = (x + y\, ,\, e^{-\lambda y}u + v)$$

which means that $S(\lambda) = \mathbb{R} \ltimes \mathbb{C}$ is a semidirect product with minimal normal subgroup $\mathbb{C}$, if $\lambda_2 \neq 0$. With $(x_1, x_2 + ix_3) = (x_1, x_2, x_3) \in S(\lambda)$ the b_j are given by

$$b_1 = \partial_1 - (\lambda_1 x_2 - \lambda_2 x_3)\partial_2 - (\lambda_2 x_2 + \lambda_1 x_3)\partial_3 , \; b_2 = \partial_2 , \; b_3 = \partial_3$$

and the exponential mapping in the complex coordinates $\xi_2 + i\xi_3$ by

(3) $$E(\xi_1 b_1 + \xi_2 b_2 + \xi_3 b_3) = (\xi_1, \tilde{e}(-\lambda\xi_1)(\xi_2 + i\xi_3)) .$$

We see that $S(\lambda)$ is exponential if and only if $\lambda_1 \neq 0$, hence $S(i)$ is (up to isomorphisms) the only nonexponential simply connected solvable Lie group in dimensions $\leqq 3$. In this case the complement of the image of E is the set of all elements $(2\pi k, 0) \in S(i)$ with $k \in \mathbb{Z} \setminus \{0\}$ and $E(2\pi k b_1 + x) = E(2\pi k b_1) = (2\pi k, 0)$ for all $k \in \mathbb{Z} \setminus \{0\}$, $x \in \mathbb{R}\, b_2 \oplus \mathbb{R}\, b_3$.

It turns out that the absence of $S(i)$, resp. $\mathfrak{s}(i)$ is responsible for a simply connected solvable group G to be exponential:

Theorem 1. *Let G be a connected, simply connected solvable Lie group of dimension n and let $\mathfrak{g}$ be its Lie algebra. The following conditions are equivalent:*
(i) *No factor group of G has closed subgroups isomorphic with $S(i)$.*
(ii) *No ad x, $x \in \mathfrak{g}$, has non zero purely imaginary eigenvalues.*
(iii) *E is injective.*
(iv) *E is surjective.*

(v) *E is a diffeomorphism from $\mathfrak{g}$ onto G.*

Proof. If (i) does not hold then $\mathfrak{g}$ contains an ideal $\mathfrak{h}$ and elements $x, y, z \notin \mathfrak{h}$ with

$$[x, y] \equiv -z (\mathrm{mod}\ \mathfrak{h}) , \ [x, z] \equiv y (\mathrm{mod}\ \mathfrak{h}) . \tag{4}$$

This means ad x $(y + iz) \equiv i(y + iz)(\mathrm{mod}\ \mathfrak{h})$, hence i is an eigenvalue of ad x. On the other hand, if i is an eigenvalue of ad x, then there is $y + iz$ in the complexification $\mathfrak{g}^{\mathbb{C}}$ of $\mathfrak{g}$ with $(\mathrm{ad}\ x - i)(y + iz) = 0$, i.e. $[x, y] = -z$, $[x, z] = y$.

Let $\mathfrak{a}$ be a maximal ideal with $z \notin \mathfrak{a}$ and $\mathfrak{b} \subset \mathfrak{g}$ an ideal with $\mathfrak{a} \subset \mathfrak{b}$ and such that $\mathfrak{b}/\mathfrak{a}$ is minimal in $\mathfrak{g}/\mathfrak{a}$. Then necessarily $z \in \mathfrak{b}$, hence $y = [x, z] \subset \mathfrak{b}$ and $(\mathbb{R}\, x \oplus \mathfrak{b})/\mathfrak{a} \cong \mathfrak{s}(i)$, which implies $J/A \cong S(i)$ for the subgroups J and A of G corresponding to $\mathbb{R}\, x \oplus \mathfrak{b}$ and $\mathfrak{a}$. Thus we proved (i)$\Leftrightarrow$(ii).

For the rest of the proof we use induction. The theorem is true for dimensions $1, 2$ and 3, so let us assume that it is true for all connected, simply connected solvable groups of dimension less than n.

Let $\mathfrak{m}$ be a nonzero minimal ideal in $\mathfrak{g}$ and M its corresponding connected normal subgroup in G. We set $\overline{\mathfrak{g}} = \mathfrak{g}/\mathfrak{m}$, resp. $\overline{G} = G/M$ for the quotients, similarly $\overline{a} = a + \mathfrak{m} \in \overline{\mathfrak{g}}$, $\overline{x} = xM \in \overline{G}$ for $a \in \mathfrak{g}$, $x \in G$. Now assume that (i) and (ii) hold.

Let $a, b \in G$ with $E(a) = E(b)$, hence $\overline{E(a)} = E(\overline{a}) = E(\overline{b})$, which implies $\overline{a} = \overline{b}$, because $\dim \overline{G} \leq n - 1$ and clearly (ii) holds also for $\overline{\mathfrak{g}}$ and $\overline{G}$. But $\overline{a} = \overline{b}$ implies that $\mathbb{R}\, a + \mathfrak{m} = \mathbb{R}\, b + \mathfrak{m} = \mathfrak{h}$ is a subalgebra and $\dim \mathfrak{h} \leq 3$. Certainly (i) is true for $\mathfrak{h}$, hence E is injective and so $E(a) = E(b)$ implies $a = b$, which means that (iii) is true for $\mathfrak{g}$ and G.

Now let $x \in G$. As $\overline{G}$ satisfies (i) and (ii) there exist $\overline{a} = a + \mathfrak{m} \in \overline{\mathfrak{g}}$ with $\overline{x} = xM = E(\overline{a})$, which means $x = E(a)y$ with $y \in M$. Again $\mathfrak{h} = \mathbb{R}\, a + \mathfrak{m}$ is a subalgebra of dimension ≤ 3, hence E is a bijection from $\mathfrak{h}$ onto the corresponding subgroup H. As $x \in H$ there exists $b \in \mathfrak{h}$ with $E(b) = x$. This shows that E is surjective, i.e. (iv) holds for $\mathfrak{g}$ and G.

To prove (iv) $\Rightarrow$ (ii) let a be an element in $\mathfrak{g}$ and $A = \mathrm{ad}\ a$. With the same notation as above we see that $E(\overline{u}) = \overline{E(u)}$ for $\overline{u} \in \overline{\mathfrak{g}}$ implies that E is surjective from $\overline{\mathfrak{g}}$ onto $\overline{G}$, hence (i) through (v) hold for $\overline{\mathfrak{g}}$ and $\overline{G}$, in particular A has no purely imaginary eigenvalues on $\mathfrak{g}/\mathfrak{m}$. Again consider the subalgebra $\mathfrak{h} := \mathbb{R}\, a + \mathfrak{m}$ with corresponding subgroup H. For $x \in H$ there exists $u \in \mathfrak{g}$ with $x = E(u)$, hence $\overline{x} = E(\overline{u})$ in $\overline{G}$. But $\overline{x} = E(t\overline{a})$ for some $t \in \mathbb{R}$, and because E is one to one on $\overline{\mathfrak{g}}$ it follows that $\overline{u} = t\overline{a}$, i.e. $u \in \mathfrak{h}$. This proves that E is surjective from $\mathfrak{h}$ to H and consequently $\mathfrak{h}$ is not isomorphic to $\mathfrak{s}(i)$, in particular A has no purely imaginary eigenvalues on $\mathfrak{m}$.

Next we show (iii)$\Rightarrow$(i): We assume that E is injective, but that there exist an ideal $\mathfrak{k}$ in $\mathfrak{g}$ and a subalgebra $\mathfrak{h}$ with $\mathfrak{h}/\mathfrak{k} \cong \mathfrak{s}(i)$, which means that there are elements a, b, c in $\mathfrak{g}$, independent modulo $\mathfrak{k}$, with $[a, b] + c$, $[a, c] - b \in \mathfrak{k}$. We know that we can choose a and b such that for the cosets $\overline{a} = a + \mathfrak{k}$, we have

$E(\overline{a}) = E(\overline{a}+\overline{b})$ in the quotient $H/K \subset G/K$. This means that there exists $q \in K$ with $E(a)q = E(a+b)$ in G. But $\mathfrak{j} := \mathbb{R}\, a \oplus \mathfrak{k}$ is a proper subalgebra of $\mathfrak{g}$ and E is injective on $\mathfrak{j}$, hence E is bijective and consequently there exists $d \in \mathfrak{j}$ with $E(d) = E(a)q = E(a+b)$. Since E is injective this implies $d = a+b$, which is a contradiction, because $a+b \notin \mathfrak{j}$.

We have proved the equivalence of conditions (i)–(iv). Evidently (v) implies (iii) and (iv), so finally we have to show that a bijective E is a diffeomorphism, i.e. that the Jacobian of E is regular in every point $x \in \mathfrak{g}$. For this we will use lemma (2). Let $\mathfrak{m}$ be a minimal non zero ideal in $\mathfrak{g}$.

If $\mathfrak{m}$ is central we may assume that $\mathfrak{m} = \mathbb{R}\, b_n$. Then, with the notation of (1) and $\xi' = (\xi_1, \xi_2, \ldots, \xi_{n-1}, 0)$, we have

$$y_n(\xi) = y_n(\xi') + \xi_n \ ,$$

hence $\frac{\partial y_n}{\partial \xi_n} = 1$. We take $\mathfrak{m}$ as last ideal of a JH–series and $\mathcal{B}$ as the corresponding JH–basis. Then we can apply induction and lemma (2): From $\dim \mathfrak{g}/\mathfrak{m} = n-1$ we see that $D_k \neq 0$ for $k < l$, while $D_l = \frac{\partial y_n}{\partial \xi_n} = 1$.

If $\mathfrak{m}$ is not central, then the centralizer $\mathfrak{c}$ is an ideal in $\mathfrak{g}$. If $\dim \mathfrak{g}/\mathfrak{c} = 2$, then also $\dim \mathfrak{m} = 2$ and we could find some $x \in \mathfrak{g}$ such that $\operatorname{ad} x|_{\mathfrak{m}}$ would have eigenvalues $\pm i$, which is excluded by (ii). Thus $\mathfrak{g} = \mathbb{R}\, a \oplus \mathfrak{c}$. Let $\{\mathfrak{h}_j\}_1^l$ be a composition series passing through $\mathfrak{c}$ and $\mathfrak{m}$, i.e. with $\mathfrak{h}_2 = \mathfrak{c}$, $\mathfrak{h}_l = \mathfrak{m}$, and let $\mathcal{B}$ be a corresponding JH–basis.

Then $\mathfrak{h}_2 = \sum_{i=2}^{n} \mathbb{R}\, b_i$, $\mathfrak{m} = \mathbb{R}\, b_n$ or $\mathfrak{m} = \mathbb{R}\, b_{n-1} \oplus \mathbb{R}\, b_n$ and $\operatorname{ad} b_1$, restricted to $\mathfrak{m}$, has an eigenvalue $\lambda = \lambda_1 + i\lambda_2$ with $\lambda_1 \neq 0$. We will only treat the case $\dim \mathfrak{m} = 2$.

Because $\mathfrak{m}$ is central in $\mathfrak{h}_2$ we see that for any $x \in \mathfrak{h}_2$ the restrictions of $\operatorname{ad}(\xi_1 b_1)$ and $\operatorname{ad}(\xi_1 b_1 + x)$ onto $\mathfrak{m}$ coincide, hence for $\xi_1 \neq 0$ the subalgebra $\mathbb{R}(\xi_1 b_1 + x) \oplus \mathfrak{m}$ is isomorphic with $\mathfrak{s}(\lambda)$ and formula (3), after applying the isomorphism $\xi_{n-1} \cdot 1 + \xi_n \cdot i \to \xi_{n-1} b_{n-1} + \xi_n b_n$ from $\mathbb{C}$ onto $\mathfrak{m}$, gives for $x = \sum_{i=2}^{n-2} \xi_i b_i$ the relation

$$E\left(\sum_{i=1}^{n} \xi_i b_i\right) = E\left(\sum_{i=1}^{n-2} \xi_i b_i\right) E((e_1(\xi_1)\xi_{n-1} - e_2(\xi_1)\xi_n)b_{n-1} + \\ + (e_2(\xi_1)\xi_{n-1} + e_1(\xi_1)\xi_n)b_n),$$

where $e_1(\xi_1)$, $e_2(\xi_1)$ are defined by $\widetilde{e}(-\lambda\xi_1) = e_1(\xi_1) + ie_2(\xi_1)$. From (1) we get

$$E\left(\sum_{i=1}^{n-2} \xi_i b_i\right) = (y_1(\xi)\,,\ y_2(\xi), \cdots, y_{n-2}(\xi)\,,\ w_1(\xi')\,,\ w_2(\xi'))$$

with $\xi' = (\xi_1, \xi_2, \ldots, \xi_{n-2}) \in \mathbb{R}^{n-2}$. Hence the last two coordinates of $E\left(\sum_{i=1}^{n} \xi_1 b_i\right)$ are given by

$$y_{n-1}(\xi) = w_1(\xi') + e_1(\xi_1)\xi_{n-1} - e_2(\xi_1)\xi_n$$
$$y_n(\xi) = w_2(\xi') + e_2(\xi_1)\xi_{n-1} + e_1(\xi_1)\xi_n \,.$$

This yields for the Jacobian of $(\xi_{n-1}, \xi_n) \to (y_{n-1}, y_n)$:

$$\left(\frac{\partial y}{\partial \xi}\right)_{n-1,n} = \begin{pmatrix} e_1(\xi_1)\,, & -e_2(\xi_1) \\ e_2(\xi_1)\,, & e_1(\xi_1) \end{pmatrix}$$

and

$$D_l = e_1(\xi_1)^2 + e_2(\xi_1)^2 = |\tilde{e}(-\lambda\xi_1)|^2 > 0$$

for all $\xi_1 \in \mathbb{R}$, because $\lambda_1 \neq 0$. This ends the proof of Theorem 1.

It is clear from theorem 1 that closed connected subgroups and simply connected quotients of exponential groups are also exponential; moreover we have

Proposition 1. *Let G be as in theorem 1. If H is a closed connected subgroup of G with corresponding subalgebra $\mathfrak{h}$, then H is exponential if and only if $H = E(\mathfrak{h})$.*

This follows immediately from theorem 1 and the functorial properties of the exponential mapping.

Let H be a closed subgroup of the Lie group G. The coset space G/H of all left cosets $\bar{x} = xH$, $x \in G$, is also a manifold. If G is solvable, connected and simply connected of dimension n and H is connected of dimension l, then G/H is diffeomorphic with $\mathbb{R}^{n-l}$. We want to describe this diffeomorphism explicitly by means of the exponential map and suitable bases in the Lie algebra $\mathfrak{g}$ of G.

Let $\mathfrak{h}$ be the Lie algebra of the closed subgroup H. Then $\mathfrak{g}/\mathfrak{h}$ is an $\mathfrak{h}$ module, hence there exists $\mathfrak{h}$ composition series of $\mathfrak{g}/\mathfrak{h}$, i.e. $\mathfrak{h}$ invariant subspaces $\mathfrak{g} = \mathfrak{g}_1 \supset \ldots \supset \mathfrak{g}_r \supset \mathfrak{g}_{r+1} = \mathfrak{h}$, such that the quotients $\mathfrak{g}_j/\mathfrak{g}_{j+1}$ are $\mathfrak{h}$ irreducible, so dim $(\mathfrak{g}_j/\mathfrak{g}_{j+1})$ is 1 or 2. Slightly incorrectly we will call such a series $\{\mathfrak{g}_j\}_{j=1,\ldots,r+1}$ an $\mathfrak{h}$ composition series of $\mathfrak{g}/\mathfrak{h}$.

Definition. A *coexponential basis for* $\mathfrak{h}$ *in* $\mathfrak{g}$ is a set $\mathcal{B}_{\mathfrak{h}} = \{b_1, b_2, \ldots, b_d\}$ of elements $b_j \in \mathfrak{g}$, independent modulo $\mathfrak{h}$, such that

(i) $\{\mathfrak{b}_j \oplus \mathfrak{h}\}_{j=1,\ldots,d}$, with $\mathfrak{b}_j = \sum_{i=j}^{d}{}^{\oplus} \mathbb{R} b_i$, is a refinement of an $\mathfrak{h}$ composition series of $\mathfrak{g}/\mathfrak{h}$ and

(ii) the mapping

$$E_{\mathfrak{h}} : (\sum_{j=1}^{d} \xi_j b_j, h) \to E(\xi_1 b_1)E(\xi_2 b_2) \cdots E(\xi_d b_d)h$$

maps $\mathfrak{b}_1 \times H$ diffeomorphically onto G.

It follows that for a coexponential basis $\mathcal{B}_{\mathfrak{h}}$ the mapping $x = \sum_1^d \xi_j b_j \to E_{\mathfrak{h}}(x, e)H \in G/H$ is a diffeomorphism of $\mathfrak{b}_1$ onto G/H.

Proposition 2. *If G is connected, simply connected solvable, then for every subalgebra $\mathfrak{h}$ of $\mathfrak{g}$ there exist coexponential bases.*

Proof. Let $\mathfrak{k}$ be a subalgebra with $\mathfrak{h} \subset \mathfrak{k} \subset \mathfrak{g}$. If $\{b_1, \ldots, b_d\}$ is a coexponential basis for $\mathfrak{k}$ in $\mathfrak{g}$ and $\{c_1, \ldots, c_k\}$ a coexponential basis for $\mathfrak{h}$ in $\mathfrak{k}$, then obviously $\{b_1, \ldots, b_d, c_1, \ldots, c_h\}$ is coexponential for $\mathfrak{h}$ in $\mathfrak{g}$. So we only need to consider the case that $\mathfrak{h}$ is a maximal proper subalgebra of $\mathfrak{g}$. If $\mathfrak{g}' = [\mathfrak{g}, \mathfrak{g}] \subset \mathfrak{h}$, then $\mathfrak{h}$ is an ideal and any $b \notin \mathfrak{h}$ can be taken as a coexponential basis. Otherwise $\mathfrak{g} = \mathfrak{g}' + \mathfrak{h}$. As dim $\mathfrak{g}' <$ dim $\mathfrak{g}$, induction allows us to assume that $\mathfrak{g}'$ contains a coexponential basis $\{b\}$, resp. $\{b_1, b_2\}$ for $\mathfrak{g}' \cap \mathfrak{h}$. We claim, that this basis is also coexponential for $\mathfrak{h}$ in $\mathfrak{g}$. We consider only the case $\{b_1, b_2\}$. With $B = E(\mathbb{R}b_1)E(\mathbb{R}b_2) \subset G'$ we have $G' = B(G' \cap H)$, hence $G = G'H = BH$. From $xy = x'y'$, $x, x' \in B$, $y, y' \in H$, we get $x'^{-1}x = y'y^{-1} \in G' \cap H$, hence $x = x'$, $y = y'$, because $\{b_1, b_2\}$ is coexponential for $\mathfrak{g}' \cap \mathfrak{h}$ in $\mathfrak{g}'$. Thus $E_{\mathfrak{h}}$ is bijective from $(\mathbb{R}b_1 \oplus \mathbb{R}b_2) \times H =: \mathfrak{b} \times H$ onto G. Clearly $E_{\mathfrak{h}}$ is a local diffeomorphism from a neighborhood of $\{0\} \times \{e\}$ onto a neighborhood of the unit element e in G. Let $\{b_3, \ldots, b_n\}$ be a basis of $\mathfrak{h}$, so that $x = \sum_1^n \xi_j b_j \to F(x) = E(\xi_1 b_1)E(\xi_2 b_2) \ldots E(\xi_n b_n)$ is a diffeomorphism F of a neighborhood of $0 \in \mathbb{R}^n$ onto a neighborhood of e. If $q : x \to q(x) \in \mathfrak{g}$ is a polynomial mapping of $\mathfrak{g}$ into $\mathfrak{g}$ without constant and linear terms, then also $x \to E(q(x))F(x)$ is a local diffeomorphism. Now fix an element $(x^0, h_0) = (\xi_1^0 b_1 + \xi_2^0 b_2, h_0)$, in $\mathfrak{b} \times H$. Then, since the ad y for $y \in \mathfrak{g}'$ are uniformly nilpotent, it is easy to see that there exists a $\mathfrak{g}'$ valued polynomial $q(\xi_1, \xi_2)$ without constant and linear terms, such that for $x = \xi_1 b_1 + \xi_2 b_2$, $h \in H$

$$E_{\mathfrak{h}}(x + x^0, hh^0) = E(x^0)E(q(\xi_1, \xi_2))E_{\mathfrak{h}}(x, h)h^0 \; .$$

Hence $E_{\mathfrak{h}}$ is also locally diffeomorphic in (x^0, h^0).

§ 2 Invariant measures, group algebras

The existence of Haar measures, i.e. of left invariant Radon measures in Lie groups, is a simple consequence of the existence of left invariant cotangent fields on the underlying manifold, see [5],ch. III, § 3.16. The construction shows also at once that the Haar modulus Δ_G of the Haar measure dx of the group G is given by

$$\Delta_G(x) = |\det \mathrm{Ad}\, x|^{-1}$$

where Ad x is the adjoint transformation on the Lie algebra $\mathfrak{g}$ of G, defined by $x \in G$. If G is connected, then clearly $\det \mathrm{Ad}\, x > 0$ for all $x \in G$, hence $\Delta_G(x) = \det \mathrm{Ad}\, x^{-1}$.

If G is solvable and simply connected, then the Haar measure is explicitly given by $dx = d\xi_1 d\xi_2 \dots d\xi_n$, where the ξ_j are the canonical coordinates of the second kind defined by a given M basis $\mathcal{B}$ of the Lie algebra $\mathfrak{g}$ of G. This follows immediately by induction on $\dim \mathfrak{g}$ and the following trivial fact: If the locally compact group Γ is the semidirect product of the closed subgroups A and B, i.e. $\Gamma = A \ltimes B$ with normal B, then the (left) Haar measure $d\gamma$ of Γ is the product measure of the Haar measures dx of A and dy of B, i.e. $d\gamma = dxdy$.

If H is a connected closed subgroup of G and $\mathcal{B}_\mathfrak{h} = \{b_1, b_2, \dots, b_d\}$ a coexponential basis for $\mathfrak{h}$, then $\mathfrak{b} = \sum_1^{d}{}^{\oplus} \mathbb{R} b_j$ parametrizes the coset space $\overline{G} = G/H$. It is known that $\overline{G}$ carries an essentially unique relatively G–invariant positive Radon measure $d\overline{x}$, but this measure is in general not equivalent to the Lebesgue measure on $\mathfrak{b}$. However, for particular choices of $\mathcal{B}_\mathfrak{h}$ one can describe $d\overline{x}$ explicitly and simply by use of the Lebesgue measure on $\mathfrak{b}$.

Again let G, H, $\mathfrak{g}$ and $\mathfrak{h}$ be as above. As we had mentioned, the Haar moduli Δ_G and Δ_H of G and H are given by $\Delta_G(x) = \det \mathrm{Ad}_\mathfrak{g} x^{-1}$, $\Delta_H(y) = \det \mathrm{Ad}_\mathfrak{h} y^{-1}$. We set

$$\chi_H^G(y) = \frac{\Delta_G(y)}{\Delta_H(y)} = \det \mathrm{Ad}_{\mathfrak{g}/\mathfrak{h}} y^{-1}$$

for $y \in H$. Here $\mathrm{Ad}_{\mathfrak{g}/\mathfrak{h}}(y)$ is the transformation induced by $\mathrm{Ad}_\mathfrak{g}(y)$ on the quotient space $\mathfrak{g}/\mathfrak{h}$. If $y = \exp h$ with $h \in \mathfrak{h}$, then $\mathrm{Ad}\, y = e^{\mathrm{ad}\, h}$, thus in this case

$$\chi_H^G(\exp h) = e^{-\mathrm{trace}\ \mathrm{ad}_{\mathfrak{g}/\mathfrak{h}} h} .$$

This shows that $\chi_H^G(\exp h) = 1$ whenever ad h is nilpotent, in particular

$$\chi_H^G(y) = 1 \text{ for all } y \in G' \cap H .$$

So χ_H^G has a unique extension to a continuous homomorphism, again denoted by χ_H^G, from $G'H$ into the positive reals:

$$\chi_H^G(xy) = \chi_H^G(y) \text{ for } x \in G' ,\ y \in H \ .$$

If M is a closed normal subgroup in G, then $\Delta_M = \Delta_G|_M$ and $\Delta_G(x) = \Delta_{\overline{G}}(\overline{x})\Delta_M(\text{ int } x)^{-1}$ where $\overline{G} = G/M$, $\overline{x} = xM \in \overline{G}$ and $\Delta_M(\text{int } x)$ is the Haar modulus of the restriction of the inner automorphism int $x : y \to xyx^{-1}$ on M. If $M \subset H$, $\overline{H} = H/M \subset \overline{G}$, then it follows that

$$\chi_{\overline{H}}^{\overline{G}}(\overline{x}) = \chi_H^G(x) \text{ for } x \in \overline{x} \in \overline{H} \ . \tag{5}$$

Now let $\mathfrak{g} = \mathfrak{g}_1 \supset \mathfrak{g}_2 \supset \cdots \supset \mathfrak{g}_r \supset \mathfrak{g}_{r+1} = 0$ be a composition series of $\mathfrak{g}$, so $\mathfrak{m} := \mathfrak{g}_r$ is a minimal non zero ideal in $\mathfrak{g}$. There exists a coexponential $\mathcal{B}_\mathfrak{h} = \{b_j\}_j$ for $\mathfrak{h}$ in $\mathfrak{g}$, so that $\{\mathfrak{b}_j \oplus \mathfrak{h}\}_j$ (see the definition) is a refinement of $\{\mathfrak{g}_i + \mathfrak{h}\}_i$. This means that for every i with $\mathfrak{g}_i + \mathfrak{h} \neq \mathfrak{g}_{i+1} + \mathfrak{h}$ there is a j with $b_j \in \mathfrak{g}_i$, resp. $b_j, b_{j+1} \in \mathfrak{g}_i$ and $\mathfrak{b}_j \oplus \mathfrak{h} = \mathfrak{g}_i + \mathfrak{h} = \mathbb{R}b_j \oplus (\mathfrak{g}_{i+1} + \mathfrak{h})$, resp. $= \mathbb{R}b_j \oplus \mathbb{R}b_{j+1} \oplus (\mathfrak{g}_{i+1} + \mathfrak{h})$. Let $E_\mathfrak{h}$ be the diffeomorphism from $\mathfrak{b}_1 \times H$ onto G defined by $\mathcal{B}_\mathfrak{h}$. If $\xi = (\xi_1, \xi_2, \ldots \xi_d) \in \mathbb{R}^d$ and $y \in H$, we denote the element $E_\mathfrak{h}(\sum_1^d \xi_i b_i, y)$ by (ξ, y), in particular for any function f on G we set

$$f(\xi, y) := f(E(\xi_1 b_1)E(\xi_2 b_2)\ldots E(\xi_d b_d)y) \ .$$

and denote the Lebesgue measure on $\mathfrak{b}_1$, resp. $\mathbb{R}^d$ by $d\xi$, i.e. $d\xi_1 d\xi_2 \ldots d\xi_d$. Then we have

Theorem 2. *With the above definitions and notations for every integrable function f on G, i.e. $f \in L^1(G)$, the following formula holds:*

$$\int_G f(x)\,dx = \int_{\mathbb{R}^d} \int_H f(\xi, y)\chi_H^G(y)\,dy d\xi.$$

For the *proof* we use induction on $n = \dim \mathfrak{g}$ and consider first the case in which $\mathfrak{m} := \mathfrak{g}_r \subset \mathfrak{h}$. Let $M = E(\mathfrak{m})$, hence $M \cong \mathbb{R}^l$, $l = 1$ or 2. We set $\overline{\mathfrak{g}} = \mathfrak{g}/\mathfrak{m}$, $\overline{\mathfrak{h}} = \mathfrak{h}/\mathfrak{m}$ etc. Then $\{\overline{\mathfrak{g}}_i\}_{1 \le i \le r-1}$ is a composition series of $\overline{\mathfrak{g}}$ and $\mathcal{B}_{\overline{\mathfrak{h}}} = \{\overline{b}_j\}$ with $\overline{b}_j = b_j + \mathfrak{m} \subset \overline{\mathfrak{g}}$ a coexponential basis for $\overline{\mathfrak{h}}$ in $\overline{\mathfrak{g}}$. We have $E(\mathbb{R}\overline{b}_j) = E(\mathbb{R}b_j)M$ in $\overline{G} = G/M$ and for any $\overline{f} \in L^1(\overline{G})$ the theorem holds. Let f be integrable on G. Then $\overline{f}$, defined by $\overline{f}(\overline{x}) = \int_M f(xz)dz$ for $\overline{x} = xM \in \overline{G}$, is in $L^1(\overline{G})$. Theorem 2,

(5) and Weil's formula [2], [32] yield

$$\int_G f(x)\,dx = \int_{\overline{G}} \overline{f}(\overline{x})\,d\overline{x} = \int_{\mathbb{R}^d} \int_{\overline{H}} \overline{f}(\xi, \overline{y}) \chi_{\overline{H}}^{\overline{G}}(\overline{y})\,d\overline{y} d\xi =$$
$$= \int_{\mathbb{R}^d} \int_H f(\xi, y) \chi_H^G(y) dy d\xi \,.$$

So in this case theorem 2 is proved.

If $\mathfrak{m} \not\subset \mathfrak{h}$ we set $\mathfrak{k} = \mathfrak{h} + \mathfrak{m}$. If $\mathfrak{h} \cap \mathfrak{m} = 0$, then $\mathfrak{m} = \mathbb{R}b_d$ or $\mathfrak{m} = \mathbb{R}b_{d-1} \oplus \mathbb{R}b_d$, $K = H \ltimes M$ and $\{b_d\}$, resp. $\{b_{d-1}, b_d\}$ is coexponential for $\mathfrak{h}$ in $\mathfrak{k}$, moreover $\chi_H^K(y)$ is the Haar modulus of int y^{-1}, restricted to M : $\Delta_K(y) = \Delta_H(y)\Delta_M(\text{int } y^{-1})$ for $y \in H$. This gives immediately theorem 2 for K and H. If $\mathfrak{h} \cap \mathfrak{m} \neq 0$, then the first part of our proof applies, because $\mathfrak{h} \cap \mathfrak{m}$ is a minimal ideal in $\mathfrak{k}$. So again the claim holds for K and H.

Since $\mathfrak{m} \subset \mathfrak{k}$, the theorem is true for G and K, i.e. if $\mathcal{B}_{\mathfrak{k}} = \{b_1, b_2, \ldots, b_c\}$, $c = d-1$ or $d-2$, is the coexponential basis for $\mathfrak{k}$, then

$$\int_G f(x)dx = \int_{\mathbb{R}^c} \int_K f(\xi, w) \chi_H^G(w) dw d\xi =$$
$$= \int_{\mathbb{R}^c} \int_{\mathbb{R}^{d-c}} \int_H f(\xi, \xi', y) \chi_K^G(\xi', y) \chi_H^K(y) dy d\xi' d\xi \,.$$

If $\mathfrak{m}$ is not central, in particular if $\dim \mathfrak{m} = 2$, then $\mathfrak{m} \subset \mathfrak{g}'$, hence $M \subset G'$ and χ_K^G is trivial on M. But if $\mathfrak{m}$ is central and not in $\mathfrak{g}'$, then M is a commutative direct factor of G, so again $\chi_K^G(z) = 1$ for $z \in M$. This implies $\chi_K^G(\xi', y)\chi_H^K(y) = \chi_K^G(y)\chi_H^K(y) = \chi_H^G(y)$, hence

$$\int_G f(x)dx = \int_{\mathbb{R}^c} \int_{\mathbb{R}^{d-c}} \int_H f(\xi, \xi', y) \chi_H^G(y) dy d\xi' d\xi = \int_{\mathbb{R}^d} \int_H f(\xi, y) \chi_H^G(y) dy d\xi \,.$$

So theorem 2 is proved.

It is known (see e.g. [32] ch. 3, § 3), that there exists a relatively invariant measure on the coset space $\overline{G} = G/H$. We want to express this measure explicitly in terms of the coordinates $\xi \in \mathbb{R}^d$. First we remark that we can extend χ_H^G to a continuous homomorphism χ from G into $\mathbb{R}^+$: The quotient G/G' is a real vector space with $(G'H)/G'$ as a closed subspace. Hence there exists a closed complementary subspace, i.e. a closed subgroup D in G with $G' \subset D$, $G/G' = D/G' \oplus (G'H)/G'$. It follows that $G = DH$ and that

$$\chi(xy) = \chi_H^G(y) \text{ for } x \in D\,,\ y \in H$$

defines unambigously an extension χ of χ_H^G with $\chi(x) = 1$ on D. For $f \in \mathscr{C}_0(G)$ we set

$$I_\chi(f) = \int_G f(x)\chi(x^{-1})dx$$

and as before

$$\overline{f}(\overline{x}) = \int_H f(xy)dy \text{ for } \overline{x} = xH \in \overline{G}.$$

We claim that $\overline{f} = 0$ implies $I_\chi(f) = 0$: Let e be the unit element in G. Then theorem 2 yields

$$\begin{aligned} I_\chi(f) &= \int_{\mathbb{R}^d}\int_H f(\xi, y)\chi(\xi, y)^{-1}\chi_H^G(y)dyd\xi = \\ &= \int_{\mathbb{R}^d}\int_H f(\xi, y)\chi(\xi, e)^{-1}dyd\xi = \int_{\mathbb{R}^d}\overline{f}(\overline{\xi})\chi(\xi, e)^{-1}d\xi = 0 \end{aligned}$$

with $\overline{\xi} = (\xi, e)^- = (\xi, e)H \in \overline{G}$. Since $f \to \overline{f}$ is a positive, surjective, linear mapping from $\mathscr{C}_0(G)$ onto $\mathscr{C}_0(\overline{G})$ it follows that there exists a positive Radon meassure $d_\chi\overline{x}$ on $\overline{G}$ so that

$$\int_G f(x)\chi(x)^{-1}dx = \int_{\mathbb{R}^d}\overline{f}(\overline{\xi})\chi(\xi, e)^{-1}d\xi = \int_{\overline{G}}\overline{f}(\overline{x})d_\chi\overline{x}$$

for all $f \in \mathscr{C}_0(G)$. For these f and $a \in G$ we have $\overline{f_a} = (\overline{f})_a$, where $f_a(x) = f(ax)$, $(\overline{f})_a(\overline{x}) = \overline{f}(a\overline{x}) = \overline{f}(\overline{ax})$. This implies

$$\int_{\overline{G}}\overline{f}_a(\overline{x})d_\chi\overline{x} = \int_G f(ax)\chi(x)^{-1}dx = \chi(a)\int_G f(x)\chi(x)^{-1}dx\ ,$$

hence $d_\chi(a\overline{x}) = \chi(a)^{-1}d_\chi\overline{x}$, i.e. $d_\chi\overline{x}$ is relatively invariant with modulus $\chi(a)^{-1}$. Summing up we have shown most of

Theorem 3. *Let H be a connected closed subgroup of G and let $\overline{G} = G/H$ the left coset space. There exists a coexponential basis for $\mathfrak{h}$ in $\mathfrak{g}$, which defines a global coordinate system $\mathbb{R}^d \ni \xi \to \overline{\xi} = (\xi, e)H \in \overline{G}$. In these coordinates an essentially unique relatively invariant measure $d_\chi\overline{x}$ on $\overline{G}$ is given by the formula*

$$\int_{\overline{G}} f(\overline{x})d_\chi\overline{x} = \int_{\mathbb{R}^d} f(\overline{\xi})\chi(\xi, e)^{-1}d\xi$$

in which χ is a real character of G, extending $\chi_H^G = \Delta_G|_H\Delta_H^{-1}$. For $a \in G$ one has $d_\chi a\overline{x} = \chi(a)^{-1}d_\chi\overline{x}$.

With respect to the uniqueness of $d_\chi \overline{x}$ we remark: The choice of the extension χ of χ_H^G onto G was a little arbitrary, i.e. the same proof and theorem is valid for *any* continuous homomorphism $\chi : G \to \mathbb{R}^+$ with $\chi|_H = \chi_H^G$. Clearly the different choices of χ correspond bijectively to the real characters of the factor group $G/G'H$.

Now let such a χ and a χ relatively invariant measure $d\mu$ on $\overline{G}$ be given. For $f \in \mathscr{C}_0(G)$ set $i(f) = \int\limits_{\overline{G}} (\overline{f\chi})(\overline{x}) d\mu(\overline{x})$. Then for any $a \in G$:

$$i(f_a) = \chi(a^{-1}) \int_{\overline{G}} (\overline{f\chi})(a\overline{x}) d\mu(\overline{x}) = i(f) \ ,$$

hence $i(f) = c \int\limits_G f dx$ for some $c > 0$, consequently $d\mu(\overline{x}) = c \cdot d_\chi \overline{x}$.

From now on, given a coset space $\overline{G}$ of G with elements $\overline{x}$, $\overline{y}$ etc., "the" relatively invariant measure on $\overline{G}$ will always be denoted by $d\overline{x}$ (or $d\overline{y}$ etc.). For the corresponding Banach spaces of complex valued p-integrable functions, $1 \leq p < \infty$, resp. essentially bounded functions as ususally we will write

$$L^p(G) \, , \ L^p(\overline{G}) \text{ etc., resp. } L^\infty(G), \, L^\infty(\overline{G}) \text{ etc.}$$

So $L^1(G)$ is an involutive Banach algebra with convolution

$$f \star g(x) = \int_G f(xy) g(y^{-1}) dy = \int_G f(y) g(y^{-1}x) dy$$

and involution

$$f^*(x) = \Delta_G(x)^{-1} \overline{f(x^{-1})}.$$

For the L^p norms we use the usual notation $|\cdot|_p$, so for $f \in L^1(G)$, $g \in L^p(G)$ the convolution $f \star g$ is a well defined function in $L^p(G)$ and the inequality $|f \star g|_p \leq |f|_1 |g|_p$ holds. Then for fixed $f \in L^1(G)$ the formula $\lambda_p(f)g = f \star g$, $g \in L^p(G)$, defines an operator $\lambda_p(f)$ of the Banach space $L^p(G)$, bounded by $|f|_1$ and $\lambda_p : f \to \lambda_p(f)$ is a bounded representation of $L^1(G)$ in the Banach algebra $\mathscr{L}(L^p(G))$ of bounded operators of $L^p(G)$.

More generally let us now take a quotient space $\overline{G} = G/H$ of G and a χ relatively invariant measure $d\overline{x}$ on $\overline{G}$. For $g \in L^p(\overline{G})$, $1 \leq p \leq \infty$, and $a \in G$ we set

$$(\lambda_p(a)g)(\overline{x}) = \chi(a)^{\frac{1}{p}} g(a^{-1}\overline{x}) \ .$$

Then obviously $\lambda_p(a) : g \to \lambda_p(a)g$ defines a linear isometry on $L^p(\overline{G})$, i.e. $|\lambda_p(a)g|_p = |g|_p$ for all a and g, and $\lambda_p : a \to \lambda_p(a)$ is a *continuous isometric representation* of G in $L^p(\overline{G})$, i.e. a homomorphism of G into the isometries of $L^p(\overline{G})$ for which all mappings $a \to \lambda_p(a)g$, $g \in L^p(\overline{G})$, are continuous from G

into $L^p(\overline{G})$. So for any $f \in L^1(G)$ the vector valued integrals

$$\lambda_p(f)(g) := \int_G f(x)\lambda_p(x)g\,dx \in L^p(\overline{G})$$

exist and $|\lambda_p(f)(g)|_p \leq |f|_1 |g|_p$, i.e. $\lambda_p : f \to \lambda_p(f)$ is a representation of $L^1(G)$ in the bounded operators $\mathcal{L}(L^p(\overline{G}))$. As usually we write

$$\lambda_p(f) = \int_G f(x)\lambda_p(x)dx\ . \tag{6}$$

Of particular interest is the case $p = 2$. Then the $L^2(\overline{G})$ are Hilbert spaces and the representation λ_2 of G is *unitary*, i.e. we have $\lambda_2(x^{-1}) = \lambda_2(x)^*$, the adjoint operator of $\lambda_2(x)$, for all $x \in G$. The representation λ_2 of $L^1(G)$ is a *–representation, which means that $\lambda_2(f^*) = \lambda_2(f)^*$ holds for all $f \in L^1(G)$. This is an immediate consequence of (6).

The *left regular representation* λ of $L^1(G)$, i.e. the representation λ_2 of $L^1(G)$ in $L^2(G)$, is faithful, so if we define

$$|f|_* = |\lambda(f)|$$

for $f \in L^1(G)$, where $|T|$ is the usual norm of the bounded operator T of $L^2(G)$, then $f \to |f|_*$ is a norm on $L^1(G)$, satisfying the identity $|f \star f^*|_* = |f|_*^2$.

The C^* *algebra of* G, denoted by $C^*(G)$, is the completion of $L^1(G)$ with respect to the norm $|f|_*$. Thus $C^*(G)$ is canonically isomorphic to the norm closed hull of the image $\lambda_2(L^1(G))$ in the C^* algebra $\mathcal{L}(L^2(G))$ of all bounded operators of $L^2(G)$. We remind the reader of the general definition of the group C^* algebra: $C^*(G)$ is the completion of $L^1(G)$ with respect to the maximal C^* norm and is characterized by the fact that $L^1(G)$ is dense in $C^*(G)$ and that every continuous $*$–representation of $L^1(G)$ in a Hilbert space extends uniquely to such a representation of $C^*(G)$ in the same space. $C^*(G)$ coincides with the closure of $\lambda_2(L^1(G))$ in $\mathcal{L}(L^2(G))$ if and only if G is amenable, so in particular, as in the present case, when G is solvable.

Later we will use the *adjoint algebra* $\mathcal{A}^b$ of an involutive Banach algebra $\mathcal{A}$ (see [18] or [15]). This is the subset $\mathcal{A}^b$ of all bounded operators T of the Banach *space* $\mathcal{A}$, for which there exists another bounded operator T^* of $\mathcal{A}$, such that

$$(Ta)^*b = a^*(T^*b)$$

holds for all $a, b \in \mathcal{A}$. Obviously $\mathcal{A}^b$ is a subalgebra of the algebra $\mathcal{L}(\mathcal{A})$ of bounded operators of $\mathcal{A}$. For $x \in \mathcal{A}$ let l_x be the left multiplication operator $l_x(a) = xa$ of $\mathcal{A}$. Then clearly $l_x \in \mathcal{A}^b$ with $(l_x)^* = l_{x^*}$. For our purposes it is sufficient to assume that $|l_x| = |x|$ for all $x \in \mathcal{A}$ (where $|T|$ denotes the operator norm of $T \in \mathcal{L}(\mathcal{A})$. This is condition (L^*), p 266, in [18]). Under this assumption the following facts are easy to see:

- T^* is uniquely defined for $T \in \mathcal{A}^b$ and $T \to T^*$ is an isometric involution of $\mathcal{A}^b$.
- $\mathcal{A}^b$ is closed in $\mathcal{L}(\mathcal{A})$, hence $\mathcal{A}^b$ is an involutive Banach algebra.
- The mapping $x \to l_x$ is an isometric $*$–isomorphism of $\mathcal{A}$ into $\mathcal{A}^b$, hence $\mathcal{A}^b$ can be considered as a Banach algebra extension of $\mathcal{A}$.
- In this sense $\mathcal{A}$ is a closed two sided ideal in $\mathcal{A}^b$. If $\mathcal{B}$ is another essential Banach algebra extension of $\mathcal{A}$, i.e. $\mathcal{A}$ is a two sided ideal in $\mathcal{A}$ and $x\mathcal{A} = 0$, $x \in \mathcal{B}$, implies $x = 0$, then there exists a unique homomorphism $h : \mathcal{B} \to \mathcal{A}^b$ with $h(a) = a$ for $a \in \mathcal{A}$.
- $*$–representations of $\mathcal{A}$ in Hilbert spaces extend in a functorial way to $*$–representations of $\mathcal{A}^b$.

For details and more information see [18].

§ 3 Induced representations

One of the most powerful tools in representation theory is the notion of *induced representations*. Given a group G and a subgroup H, the induction process attaches to a given representation σ of H in a functorial way a representation ind σ of G. In this paragraph we assume G to be locally compact, i.e. not necessarily Lie, and H closed in G. Furthermore, "representation" always means "continuous unitary representation". For the matter of simplicity we also assume that there exists a relatively invariant measure $d\overline{x}$ on the coset space $\overline{G} = G/H$, hence there exists a positive real character χ on G which extends χ_H^G on H and with $da\overline{x} = \chi(a)^{-1} d\overline{x}$ for $a \in G$. This allows us to define the *left regular representation* $\overline{\lambda}$ of G *in the Hilbert space* $L^2(\overline{G})$: For $a \in G$ and $f \in L^2(\overline{G})$ we set $(\overline{\lambda}(a)f)(\overline{x}) = \chi(a)^{\frac{1}{2}} f(a^{-1}\overline{x})$. Then it is clear, that $\overline{\lambda} : a \to \overline{\lambda}(a)$ is a representation of G in $L^2(\overline{G})$.

Let σ be a representation of H in a Hilbert space $\mathfrak{H}$. The space $\mathcal{F}$ of all measurable functions from G into $\mathfrak{H}$ is a G space: For $a \in G$, $f \in \mathcal{F}$ we define $\Lambda(a)f$ by $(\Lambda(a)f)(x) = f(a^{-1}x)$. Now σ defines a G invariant subspace of $\mathcal{F}$: We set

$$\mathcal{F}_\sigma = \{f \in \mathcal{F};\ f(xy) = \sigma(y)^* f(x) \text{ for all } y \in H \text{ and almost all } x \in G\}\ .$$

Then $\mathcal{F}_\sigma$ is stable under all $\Lambda(a)$ and for $f, g \in \mathcal{F}_\sigma$ we see that the function $x \to (f(x)|g(x))$ depends only on $\overline{x}$, hence we write

$$(f(x)|g(x)) = (f|g)(\overline{x})$$

and $|f(x)| = (f|f)(\overline{x})^{\frac{1}{2}} = |f|(\overline{x})$. The subspace of $\mathcal{F}_\sigma$

$$\text{ind}\ \mathfrak{H} := \{f \in \mathcal{F}_\sigma;\ \int_{\overline{G}} |f|(\overline{x})^2 d\overline{x} < \infty\}$$

is a Λ invariant Hilbert space with inner product

$$(f|g) := \int_{\overline{G}} (f|g)(\overline{x})d\overline{x}.$$

For $a \in G$, $f \in \operatorname{ind} \mathfrak{H}$ we find

$$|\Lambda(a)f|^2 = \int_{\overline{G}} |f|(a^{-1}\overline{x})^2 d\overline{x} = \chi(a)^{-1}|f|^2 ,$$

hence, the mapping

$$(\operatorname{ind}_H^G \sigma)(a) : f \to \chi(a)^{\frac{1}{2}}\Lambda(a)f \tag{7}$$

defines a unitary operator $(\operatorname{ind}_H^G \sigma)(a)$ of the Hilbert space $\operatorname{ind} \mathfrak{H}$. Now obviously the following definition makes sense.

Definition. The mapping

$$\operatorname{ind}_H^G \sigma : x \to (\operatorname{ind}_H^G \sigma)(x) ,$$

where $(\operatorname{ind}_H^G \sigma)(x)$ is the unitary operator defined by (7), is called the *induced representation* of σ on G.

So the left regular representation on $L^2(\overline{G})$, defined above, is equal to $\operatorname{ind}_H^G 1_H$, where 1_H is the trivial representation of H.

We get particularly simple expressions for $\operatorname{ind}_H^G \sigma$, if G is a semidirect product: $G = R \ltimes H$. In this case we can identify $\overline{G}$ with R and it is $\chi(x) \equiv 1$. Let $\mathcal{L} = L^2(R, \mathfrak{H})$ be the Hilbert space of $\mathfrak{H}$ valued square integrable functions on R. For $\xi \in \mathcal{L}$ we define a function $\Omega\xi$ on G by

$$(\Omega\xi)(r, y) = \sigma(y)^*\xi(r) \text{ for } (r, y) \in R \ltimes H .$$

Then obviously $\Omega\xi(xy) = \sigma(y)^*\Omega\xi(x)$ for all $x \in G$, $y \in H$, moreover $\Omega\xi$ is measurable and for $x = (r, y)$ we have $|\Omega\xi|(\overline{x}) = |\xi(r)|$, hence $\Omega\xi \in \operatorname{ind} \mathfrak{H}$ and $|\Omega\xi| = |\xi|$. It follows that $\Omega : \xi \to \Omega\xi$ maps $\mathcal{L}$ isometrically into $\operatorname{ind} \mathfrak{H}$. Let f be a continuous function in $\operatorname{ind} \mathfrak{H}$. Then $\xi := f|_R$, is continuous on R and $|\xi(r)| = |f(r, e)| = |f|((r, e)^-)$ implies $|\xi| = |f|$, consequently $\xi \in \mathcal{L}$. Clearly $f = \Omega\xi$, from which it follows easily that Ω is onto.

Let $y \to y^t$ be the automorphism of H, defined by $t \in R$, so that the product in G is given by $(r, y)(t, z) = (rt, y^t z)$. For $\xi \in \mathcal{L}$ and $a = (t, u) \in G$ we get

$$(\operatorname{ind}_H^G \sigma(a))\Omega\xi(r, y) = \Omega\xi(t^{-1}r, (u^{t^{-1}r})^{-1}y) = \sigma(y)^*\sigma(u^{t^{-1}r})\xi(t^{-1}r) .$$

Hence, if we set

$$(8) \qquad (\tilde{\sigma}(t,u)\xi)(r) := \sigma(u^{t^{-1}r})\xi(t^{-1}r)$$

we see that

$$\tilde{\sigma}(a) = \Omega^* \mathrm{ind}_H^G \sigma(a)\Omega$$

for all $a \in G$. This means that $\tilde{\sigma} : a \to \tilde{\sigma}(a)$ is a representation of G in $L^2(R, \mathfrak{H})$, equivalent to $\mathrm{ind}_H^G \sigma$:

Proposition 3. *The induced representation* $\mathrm{ind}_H^G \sigma$ *of* G *is equivalent to the representation* $\tilde{\sigma}$ *of* G *in* $L^2(R, \mathfrak{H})$, *defined by (8).*

We return to the general situation of an arbitrary locally compact group G, a closed subgroup H and a relatively invariant measure $d\overline{x}$ von $\overline{G} = G/H$. We leave the proof of the following useful fact as an easy exercise to the reader:

(9) *For* $u \in \mathscr{C}_0(G)$ *and* $\xi \in \mathfrak{H}$ *define the function* $u\#\xi$ *by*

$$u\#\xi(x) = \int_H u(xy)\sigma(y)\xi dy\ .$$

The mapping $u \otimes \xi \to u\#\xi$ *defines a linear injection from the tensor product* $\mathscr{C}_0(G) \otimes \mathfrak{H}$ *onto a dense subspace of* ind $\mathfrak{H}$.

One of the basic problems in the representation theory of G is to decide whether a given representation π of G is induced from a proper subgroup. There is an easy necessary condition. Assume that $\pi = \mathrm{ind}_H^G \sigma$ for some representation σ of H. Then we can canonically identify the Banach space $L^\infty(\overline{G})$ with a von Neumann subalgebra of $\mathscr{L}(\mathfrak{H}(\pi))$, $\mathfrak{H}(\pi) = \mathrm{ind}\ \mathfrak{H}(\sigma)$: The bounded measurable function $q \in L^\infty(\overline{G})$ defines an operator $Q \in \mathscr{L}(\mathfrak{H}(\pi))$ by virtue of the definition

$$(Qf)(x) = q(\overline{x})f(x)\,,\ x \in G\,,$$

for $f \in \mathfrak{H}(\pi)$. The mapping

$$\delta : q \to \delta(q) = Q \in \mathscr{L}(\mathfrak{H}(\pi))$$

is a *–algebra isomorphism of $L^\infty(\overline{G})$ onto a commutative von Neumann subalgebra $\mathfrak{Q}$ of $\mathscr{L}(\mathfrak{H}(\pi))$. Now $L^\infty(\overline{G})$ is a G space: For $a \in G$, $q \in L^\infty(\overline{G})$ we define $q^a = \lambda_\infty(a)^{-1}q$, i.e. $q^a(\overline{x}) = q(a\overline{x})$. With this notation we can write

$$(10) \qquad \delta(q^a) = \pi(a)^*\delta(q)\pi(a)\ .$$

This identity is an immediate consequence of the definition of $\pi = \mathrm{ind}_H^G \sigma$.

When we restrict δ to the C^* algebra $\mathcal{C}_\infty(\overline{G})$ of continuous functions on $\overline{G}$, vanishing at infinity, we obtain an isomorphism onto a C^* subalgebra $\mathfrak{D}$ of $\mathfrak{Q}$, such that also (10) holds. We remark that $\mathfrak{Q}$ is the closure of $\mathfrak{D}$ in the weak operator topology.

Definition. Let H be a closed subgroup of the locally compact group G. A *system of imprimitivity* for (G, H) is a pair (δ, π), where
(i) δ is a faithful $*$–representation of $\mathcal{C}_\infty(\overline{G})$ in a Hilbert space $\mathfrak{H}$, i.e. a $*$–isomorphism of $\mathcal{C}_\infty(\overline{G})$ into $\mathcal{L}(\mathfrak{H})$,
(ii) π is a continuous unitary representation of G in $\mathfrak{H}$,
(iii) the relation (10) holds for all $q \in \mathcal{C}_\infty(\overline{G})$ and all $a \in G$.

Two systems (δ_1, π_1) and (δ_2, π_2) are *unitarily equivalent* if there exists a unitary operator U from the space $\mathfrak{H}_1$ of (δ_1, π_1) onto the space $\mathfrak{H}_2$ of (δ_2, π_2) such that $U\delta_1 = \delta_2 U$ and $U\pi_1 = \pi_2 U$.

The system $(\delta, \mathrm{ind}_H^G \sigma)$, constructed above from the representation σ of H, will be called the *imprimitivity system* of (G, H) *induced* by σ and we say that an imprimitivity system is induced, if it is in this sense induced by some representation σ of H.

The most fundamental result in this context is

Theorem 4 (the imprimitivity theorem). *Every system of imprimitivity* (δ, π) *for* (G, H) *is unitarily equivalent to an induced system.*

The *proof* follows essentially Blattner's ideas [4], with the simplification by Müller–Römer [24]. $\mathcal{C}_\infty(\overline{G})$ is in a natural way a G algebra, i.e. there is a canonical homomorphism of G into the automorphisms of the C^* algebra $\mathcal{C}_\infty(\overline{G})$, defined by $f \to f^a$, $f^a(\overline{x}) = f(a\overline{x})$ for $a \in G$. The corresponding twisted L^1 algebra $L^1(G, \mathcal{C}_\infty(\overline{G}))$ is called the *imprimitivity algebra* of (G, H) and it is easy to see that the $*$–representations of $L^1(G, \mathcal{C}_\infty(\overline{G}))$ are in a canonical $1:1$ correspondence with the imprimitivity systems for (G, H): The representation Λ, defined by (δ, π), is given by

$$\Lambda(f) = \int_G \pi(x)\delta(f(x))dx \ .$$

It is certainly sufficient to prove the theorem for cyclic representations. So let us assume that ε is a cyclic vector in the Hilbert space $\mathfrak{H}$ of (δ, π), resp. of Λ, i.e. that the $\Lambda(f)\varepsilon$, $f \in L^1(G, \mathcal{C}_\infty(\overline{G}))$, span a dense subspace of $\mathfrak{H}$.

Clearly the space $\mathcal{C}_0(G \times \overline{G})$ of compactly supported continuous functions on the product space $G \times \overline{G}$ can be identified with a dense subspace of $L^1(G, \mathcal{C}_\infty(\overline{G}))$. As usually $f \in \mathcal{C}_0(G \times \overline{G})$ defines the function $x \to f(x) \in \mathcal{C}_\infty(\overline{G})$ with $f(x)(\overline{y}) = f(x, \overline{y})$. The explicit expression for the convolution of two functions

$f, g \in \mathscr{C}_0(G \times \overline{G})$ is then given by

$$f \star g(x, \overline{y}) = \int_G f(xt, t^{-1}\overline{y})g(t^{-1}, \overline{y})dt$$

and the involution by

$$f^*(x, \overline{y}) = \Delta_G(x^{-1})\overline{f(x^{-1}, x\overline{y})} .$$

It follows immediately that $\mathscr{C}_0(G \times \overline{G})$ is a dense $*$–subalgebra of $L^1(G, \mathscr{C}_\infty(\overline{G}))$.

We will realize $\mathfrak{H}$ as a function space on $G \times \overline{G}$. For this we define a Radon measure $d\lambda$ on $G \times \overline{G}$ by

$$\int_{G\times\overline{G}} f d\lambda = \int_{G\times\overline{G}} f(x, \overline{y})d\lambda(x, \overline{y}) := (\Lambda(f)\varepsilon|\varepsilon) .$$

Then $\mathscr{C}_0(G \times \overline{G})$ becomes a pre–Hilbert space with inner product

$$(f|g)_\lambda := \int g^* \star f d\lambda = (\Lambda(f)\varepsilon|\Lambda(g)\varepsilon) , \tag{11}$$

in particular, $f \to \Lambda(f)\varepsilon$ is unitary from $\mathfrak{H}' := \mathscr{C}_0(G \times \overline{G})/\mathscr{N}$, with $\mathscr{N} = \{f; \Lambda(f)\ \varepsilon\ =\ 0\ \}$, into $\mathfrak{H}$. If we define (δ', π') on $\mathfrak{H}'$ by

$$(\delta'(u)f)(x, \overline{y}) = u(x\overline{y})f(x, \overline{y}) \text{ for } u \in \mathscr{C}_\infty(\overline{G})$$
$$(\pi'(a)f)(x, \overline{y}) = f(a^{-1}x, \overline{y}) \text{ for } a \in G ,$$

we see that (δ', π') and (δ, π) are unitarily equivalent, hence we will assume that $\mathfrak{H}$ is the completion of $\mathfrak{H}'$ and $\delta = \delta'$, $\pi = \pi'$.

Our next step is to move back from $G \times \overline{G}$ to the product space $G \times G$: By

$$Tf(x, \overline{y}) = \int_H f(x, yz)dz$$

we define the usual positive linear mapping T from $\mathscr{C}_0(G \times G)$ onto $\mathscr{C}_0(G \times \overline{G})$. The mapping T has a right inverse: Let β be a Bruhat function for H (see [32]), i.e. a nonnegative continuous function on G, such that for any compact $C \subset G$ the restriction $\beta|_{CH}$ has compact support and $T\beta = 1$. Then clearly βf, defined by $(\beta f)(x, y) = \beta(y)f(x, \overline{y})$, is in $\mathscr{C}_0(G\times G)$ for all $f \in \mathscr{C}_0(G\times\overline{G})$ and $T' : f \to \beta f$ is the desired right inverse.

We use T to define a Radon measure μ on $G \times G$: For $(x, y) \in G \times G$ we set

$$\theta(x, y) = (y^{-1}, y^{-1}x^{-1}), \quad \Delta(x, y) = \Delta_G(xy) ,$$

hence θ is a homeomorphism of $G \times G$ with $\theta^{-1}(x, y) = (y^{-1}x, x^{-1})$ and Δ is positive and continuous on $G \times G$. For $u \in \mathscr{C}_0(G \times G)$ we set

$$\int_{G\times G} u(x, y) d\mu(x, y) := \int_{G\times\overline{G}} T(\Delta^{-1} u \circ \theta)(x, \overline{y}) d\lambda(x, \overline{y}) .$$

It is evident that this formula defines a measure μ and that we also can write

$$(12) \qquad \int_{G\times\overline{G}} Tw(x, \overline{y}) d\lambda(x, \overline{y}) = \int_{G\times G} w(y^{-1}x, x^{-1}) \Delta_G(y)^{-1} d\mu(x, y)$$

for $w \in \mathscr{C}_0(G \times G)$.

Now we return to (11). One computes easily that

$$g^* \star f(x, \overline{y}) = \int_G f(tx, \overline{y}) \overline{g(t, x\overline{y})} dt ,$$

hence, with

$$w_t(x, y) = \beta(txy) f(tx, \overline{y}) \overline{g(t, x\overline{y})}$$

we defined $w_t \in \mathscr{C}_0(G \times G)$, for which $g^* \star f(x, \overline{y}) = \int_G Tw_t(x, \overline{y}) dt$. It follows from (12) that

(13)

$$\int_{G\times\overline{G}} Tw_t(x, \overline{y}) d\lambda(x, \overline{y}) = \int_{G\times G} \beta(ty^{-1}) f(ty^{-1}x, \overline{x^{-1}}) \overline{g(t, \overline{y^{-1}})} \Delta_G(y^{-1}) d\mu(x, y).$$

The integrand on the right hand side is a compactly supported continuous function in $(t, x, y) \in G \times G \times G = G^3$, hence it is integrable with respect to the product measure $d\mu(x, y)dt$ on G^3. We apply Fubini, integrate first over t, substitute $t \to ty$ and again change the order of integration. The result is

$$\int_G \beta(t) \left(\int_{G\times G} f(tx, \overline{x^{-1}}) \overline{g(ty, \overline{y^{-1}})} d\mu(x, y) \right) dt =: \int_G \beta(t) (f^\#(t) | g^\#(t))_\mu dt$$

where we have set

$$f^\#(t)(x) = f(tx, \overline{x^{-1}}) \text{ for } f \in \mathscr{C}_0(G \times \overline{G}) ,$$

hence $f^\#(t) \in \mathscr{C}_0(G)$ and $f^\# : t \to f^\#(t)$ is a continuous mapping from G into $\mathscr{C}_0(G)$. Furthermore,

$$(14) \qquad (u|v)_\mu := \int_{G\times G} u(x) \overline{v(y)} d\mu(x, y)$$

defines an inner product on $\mathscr{C}_0(G)$.

Similar arguments show that we can also integrate the left hand side of (13) over t. The result, using Fubini and (11), is the identity

$$(f|g)_\lambda = \int_G \beta(t)(f^\#(t)|g^\#(t))_\mu dt. \tag{15}$$

We claim now that (14) is positive semidefinite on $\mathscr{C}_0(G)$, hence it defines a Hilbert space $\mathfrak{L}$, and

$$(\sigma(z)u)(x) = \chi(z)^{-\frac{1}{2}}u(z^{-1}x), \quad z \in H\,,\ x \in G \tag{16}$$

defines a unitary representation σ of H in $\mathfrak{L}$. Furthermore, we will show that (δ, π) is induced from σ. This will finish the proof of theorem 4.

By definition of $d\mu$ and θ we see that for $z \in H$

$$(\sigma(z)u|\sigma(z)v)_\mu =$$
$$= \int_{G\times\overline{G}}\int_H \Delta_G(xyh)^{-1}\Delta_G(z)^{-1}\Delta_H(x)u((yhz)^{-1})v((xyhz)^{-1})dhd\lambda(x,\overline{y}).$$

Substituting hz^{-1} for h in the inner integral gives $(\sigma(z)u|\sigma(z)v)_\mu = (u|v)_\mu$.

Now we take $f \in \mathscr{C}_0(G\times\overline{G})$ and $q \in \mathscr{C}_0(\overline{G})$. Let $q \times f$ be the function $(x, \overline{y}) \to q(x\overline{y})f(x, \overline{y})$, hence $q \times f \in \mathscr{C}_0(G\times\overline{G})$ and $(q\times f)^\#(x)(y) = q(\overline{x})f^\#(x)(y)$. With $q \times f$ both for f and g in (15) we get

$$\int_G \beta(x)q(\overline{x})(f^\#(x)|f^\#(x))_\mu dx \geq 0 \tag{17}$$

for all $q \in \mathscr{C}_0(\overline{G})$. For $z \in H$ we have

$$f^\#(xz)(y) = f(xzy, \overline{y^{-1}}) = f(xzy, \overline{(zy)^{-1}}) = f^\#(x)(zy)\ ,$$

hence $f^\#(xz) = \chi(z)^{\frac{1}{2}}\sigma(z^{-1})(f^\#(x))$ and

$$(f^\#(xz)|f^\#(xz))_\mu = \chi(z)^{-1}(f^\#(x)|f^\#(x))_\mu. \tag{18}$$

As (17) implies $\int_H \beta(xz)(f^\#(xz)|f^\#(xz))_\mu dy \geq 0$ for all $x \in G$ (observe that $(f^\#(x)|f^\#(x))$ is continuous in x!), (17) and (18) yield

$$(f^\#(x)|f^\#(x))_\mu \geq 0 \text{ for all } f \in \mathscr{C}_0(G\times\overline{G}) \text{ and all } x \in G\ .$$

For $u \in \mathscr{C}_0(G)$ with supp $u = K$, we can find $v \in \mathscr{C}_0(\overline{G})$, such that $v(\overline{y}) = 1$ for all $y^{-1} \in K$. We set $f(x, \overline{y}) = u(x)v(\overline{y})$. Then $f^\#(e)(y) = f(y, \overline{y^{-1}}) = u(y)v(\overline{y^{-1}}) = u(y)$, i.e. $f^\#(e) = u$. It follows that $(u|u) \geq 0$, hence the inner product (14) is positive semidefinite and (16) defines a continuous unitary representation of H in $\mathfrak{L}$.

Let $\pi^\sigma = \mathrm{ind}_H^G\sigma$. The space $\mathfrak{H}^\sigma$ of π^σ consists of measurable functions $\varphi : G \to \mathfrak{L}$ with $\varphi(xz) = \sigma(z)^*\varphi(x)$ for $z \in H$ and almost all $x \in G$. The representation of $\mathscr{C}_\infty(\overline{G})$ in $\mathfrak{H}^\sigma$ will be denoted by δ^σ, i.e. $(\delta^\sigma(q)f)(x) = q(\overline{x})f(x)$ for $q \in \mathscr{C}_0(\overline{G})$, $f \in \mathfrak{H}^\sigma$. Now if $f \in \mathscr{C}_0(G \times \overline{G})$, then $f^\#(x) \in \mathfrak{L}$ for all $x \in G$, moreover we had seen that $f^\#(xz) = \chi(z)^{\frac{1}{2}}\sigma(z)^* f^\#(x)$ hence, the function $\Omega f : G \to \mathfrak{L}$, defined by

$$(\Omega f)(x) = \chi(x)^{\frac{1}{2}} f^\#(x)$$

is continuous from G into $\mathfrak{L}$ and satisfies the identity $(\Omega f)(xz) = \sigma(z)^*(\Omega f)(x)$ for $z \in H$. We remark that this is the first time where we used the assumption that $\chi = \frac{\Delta_G|_H}{\Delta_H}$ extends to a real character of g. It is easy to see how this has to be modified in the general case.

Now recall that, given the Bruhat function β for H, a χ relatively invariant measure $d\overline{x}$ on $\overline{G}$ is given by $\int_{\overline{G}} F(\overline{x})d\overline{x} = \int_G F(\overline{x})\beta(x)\chi(x)^{-1}dx$, see [32], ch. 8, § 1. Hence we derive from (15) for $f, g \in \mathscr{C}_0(G \times \overline{G})$:

$$\begin{aligned}\int_{\overline{G}}(\Omega f|\Omega g)_\mu(\overline{x})d\overline{x} &= \int_G (\Omega f(x)|\Omega g(x))_\mu \beta(x)\chi(x)^{-1}dx = \\ &= \int_G \beta(x)(f^\#(x)|g^\#(x))dx = (f|g)_\mu .\end{aligned}$$

This means that $\Omega : f \to \Omega f$ defines a unitary operator from $\mathfrak{H}$ into $\mathfrak{H}^\sigma$. To show that Ω is onto, we apply (9): the elements $u\#\varphi$ with u and φ in $\mathscr{C}_0(G)$ span a dense subspace in $\mathfrak{H}^\sigma$. On the other hand u and φ define a function $f \in \mathscr{C}_0(G \times \overline{G})$ by

$$f(x, \overline{y}) = \int_H u(xyz)\varphi(z^{-1}y^{-1})\chi(xyz)^{\frac{1}{2}}dz .$$

Hence $f \in \mathfrak{H}$ and

$$\begin{aligned}(\Omega f)(x, y) &= \chi(x)^{\frac{1}{2}} f^\#(x)(y) = \chi(x)^{\frac{1}{2}} f(xy, \overline{y^{-1}}) = \\ &= \chi(x)^{\frac{1}{2}} \int_H u(xz)\varphi(z^{-1}y)\chi(xz)^{\frac{1}{2}}dz = (u\#\varphi)(x, y) ,\end{aligned}$$

i.e. $\Omega f = u\#\varphi$. It follows that $\Omega\mathfrak{H}$ is dense in, hence equal to $\mathfrak{H}^\sigma$.

For $a \in G$ and $f \in \mathscr{C}_0(G \times \overline{G})$ we have

$$\begin{aligned}\Omega(\pi(a)f)(x)(y) &= \chi(x)^{\frac{1}{2}}(\pi(a)f)^\#(x, y) = \chi(x)^{\frac{1}{2}}(\pi(a)f)(xy, \overline{y^{-1}}) = \\ &= \chi(a)^{\frac{1}{2}}\chi(a^{-1}x)^{\frac{1}{2}} f(a^{-1}xy, \overline{y^{-1}}) = \chi(a)^{\frac{1}{2}}(\Omega f)(a^{-1}x)(y) ,\end{aligned}$$

thus, after (7):

$$\Omega\pi(a) = (\mathrm{ind}_H^G\sigma)(a)\Omega .$$

Similarly we see for $q \in \mathscr{C}_\infty(\overline{G})$:

$$(\Omega\delta^\sigma(q)f)(x)(y) = q(x\overline{y})(\Omega f)(x)(y) \ ,$$

which means $\Omega\delta^\sigma(q) = \delta(q)\Omega$. Hence (δ, π) and $(\delta^\sigma, \pi^\sigma)$ are unitarily equivalent and theorem 4 is proved.

Given two equivalent representations σ_1 and σ_2 of H it is clear that also $(\delta^{\sigma_1}, \pi^{\sigma_1})$ and $(\delta^{\sigma_2}, \pi^{\sigma_2})$ are equivalent. As a matter of fact a much stronger result is true. The spaces $\mathfrak{H}$ of impimitivity systems (δ, π) for (G, H) are by definition G and $\mathscr{C}_\infty(\overline{G})$ modules, hence it is natural to define:

Definition. Let (δ_1, π_1) and (δ_2, π_2) be systems of imprimitivity for G, H and $\mathfrak{H}_1$, $\mathfrak{H}_2$ the corresponding Hilbert spaces. An *intertwining operator* of the two systems is a bounded operator $T \in \mathscr{L}(\mathfrak{H}_1, \mathfrak{H}_2)$ with $T\pi_1(x) = \pi_2(x)T$, $T\delta_1(f) = \delta_2(f)T$ for all $x \in G$, $f \in \mathscr{C}_\infty(\overline{G})$. The space of all these intertwining operators is denoted by

$$\mathscr{L}_H(\pi_1, \pi_2) \ .$$

If $H = G$, then $\mathscr{L}_G(\pi_1, \pi_2) =: \mathscr{L}(\pi_1, \pi_2)$ is just the space of ordinary intertwining operators of π_1 and π_2. If (δ_3, π_3) is a third system of imprimitivity, then there is a canonical bilinear product $\mathscr{L}_H(\pi_2, \pi_3) \times \mathscr{L}_H(\pi_1, \pi_2) \to \mathscr{L}_H(\pi_1, \pi_3)$, in particular

$$\mathscr{L}_H(\pi) := \mathscr{L}_H(\pi, \pi)$$

is a von Neumann subalgebra of $\mathscr{L}(\mathfrak{H}(\pi))$.

We consider again the representations σ_1 and σ_2 of H and the systems $(\delta^{\sigma_1}, \pi^{\sigma_1})$, $(\delta^{\sigma_2}, \pi^{\sigma_2})$, but now we denote the space of σ_i by $\mathfrak{H}_i$ and the space of $\pi^{\sigma_i} = \operatorname{ind}_H^G \sigma_i$ by $\operatorname{ind} \mathfrak{H}_i$. Let $\{T(\overline{x})\}_{\overline{x} \in \overline{G}}$ be a uniformly bounded measurable family of operators in $\mathscr{L}(\sigma_1, \sigma_2)$, i.e. $\|T\| := \sup_{\overline{x}} |T(\overline{x})| < \infty$ and $\overline{x} \to T(\overline{x})\xi$ is measurable from $\mathfrak{H}_1$ into $\mathfrak{H}_2$ for very $\xi \in \mathfrak{H}_1$. Then the mapping $\widetilde{T}$, defined by

$$\widetilde{T} : f \to Tf \quad \text{with} \quad (Tf)(x) = T(\overline{x})f(x)$$

is a bounded operator from $\operatorname{ind} \mathfrak{H}_1$ into $\operatorname{ind} \mathfrak{H}_2$ which commutes with the δ^{σ_i}.

Now any $A \in \mathscr{L}(\sigma_1, \sigma_2)$ defines the constant family $A(\overline{x}) = A$ for all $\overline{x} \in \overline{G}$ and in this case $\widetilde{A}$ obviously commutes also with the π^{σ_i}, hence $\widetilde{A} \in \mathscr{L}_H(\pi^{\sigma_1}, \pi^{\sigma_2})$.

Theorem 5. *The mapping $A \to \widetilde{A}$ is a linear isomorphism from $\mathscr{L}(\sigma_1, \sigma_2)$ onto $\mathscr{L}_H(\pi^{\sigma_1}, \pi^{\sigma_2})$. If σ_3 is a third representation of H, $A \in \mathscr{L}(\sigma_2, \sigma_3)$, $B \in \mathscr{L}(\sigma_1, \sigma_2)$, hence $AB \in \mathscr{L}(\sigma_1, \sigma_3)$, then $\widetilde{AB} = \widetilde{A}\widetilde{B}$; in particular if $\sigma_1 = \sigma_2 = \sigma$, then $A \to \widetilde{A}$ is an isomorphism of the von Neumann algebra $\mathscr{L}(\sigma)$ onto the von Neumann algebra $\mathscr{L}_H(\pi^\sigma)$.*

Proof. The only nontrivial fact is the surjectivity of the mapping $A \to \widetilde{A}$. To prove this we consider the mappings Θ_u^i from $\mathfrak{H}_i$ into ind $\mathfrak{H}_i$, which for $u \in \mathscr{C}_0(G)$ we had defined in (9) by

$$\Theta_u^i : \xi \to u\#\xi \in \text{ind } \mathfrak{H}_i .$$

Θ_u^i is a linear operator from $\mathfrak{H}_i$ into ind $\mathfrak{H}_i$, bounded by $(\int_{\overline{G}} |\overline{u}(\overline{x})|^2 d\overline{x})^{\frac{1}{2}}$, $\overline{u}(\overline{x}) = \int_H u(xy)dy$. It follows that the adjoint Θ_u^{i*} maps ind $\mathfrak{H}_i$ into $\mathfrak{H}_i$.

An easy computation shows that

$$\Theta_u^{i*} f = \int_G \overline{u(x)}\chi(x)^{-1} f(x)dx$$

thus for $u, v \in \mathscr{C}_0(G)$:

$$(\Theta_u^i \Theta_v^{i*} f)(x) = \int_H \int_G u(xz)\overline{v(y)}\chi(y)^{-1} f(yz^{-1})dydz .$$

In the inner integral we substitute yxz for y. Observing that $f(yx) = \chi(y)^{\frac{1}{2}}(\text{ind}_H^G \sigma^i(y^{-1})f)(x)$ and after changing the order of integration we obtain

$$\begin{aligned} \Theta_u^i \Theta_v^{i*} f(x) &= \int_G \left(\int_H u(xz)\overline{v(yxz)}(\chi^{-1}\Delta_G(xz)dz \right) \chi(y)^{-\frac{1}{2}} \left(\text{ind}_H^G \sigma^i(y^{-1})f\right)(x)dy \\ &= \int_G W(\overline{x}, y) \left(\text{ind}_H^G \sigma^i(y^{-1})f\right)(x)dy , \end{aligned}$$

where

$$W(\overline{x}, y) = \int_H u(xz)\overline{v(yxz)}(\chi^{-1}\Delta_G)(xz)dz\chi(y)^{-\frac{1}{2}}$$

clearly defines a function $W \in \mathscr{C}_0(\overline{G} \times G)$. So formally we can write

$$\Theta_{u,v}^i := \Theta_u^i \Theta_v^{i*} = \int_G W(\overline{x}, y)\pi^{\sigma_i}(y^{-1})dy .$$

Now let S be in $\mathscr{L}_H(\pi^{\sigma_1}, \pi^{\sigma_2})$. Since S commutes with the π^σ's and the δ's, we see that

$$S\Theta_{u,v}^1 = \Theta_{u,v}^2 S$$

for all $u, v \in \mathscr{C}_0(G)$. The elements $\Theta_u^{i*} f$ with $u \in \mathscr{C}_0(G)$, $f \in$ ind $\mathfrak{H}_i$ span a dense subspace $\mathfrak{H}_i^0$ of $\mathfrak{H}_i$. Let $\alpha = \sum_{j=1}^m \Theta_{u_j}^{1*} f_j \in \mathfrak{H}_1^0$. For an operator X : ind $\mathfrak{H}_1 \to$ ind $\mathfrak{H}_i$, $i = 1, 2$, we set

$$\alpha_X = \sum_{j=1}^m \Theta_{u_j}^{i*} X f_j ,$$

hence $\alpha_X \in \mathfrak{H}_i$. Now we fix S in $\mathcal{L}_H(\pi^{\sigma_1}, \pi^{\sigma_2})$. Then

$$|\alpha_S|^2 = \sum_{j,k}(\Theta^{2*}_{u_j} Sf_j | \Theta^{2*}_{u_k} Sf_k) = \sum_{j,k}(\Theta^1_{u_k,u_j} S^*Sf_j | f_k) \ ,$$

since S^*S commutes with the $\Theta^1_{u,v}$. Clearly $R := S^*S$ is a positive operator in $\mathcal{L}_H(\pi^{s_1})$ and we can assume that $0 \leq R \leq \mathbf{1}$. Then also $T := (\mathbf{1} - R)^{\frac{1}{2}} \in \mathcal{L}_H(\pi^{\sigma_1})$ and $\mathbf{1} = S^*S + T^2$, hence $|\alpha_T|^2 = \sum_{j,k}(\Theta^1_{u_k,u_j} T^2 f_j | f_k)$ and consequently

$$|\alpha|^2 = |\alpha_S|^2 + |\alpha_T^2| \geq |\alpha_S|^2 \ ,$$

in particular $\alpha_S = 0$ if $\alpha = 0$. It follows that

$$A : \alpha \to \alpha_S$$

defines an operator $A \in \mathcal{L}(\sigma_1, \sigma_2)$, bounded by 1.

By definition of A the identity $(A\Theta^{1*}_{u\chi} f | \xi) = (\Theta^{2*}_{u\chi} Sf | \xi)$ holds for all $u \in \mathcal{C}_0(G)$, $\xi \in \mathfrak{H}_2$. This means

$$\int \overline{u(x)}(A(f(x))|\xi)dx = \int \overline{u(x)}((Sf)(x)|\xi)dx$$

and implies $Sf(x) = A(f(x))$ for all $x \in G$ if f is continuous. For $y \in H$ we get

$$\sigma_2(y)A(f(x)) = \sigma_2(y)(Sf)(x) = Sf(xz^{-1}) = A(f(xy^{-1}) = A\sigma_1(y)(f(x))$$

hence $A \in \mathcal{L}(\sigma_1, \sigma_2)$ and $S = \widetilde{A}$.

Theorem 5 has the following important

Corollary. *The systems of imprimitivity $(\delta^{\sigma_1}, \pi^{\sigma_1})$ and $(\delta^{\sigma_2}, \pi^{\sigma_2})$ are unitarily equivalent if and only if σ_1 and σ_2 are unitarily equivalent.*

For the proof of our next theorem it is convenient to realize the induced representation $\operatorname{ind}_H^G \sigma$ in the Hilbert space $L^2(\overline{G}, \mathfrak{H})$, with $\mathfrak{H} = \mathfrak{H}(\sigma)$. Basically this is done as in the case of semidirect products, see proposition (3). In that case $\overline{G}$ can be identified with the subgroup R and $\overline{x} = (r, y)H \to r$ is a continuous section from $\overline{G}$ into $G = R \ltimes H$. In general such sections do not exist, however, E.T. Kehlet proved in [16]:

For every locally compact group G and closed subgroup H there exists a locally bounded Baire cross section from $\overline{G} = G/H$ into G, i.e. a mapping $s : \overline{G} \to G$ with the following properties: (i) $s(\overline{x}) \in \overline{x}$, i.e. $x^{-1}s(\overline{x}) \in H$ for all $x \in G$, with $\overline{x} = xH \in \overline{G}$. (ii) If $\overline{K}$ is compact in $\overline{G}$, then $s(\overline{K})$ has compact closure in G. (iii) If $B \subset G$ is a Baire set, then its inverse image $s^{-1}(B)$ is a Baire set in $\overline{G}$.

As $x \to \overline{x}$ is continuous and surjective, it follows that $x \to s(\overline{x})^{-1}$, hence

$$h : x \to h(x) := s(\overline{x})^{-1}x \in H$$

is Baire. Consequently

$$S : (\overline{x}, y) \to s(\overline{x})y \in G$$

is a bijective Baire mapping with Baire inverse

$$S^{-1} : z \to (s(\overline{x}), h(x)) \in \overline{G} \times H \ ,$$

hence G and $\overline{G} \times H$ are Baire isomorphic via S. Now suppose that H is normal. Then $d\overline{x}$ is the Haar measure on $\overline{G}$ and we may assume that $\chi = 1$, thus $\int\limits_G f(x)dx = \int\limits_{\overline{G}}\int\limits_H f(xy)dyd\overline{x}$ for all $f \in L^1(G)$. As $\int\limits_H f(xy)dy = \int\limits_H f(s(\overline{x})y)dy$ we can write

$$\int_G f(x)dx = \int_{\overline{G}}\int_H f(S(\overline{x}, y))dyd\overline{x} \ .$$

Now let σ be a unitary representation of the *normal* closed subgroup H in the Hilbert space $\mathfrak{H}$. If $f \in \operatorname{ind} \mathfrak{H}$ is continuous in $\overline{x}$, then $Uf : \overline{x} \to f(s(\overline{x}))$ is a Baire, hence $d\overline{x}$ measurable function with $|Uf(\overline{x})| = |f(x)| = |f|(\overline{x})$, which shows that $U : f \to Uf$ is an isometric linear operator from $\operatorname{ind} \mathfrak{H}$ into $L^2(\overline{G}, \mathfrak{H})$. On the other hand, if $g \in L^2(\overline{G}, \mathfrak{H})$ is continuous on $\overline{G}$, then $f : x \to \sigma(h(x))^* g(\overline{x})$ is measurable on G and the identity $f(xy) = \sigma(y)^* f(x)$ holds for all $x \in G$, $y \in H$. Evidently $Uf = g$ and consequently U is a unitary operator from $\operatorname{ind} \mathfrak{H}$ onto $L^2(\overline{G}, \mathfrak{H})$.

Let $a \in G$, $\overline{x} \in \overline{G}$ and $g \in L^2(\overline{G}, \mathfrak{H})$. We set

$$\eta(a, \overline{x}) = s(\overline{x})^{-1}as(a^{-1}\overline{x}) \in H$$

and define $\tilde{\pi}(a)g \in L^2(\overline{G}, \mathfrak{H})$ by

$$(\tilde{\pi}(a)g)\,(\overline{x}) = \sigma(\eta(a, \overline{x}))g(a^{-1}\overline{x}) \ . \tag{19}$$

Then a straightforward calculation shows that for every $f \in \operatorname{ind} \mathfrak{H}$ the identity

$$(U\pi(a)f)(\overline{x}) = (\tilde{\pi}(a)Uf)(\overline{x})$$

holds. Hence

$$\tilde{\pi}(a) = U\pi(a)U^* \ .$$

This means:

(20) *The formula* (19) *defines a unitary representation* $\tilde{\pi}$ *of* G *in* $L^2(\overline{G}, \mathfrak{H})$, *which is equivalent to* $\mathrm{ind}_H^G \sigma$.

For $y \in H$ clearly $y^{-1}\overline{x} = \overline{x}$ for all $\overline{x} \in \overline{G}$, hence $\eta(y, \overline{x}) = s(\overline{x})^{-1} y s(\overline{x}) =: y^{\overline{x}}$ and

$$(21) \qquad (\tilde{\pi}(y)g)(\overline{x}) = \sigma\left(y^{\overline{x}}\right) g(\overline{x}) ,$$

which means that $\tilde{\pi}|_H$ is the direct integral of the conjugate representations ${}^{\overline{x}}\sigma$: $y \to \sigma\left(y^{\overline{x}}\right)$. Observe that ${}^{\overline{x}}\sigma$ depends on the section s, however for fixed $\overline{x} \in \overline{G}$ different sections clearly give equivalent representations, in particular

$$\mathfrak{k}_{\overline{x}} := \ker_{C^*(H)} {}^{\overline{x}}\sigma \subset C^*(H)$$

depends only on $\overline{x}$.

Theorem 6. *Let* H *be a closed normal subgroup of* G *and* σ *an irreducible unitary representation of* H *in the Hilbert space* $\mathfrak{H}$. *If* $\mathfrak{k}_{\overline{x}} \neq \mathfrak{k}_{\overline{y}}$ *for* $\overline{x}, \overline{y} \in \overline{G}$, $\overline{x} \neq \overline{y}$, *then* $\mathrm{ind}_H^G \sigma$ *is irreducible.*

Proof (see [19], Satz 7). We realize $\pi := \mathrm{ind}_H^G \sigma$ in $\mathfrak{H}_\pi := L^2(\overline{G}, \mathfrak{H})$. As before the multiplications with functions in $L^\infty(\overline{G})$ form a commutative von Neumann algebra $\mathfrak{Q}$ in $\mathcal{L}(\mathfrak{H}_\pi)$. Our first step is to prove that the commutant $\pi(H)'$ of the image $\pi(H)$ of H in $\mathcal{L}(\mathfrak{H}_\pi)$ is contained in the commutant $\mathfrak{Q}'$ of $\mathfrak{Q}$:

$$(22) \qquad \pi(H)' \subset \mathfrak{Q}'$$

or, equivalently: $T \notin \mathfrak{Q}'$ implies $T \notin \pi(H)'$. We identify functions $\Phi \in L^\infty(\overline{G})$ with their corresponding operators, in particular we consider the characteristic function χ_A of a measurable set $A \subset \overline{G}$ also as an element in $\mathfrak{Q}$. As these χ_A with relatively compact open A generate a weakly dense subalgebra of $\mathfrak{Q}$, there exists such an A_0 with $\chi_{A_0} T \neq T \chi_{A_0}$, A_0 not dense in $\overline{G}$. Hence $\chi_{A_0} T \chi_{A_0'} \neq \chi_{A_0'} T \chi_{A_0}$ for the complement $A_0' = \complement A_0$. If e.g. $\chi_{A_0} T \chi_{A_0'} \neq 0$ there exists another open $B_0 \subset A_0'$ with compact closure B, disjoint from the closure A of A_0, with $\chi_A T \chi_B \neq 0$. As $\mathfrak{Q} \subset \pi(H)'$, we may assume that

$$T = \chi_A T \chi_B \neq 0 ,$$

and moreover that $\chi_{A_1} T \chi_{B_1} \neq 0$ for all relatively open non empty subsets $A_1 \subset A$ and $B_1 \subset B$.

For $p \in A$ and $q \in B$ our assumption yields $\mathfrak{k}_p \neq \mathfrak{k}_q$. Let us assume that $\mathfrak{k}_p \not\subset \mathfrak{k}_q$. Then there exists a positive $a \in C^*(H)$ with $|{}^p\sigma(a)| > 1$ and ${}^q\sigma(a) = 0$. As $\overline{x} \to {}^{\overline{x}}\sigma(a)$ is continuous, $|{}^{\overline{x}}\sigma(a)| > 1$, resp. $|{}^{\overline{x}}\sigma(a)| < \frac{1}{2}$ describe neighborhoods of p, resp. q, hence we can assume that

$$|{}^{\overline{x}}\sigma(a)| > 1 \qquad \text{for all } \overline{x} \in A$$
$$|{}^{\overline{x}}\sigma(a)| < \frac{1}{2} \qquad \text{for all } \overline{x} \in B\ .$$

Let $\mathcal{A}$ be the C^* subalgebra of $C^*(H)$, generated by a. Then $\mathcal{A} \cong \mathcal{C}_\infty(Z)$ with $Z = \widehat{\mathcal{A}}$ the spectrum of $\mathcal{A}$. For $\overline{x} \in \overline{G}$ let $Z_{\overline{x}}$ be the hull in Z of the kernel of the restriction of ${}^{\overline{x}}\sigma$ onto $\mathcal{A}$. If $\hat{a} \in \mathcal{C}_\infty(Z)$ is the Gelfand transform of a, then all $Z_{\overline{x}}$, $\overline{x} \in B$, are contained in the closed set $Y = \{\zeta \in Z; \hat{a}(\zeta) \leq \frac{1}{2}\}$. On the other hand, every $Z_{\overline{x}}$ for $\overline{x} \in A$ meets the compact set $X = \{\zeta \in Z; \hat{a}(\zeta) \geq 1\}$. We choose $h \in \mathcal{C}_\infty(Z)$ with $0 \leq h \leq 1$, $h(\zeta) = 1$ for $\zeta \in X$, $h(\zeta) = 0$ for $\zeta \in Y$. Let b be the element in $\mathcal{A}$ with $\hat{b} = h$ and $c = ab$. Then obviously ${}^{\overline{x}}\sigma(c) = {}^{\overline{x}}\sigma(a) \neq 0$ for $\overline{x} \in A$, ${}^{\overline{x}}\sigma(c) = 0$ for $\overline{x} \in B$, hence

$$c \in \mathfrak{k} := \bigcap_{\overline{x} \in B} \mathfrak{k}_{\overline{x}}, \qquad {}^{\overline{x}}\sigma(c) \neq 0 \text{ for } \overline{x} \in A\ .$$

From (21) it follows that $(\pi(S)g)(\overline{x}) = {}^{\overline{x}}\sigma(S)g(\overline{x})$ for all $S \in C^*(H)$, which implies $\pi(\mathfrak{k})\chi_B\mathfrak{H}_\pi = 0$, i.e. $\chi_B\pi(\mathfrak{k}) = \pi(\mathfrak{k})\chi_B = 0$. On the other hand ${}^{\overline{x}}\sigma$ acts irreducibly on $\mathfrak{H}$ and is nontrivial on $\mathfrak{k}$ for all $\overline{x} \in A$. Hence ${}^{\overline{x}}\sigma(\mathfrak{k})\mathfrak{H} = \mathfrak{H}$ for $\overline{x} \in A$ and consequently $\chi_A\pi(\mathfrak{k})\mathfrak{H}_\pi = \pi(\mathfrak{k})\chi_A\mathfrak{H}_\pi = \chi_A\mathfrak{H}_\pi$. If T, hence also T^* would commute with $\pi(H)$, we would have $TT^*\pi(\mathfrak{k}) = T\pi(\mathfrak{k})T^* = T\chi_B\pi(\mathfrak{k})T^* = 0$. But $TT^* = TT^*\chi_A$ implies $TT^*\mathfrak{H}_\pi = TT^*\chi_A\mathfrak{H}_\pi = TT^*\pi(\mathfrak{k})\chi_A\mathfrak{H}_\pi = 0$, which contradicts $T \neq 0$. It follows that $T \notin \pi(H)'$, so (22) is proved.

Next suppose $S \in \pi(H)'$. We claim that then $S \in \mathfrak{Q}$. It is easy to see that if suffices to show $\chi_K S \in \mathfrak{Q}$ for all compact sets K in $\overline{G}$. So assume that $S = \chi_K S$ for some compact $K \subset \overline{G}$, hence also $S = S\chi_K$, because S commutes with $\mathfrak{Q}$.

The subspace $\chi_K\mathfrak{H}_\pi = L^2(K, \mathfrak{H})$ contains the constant functions $\overline{x} \to \xi \in \mathfrak{H}$, which means essentially, that we can consider $\mathfrak{H}$ as a closed subspace of $\chi_K\mathfrak{H}_\pi$. Let us assume that there exists an element $\theta \in \mathfrak{H}$ such that $\varphi = S\theta$ is *not* of the form $f \otimes \theta$ with $f \in L^2(K)$ and $\xi \in \mathfrak{H}$. This means that there exists a measurable subset $M \subset K$ with positive measure $|M|$, such that $\varphi(\overline{x}) \notin \mathbb{C}\theta$ for $\overline{x} \in M$. Because φ is measurable, M contains a compact subset C, still with $|C| > 0$, such that the restriction $\varphi|_C$ is continuous, moreover we can assume that every non empty relatively open subset in C also has positive measure. We fix an element $p \in C$. Then $\varphi(p)$ and θ are linearly independent, hence, because ${}^p\sigma$ acts irreducibly on $\mathfrak{H}$, it follows from Kadison's transitivity theorem that there exists $z \in C^*(H)$ with ${}^p\sigma(z)\theta = 0$ and $|{}^p\sigma(z)\varphi(p)| > 1$. Then

$$W = \{w \in C\,;\ |{}^w\sigma(z)\theta| < \frac{1}{2|S|}\,,\ |{}^w\sigma(z)\varphi(w)| > 1\}$$

is a neighborhood of p in C with $w := |W| > 0$. It follows that

$$0 < w \leq \int_W |^{\overline{x}}\sigma(z)\varphi(\overline{x})|^2 d\overline{x} = |\chi_W \pi(z) S\theta|^2 = |S\chi_W \pi(z)\theta|^2 \leq$$
$$\leq |S|^2 \int_W |^{\overline{x}}\sigma(z)\theta|^2 d\overline{x} \leq |S|^2 w \frac{1}{4|S|^2} = \frac{1}{4} w \ ,$$

which is impossible. So necessarily $S\theta(\overline{x}) \in \mathbb{C}\theta$ for $d\overline{x}$ almost all $\overline{x} \in \overline{G}$, which means $S\theta = h \otimes \theta$, $f \in L^2(\overline{G})$. If $\xi \in \mathfrak{H}$ is independent of θ and $S\xi = g \otimes \xi$, then $S(\theta + \xi) = f \otimes (\theta + \xi)$ for some $h \in L^2(\overline{G})$, hence $f = h = g$, i.e. $S\xi = f \otimes \xi$ for *all* $\xi \in \mathfrak{H}$. For $u \in L^\infty(\overline{G})$ we get $Su\xi = uS\xi = (uf) \otimes \xi$, hence $|uf|_2|\xi| = |Su\xi| \leq |S|\,|u\xi| = |S|\,|u|_2\,|\xi|$, which implies $|f|_\infty \leq |S|$, $f \in L^\infty$, and consequently – we omit the details – $S = f \in \mathfrak{Q}$.

Finally let the bounded operator R commute with $\pi(G)$. Then $R \in \pi(H)'$, hence $R = r \in \mathfrak{Q}$, i.e. $Rf(\overline{x}) = r(\overline{x})f(\overline{x})$ and for $a \in G$ (18) gives

$$(\pi(a)Rf)(\overline{x}) = \sigma(\eta(a,\overline{x}))r(a^{-1}\overline{x})f(a^{-1}\overline{x}) =$$
$$= (R\pi(a)f)(\overline{x}) = r(\overline{x})\sigma(\eta(a,\overline{x}))f(a^{-1}\overline{x})$$

which means $r(\overline{x}) = r(a^{-1}\overline{x})$ for all $a \in G$ and, depending on a, almost all $\overline{x} \in \overline{G}$. Because G acts transitively on $\overline{G}$, the usual arguments show that r has to be constant, i.e. $R = r\mathbf{1}$, $r \in \mathbb{C}$. This proves theorem 6.

Our next result gives a sufficient condition that an irreducible representation of the group G is induced from a proper subgroup.

Suppose that G contains a closed normal commutative subgroup A with character group $\widehat{A}$. We write $\langle x, \xi \rangle$ for the canonical pairing on $A \times \widehat{A}$, hence $\xi \in \widehat{A}$ is the character $x \to \langle x, \xi \rangle$. For $x \in A$, $z \in G$ we write as usually $x^z = z^{-1}xz$ and define ${}^z\xi \in \widehat{A}$ by $\langle x, {}^z\xi \rangle = \langle x^z, \xi \rangle$. Thus $\widehat{A}$ is a locally compact G space with action $(z, \xi) \to {}^z\xi$. Let $\tilde{\xi}$ denote the closure of the orbit ${}^G\xi$ in $\widehat{A}$ and $\mathfrak{B} = \mathfrak{B}(A)$ the set of all $\tilde{\xi}$, $\xi \in \widehat{A}$. We provide $\mathfrak{B}$ with the finest topology for which the surjective mapping $\xi \to \tilde{\xi}$ is continuous and claim: The mapping $\xi \to \tilde{\xi}$ is also open: Let $U \subset \widehat{A}$ be open. As $\widetilde{{}^z\xi} = \tilde{\xi}$ we may assume that U is G–invariant. Then the closed complement $U' = \complement U$ is also G invariant, hence $\vartheta \notin U$ implies $\tilde{\vartheta} \subset U'$ and consequently $\tilde{\xi} \neq \tilde{\vartheta}$ for every $\xi \in U$. This means that U is the inverse image of its image $\widetilde{U}$ in $\mathfrak{B}$ and implies that $\widetilde{U}$ is open.

Now we will also assume that A is second countable, i.e. A, hence $\widehat{A}$ has a countable basis of open sets. Then also $\mathfrak{B}$ is second countable, moreover $\mathfrak{B}$ is a T_0–space: $\tilde{\vartheta} \in \mathfrak{B}$ is in the closed hull of $\tilde{\xi}$ if and only if $\vartheta \in \tilde{\xi}$, thus $\tilde{\vartheta}$ and $\tilde{\xi}$ have the same hull in $\mathfrak{B}$ if and only if $\tilde{\vartheta} = \tilde{\xi}$.

Let π be a unitary representation of G and let π_A be its restriction to A. Then π_A defines a representation of $C^*(A)$ and as $C^*(A)$ is isomorphic to $\mathscr{C}_\infty(\widehat{A})$ we get a representation ρ of $\mathscr{C}_\infty(\widehat{A})$ in $\mathfrak{H}$ with $\rho(\hat{f}) = \pi_A(f)$ for the Gelfand transform $\hat{f}$ of the element $f \in C^*(A)$. G acts on $L^1(A) \subset C^*(A)$ by $f \to f^z$ with $f^z(y) = \Delta_A(z)^{-1} f(zyz^{-1})$ where Δ_A denotes the modulus of the automorphism $y \to y^z$ of A. It follows immediately that $\pi_A(f^z) = \pi(z)^* \pi_A(f) \pi(z)$ for all $f \in L^1(A)$ and consequently for all $f \in C^*(A)$. As the transpose of $f \to f^z$ on $\mathscr{C}_\infty(\widehat{A})$ is given by $g \to g^z$ with $g^z(\xi) = g({}^z\xi)$, we find that

$$\rho(g^z) = \pi(z)^* \rho(g) \pi(z) \tag{23}$$

for all $g \in \mathscr{C}_\infty(\widehat{A})$, $z \in G$.

Let ω_π be the hull of the kernel of ρ in $\widehat{A}$; hence ω_π is G invariant and the smallest closed subset of $\widehat{A}$ such that ρ factors over the quotient algebra $\mathscr{C}_\infty(\widehat{A})/k(\omega_\pi) \cong \mathscr{C}_\infty(\omega_\pi)$, with $k(\omega_\pi)$ the ideal of all $f \in \mathscr{C}_\infty(\widehat{A})$ vanishing on ω_π. The following fact is basic:

(24) *If π is irreducible then $\omega_\pi \in \mathfrak{B}$, i.e. there exists some $\chi \in \widehat{A}$, such that the orbit ${}^G\chi$ is dense in ω_π.*

Remark. This is a special case of a result of Guichardet, see [13].

Proof. By Stone's theorem there exists a projection valued Borel measure P on $\widehat{A}$ such that $\rho(g) = \int_{\widehat{A}} g(\xi) dP(\xi)$ for $g \in \mathscr{C}_\infty(\widehat{A})$. For a Borel set $X \subset \widehat{A}$ the formula (23) implies $\pi(z)^* P(X) \pi(z) = P({}^{z^{-1}}X)$ for all $z \in G$, with ${}^{z^{-1}}X = \{{}^{z^{-1}}\xi ; \xi \in X\}$. It follows that for a G invariant Borel set X the projection $P(X)$ commutes with π, hence in this case $P(X) = 0$ or $\mathbf{1}$. Now the mapping $\xi \to \tilde{\xi}$ is continuous and open from $\widehat{A}$ onto $\mathfrak{B}$, so P defines also a projection valued measure on $\mathfrak{B}$, denoted by $\widetilde{P}$: If X is the inverse image of $\widetilde{X} \subset \mathfrak{B}$, then $\widetilde{P}(\widetilde{X}) = P(X)$. Evidently these inverse images are all G invariant, hence $\widetilde{P}(Z) = 0$ or $\mathbf{1}$ for all Borel sets $Z \subset \mathfrak{B}$. This fact together with the identity $\widetilde{P}(U \cap V) + \widetilde{P}(U \cup V) = \widetilde{P}(U) + \widetilde{P}(V)$ implies that $\widetilde{P}(U)\widetilde{P}(V) = \widetilde{P}(U \cap V)$ for all Borel sets $U, V \subset \mathfrak{B}$.

Now let $\{U_j\}$ be a countable basis of the open sets in $\mathfrak{B}$. We set

$$S_j = U_j \ \text{ if } \ \widetilde{P}(U_j) = \mathbf{1}, S_j = \complement U_j \quad \text{if } \ \widetilde{P}(U_j) = 0 \ .$$

It follows that $\widetilde{P}(S_1 \cap S_2 \cap \cdots \cap S_k) = \prod_1^k \widetilde{P}(S_j) = \mathbf{1}$ for all $k \in \mathbb{N}$, hence $\widetilde{P}(S) = \mathbf{1}$ for $S = \bigcap_{k=1}^{\infty} S_k$ and consequently $S \neq \emptyset$. Let $\tilde{\chi} \in S$.

If $\tilde{\xi} \in \mathfrak{B}$ and $\tilde{\xi} \neq \tilde{\chi}$, then, since $\mathfrak{B}$ is T_0, there exists U_j either with $\tilde{\chi} \in U_j$, $\tilde{\xi} \notin U_j$ or with $\tilde{\xi} \in U_j$, $\tilde{\chi} \notin U_j$. This implies $\tilde{\chi} \in S_j$, $\tilde{\xi} \notin S_j$, hence $\tilde{\xi} \notin S$, which proves that $S = \{\tilde{\chi}\}$. For our measure P this means that $P(\tilde{\chi}) = \mathbf{1}$, thus $\rho(g) = \int_{\tilde{\chi}} g(\xi) dP(\xi)$ and $k(\omega_\pi)$ contains the kernel $k(\tilde{\chi})$ of the orbit closure $\overline{{}^G\chi} = \tilde{\chi}$. It follows that $\omega_\pi \subset \tilde{\chi}$. On the other hand ω_π is closed and G–invariant, hence its image $\tilde{\omega}$ in $\mathfrak{B}$ is closed and $\widetilde{P}(\tilde{\omega}) = P(\omega_\pi) = \mathbf{1}$, which implies $\tilde{\chi} \in \tilde{\omega}$ consequently $\chi \in \omega_\pi$, $\tilde{\chi} \subset \omega_\pi$ and finally $\tilde{\chi} = \omega_\pi$. So the orbit ${}^G\chi$ is dense in ω_π.

Our next assumption is most restrictive, however in the cases which interest us it will always be satisfied: As above we assume that π is irreducible and A second countable. Then, by (24), there exists $\chi_0 \in \widehat{A}$, such that the orbit $\Gamma := {}^G\chi_0$ is dense in the hull $\omega_\pi \subset \widehat{A}$. Let

$$F := \{x \in G; {}^x\chi_0 = \chi_0\}$$

be the closed stabilizer of χ_0 in G. Then we suppose:

(25) *Γ is locally closed in $\widehat{A}$, i.e. open in its closure ω_π, and $x \to {}^x\chi_0 \in \Gamma$ defines a homeomorphism from the coset space $\overline{G} = G/F$ onto Γ.*

We remark that the mapping $\overline{G} \to \Gamma$ is always continuous, thus a Borel isomorphism in case that G e.g. is σ–compact. But because in our applications the bicontinuity will be obvious we content ourselves with the stronger assumption. As we had seen, the representation π determines the representation ρ of $\mathscr{C}_\infty(\widehat{A})$ and in turn – we use the same letter ρ – on $\mathscr{C}_\infty(\omega_\pi)$. Now Γ is open in ω_π, hence $\mathscr{C}_\infty(\Gamma)$ is contained in $\mathscr{C}_\infty(\omega_\pi)$ and we can restrict ρ onto $\mathscr{C}_\infty(\Gamma)$. Finally the homeomorphism $\eta : \overline{x} = xF \to {}^x\chi_0$ from $\overline{G}$ onto Γ gives an isomorphism $\eta^* : \mathscr{C}_\infty(\Gamma) \to \mathscr{C}_\infty(\overline{G})$ and a representation $\Delta = \rho \circ \eta^{*-1}$ of $\mathscr{C}_\infty(\overline{G})$ in $\mathfrak{H}$ with $\Delta(f \circ \eta) = \rho(f)$ for $f \circ \eta \in \mathscr{C}_\infty(\overline{G})$. Because G acts on $\mathscr{C}_\infty(\overline{G})$ by $g^a(\overline{x}) = g(a\overline{x})$, $a \in G$, $\overline{x} \in \overline{G}$, it is now obvious that (Δ, π) is a system of imprimitivity for (G, F): Condition (10) follows from (23), and the nondegeneracy of Δ is a consequence of the fact, that the kernel of ρ, restricted to $\mathscr{C}_\infty(\Gamma)$, is π–invariant, hence 0, because π is irreducible. We collect the results of our discussion:

(26) *Let A be a closed normal second countable abelian subgroup of G, $\widehat{A}$ the dual group of A, π an irreducible representation of G, ρ the representation of $\mathscr{C}_\infty(\widehat{A})$, which is defined by the restriction of π onto A and let $\omega_\pi \subset \widehat{A}$ be the hull of the kernel of ρ in $\widehat{A}$, i.e. the set of common zeros of all $f \in \mathscr{C}_\infty(\widehat{A})$ with $\rho(f) = 0$. Then there exists $\chi_0 \in \widehat{A}$ such that the orbit $\Gamma = {}^G\chi_0$ is dense in ω_π. Let F be the stabilizer of χ_0; hence $\overline{x} = xF \to {}^x\chi_0$ maps $\overline{G} = G/F$ one to one onto Γ. If this mapping is a homeomorphism and if Γ is locally closed in $\widehat{A}$, i.e. is open in ω_π, then π is induced from some irreducible representation σ of F, i.e. $\pi = \mathrm{ind}_F^G\sigma$, $\sigma \in \widehat{F}$.*

To finish this paragraph we mention two simple facts on induced representations. The first is the theorem about *induction in stages*: Let H and L be closed subgroups of G with $L \subset G$ and let τ be a unitary representation of L. Then the induced representations $\mathrm{ind}_L^G\tau$ and $\mathrm{ind}_H^G(\mathrm{ind}_L^H\tau)$ of G are unitarily equivalent.

The second remark is stated in

(27) *Let σ be a unitary representation of G in the Hilbert space $\mathfrak{H}$, let σ_0 be the restriction of σ onto the closed subgroup H and let τ be a unitary representation of H in $\mathfrak{H}_\tau$. The induced representation $\operatorname{ind}_H^G(\tau \otimes \sigma_0)$ is unitarily equivalent to the tensor product representation $(\operatorname{ind}_H^G \tau) \otimes \sigma$ of G in $(\operatorname{ind}\mathfrak{H}_\tau) \otimes \mathfrak{H}$.*

Proof. The space of $\tau \otimes \sigma_0$ is $\mathfrak{H}_\tau \otimes \mathfrak{H}$, hence the elements φ in ind $(\mathfrak{H}_\tau \otimes \mathfrak{H})$ satisfy $\varphi(xy) = (\tau(y)^* \otimes \sigma(y)^*)\varphi(x)$ for $x \in G$, $y \in H$. It is easy to see that (ind $\mathfrak{H}_\tau) \otimes \mathfrak{H}$, i.e. the space of $(\operatorname{ind}_H^G \tau) \otimes \sigma$, can be identified with the Hilbert space of functions $\theta : G \to \mathfrak{H}_\tau \otimes \mathfrak{H}$ with $\theta(xy) = (\tau(y)^* \otimes \mathbf{1})\theta(x)$ for $x \in G$, $y \in H$; the identifying mapping Φ from $(\operatorname{ind}\mathfrak{H}_\tau \otimes \mathfrak{H})$ onto this space is given by $\Phi(\psi \otimes \xi)(x) = \psi(x) \otimes \xi$ for $\psi \otimes \xi \in$ (ind $\mathfrak{H}_\tau) \otimes \mathfrak{H}$.

Now let $\theta \in$ (ind $\mathfrak{H}_\tau) \otimes \mathfrak{H}$. We define the function $T\theta : G \to \mathfrak{H}_\tau \otimes \mathfrak{H}$ by

$$(T\theta)(x) = (\mathbf{1} \otimes \sigma(x)^*)\theta(x) \ .$$

Then we have for $x \in G$, $y \in H$

$$\begin{aligned}(T\theta)(xy) &= (\mathbf{1} \otimes \sigma(y)^*\sigma(x)^*)(\tau(y)^* \otimes \mathbf{1})\theta(x) = \\ &= (\tau(y) \otimes \sigma(y))^*(T\theta)(x) \ .\end{aligned}$$

Since $|T\theta(x)| = |\theta(x)|$ we see that $T\theta \in$ ind $(\mathfrak{H}_\tau \otimes \mathfrak{H})$, hence T maps (ind $\mathfrak{H}_\tau) \otimes \mathfrak{H}$ unitarily onto ind $(\mathfrak{H}_\tau \otimes \mathfrak{H})$. Finally, for $a \in G$ and the modulus χ of the measure $d\overline{x}$ on G/H we get

$$\begin{aligned}&\left(\operatorname{ind}_H^G(\tau \otimes \sigma_0)(a)T\theta\right)(x) = \chi(a)^{\frac{1}{2}}(\mathbf{1} \otimes \sigma(a^{-1}x))\theta(a^{-1}x) = \\ &= (\mathbf{1} \otimes \sigma(x)^*)(\mathbf{1} \otimes \sigma(a))\left\{((\operatorname{ind}_H^G \tau)(a) \otimes \mathbf{1})\theta\right\}(x) = \\ &= (\mathbf{1} \otimes \sigma(x)^*)\left\{(\operatorname{ind}_H^G \tau(a) \otimes \sigma(a))\theta\right\}(x) = (T(\operatorname{ind}_H^G \tau(a) \otimes \sigma(a))\theta)(x)\end{aligned}$$

which implies $\operatorname{ind}_H^G(\tau \otimes \sigma_0)(a)T = T(\operatorname{ind}_H^G \tau(a) \otimes \sigma(a))$ and proves (27).

In the special case τ = trivial representation, hence $\tau \otimes \sigma_0 = \sigma_0$, $\operatorname{ind}_H^G \tau = \lambda_H$ = the regular representation in $L^2(G/H)$ we conclude that $\operatorname{ind}_H^G \sigma_0$ is unitarily equivalent to $\lambda_H \otimes \sigma$. This yields

(28) *With the notation of (27) let G be an exponential group and H a proper connected closed subgroup. The induced representation $\operatorname{ind}_H^G \sigma_0$ is reducible.*

Proof. It suffices to prove the reducibility of λ_H in the case that H is a maximal proper connected subgroup. Let $\mathfrak{h}$ be the subalgebra of $\mathfrak{g}$, corresponding to H. We use induction on $n = \dim \mathfrak{g}$. If $\mathfrak{m}$ is an ideal in $\mathfrak{g}$ with $0 \neq \mathfrak{m} \subset \mathfrak{h}$, then $M = E(\mathfrak{m})$ is normal in G and λ_H factors over $\lambda_{\dot{H}}$, $\dot{H} = H/M \subset \dot{G} = G/M$. So λ_H is reducible if and only if $\lambda_{\underset{H}{\bullet}}$ is reducible, which allows us to assume that $\mathfrak{h}$ contains no nonzero ideal of $\mathfrak{g}$. Let $\mathfrak{a}$ be a minimal nonzero ideal in $\mathfrak{g}$, hence $\mathfrak{a}' = 0$, $\dim \mathfrak{a} \leq 2$ and $\mathfrak{g} = \mathfrak{a} + \mathfrak{h}$. If $\mathfrak{c}$ is the centralizer of $\mathfrak{a}$ in $\mathfrak{g}$, then $\mathfrak{c} \cap \mathfrak{h}$ is an ideal, contained in $\mathfrak{h}$. It follows that $\mathfrak{c} \cap \mathfrak{h} = 0$ and as $\mathfrak{a}' = 0$, i.e. $\mathfrak{a} \subset \mathfrak{c}$, we see that $\mathfrak{a} = \mathfrak{c}$, $\mathfrak{g} = \mathfrak{a} \oplus \mathfrak{h}$ and $G = H \ltimes A$ is the semidirect product of H and $A = E(\mathfrak{a})$; moreover, $\mathfrak{h}$, resp. H, acts faithfully on $\mathfrak{a}$, resp. on A, which implies that $\dim \mathfrak{h} = 1$, i.e. $H = \mathbb{R}$. So finally we see that only the following two cases are left over: $\dim \mathfrak{a} = 1$: $G = \mathbb{R} \ltimes \mathbb{R}$, the "$ax + b$"–group with $H = \mathbb{R} \ltimes \{0\}$ and action $x \to x^t = e^{-t}x$ on $A = \{0\} \ltimes \mathbb{R}$, $t \in H$; $\dim \mathfrak{a} = 2$: $G = \mathbb{R} \ltimes \mathbb{C}$ with action $z \to z^t = e^{-qt}z$, $q = 1 + \lambda i$, $\lambda \neq 0$, on $A = \{0\} \ltimes \mathbb{C} \cong \mathbb{C}$ for $t \in H = \mathbb{R} \ltimes \{0\}$. Let us compute λ_H in the second case. Here $z \to (0, z)H = \{(t, e^{-qt}z)\,;\, t \in \mathbb{R}\}$ identifies $\mathbb{C}$ with the homogeneous space G/H. For $(t, w) \in G$ we have $(t, w)(0, z)H = (t, w + z)H = (0, e^{qt}(w + z))H$, hence G acts on $\mathbb{C}$ by $(t, w)z = e^{qt}(w+z)$. This implies that the Lebesgue measure $dz \wedge d\bar{z} = dxdy$ on $\mathbb{C}$ is relatively invariant with modulus $\chi_H^G(t, w) = e^{2t}$ and that λ_H acts on $L^2(\mathbb{C})$ by

$$(\lambda_H(t, w)f)(z) = e^{-t}f(e^{-qt}z - w)\ .$$

The Fourier transform of $L^2(\mathbb{C})$ defines an equivalent representation $\hat{\lambda}_H$ on $L^2(\mathbb{C})$ given by

$$(\hat{\lambda}_H(t, w)g)(z) = e^t e^{i(w, e^{\bar{q}t}z)} g(e^{\bar{q}t}z)$$

with euclidean inner product $(u, v) = \operatorname{Im} u\bar{v}$ on $\mathbb{C}$. If follows that $\hat{\lambda}_H$ commutes with multiplication operators $g \to hg$ on $L^2(\mathbb{C})$, defined by functions $h \in L^\infty(\mathbb{C})$, which are constant on all spirals $\{e^{\bar{q}t}z\,;\, t \in \mathbb{R}\}$, $z \in \mathbb{C}^\times$. The algebra of these functions is isomorphic to $L^\infty(\mathbb{T})$, hence $\hat{\lambda}_H$ and λ_H are reducible.

For $\dim \mathfrak{a} = 1$ the representation $\hat{\lambda}_H$ acts on $L^2(\mathbb{R})$ and is given by

$$(\hat{\lambda}_H(t, x)g)(y) = e^{\frac{1}{2}t}e^{ie^t xy}g(e^t y)\ ,$$

hence the subspace of all $g \in L^2(\mathbb{R})$ with $g(y) = 0$ for $y \leq 0$ is invariant, so again λ_H is reducible. Let us remark that we really used the fact that G is exponential. For $G = \mathbb{C} \ltimes \mathbb{C}$ with action $z \to e^u z$ for $u \in H = \mathbb{C} \ltimes \{0\}$ the representation λ_H is irreducible.

§ 4 The dual of an exponential group

We return in this paragraph to exponential groups, so G will always denote a connected, simply connected solvable exponential Lie group and $\mathfrak{g}$ its Lie algebra. Our notation will be the same as in §1.

Recall that a representation π of a locally compact group Γ is called *monomial*, if it is induced from a one dimensional representation χ, i.e. from an abelian character of some closed subgroup H of Γ : $\pi = \mathrm{ind}_H^\Gamma \chi$. Our first basic result is an easy consequence of (26):

(29) *Every irreducible unitary representation of the exponential group G is monomial.*

Proof. We proceed by induction on $n = \dim G$. The theorem is true for $n = 1$, so we suppose $n > 1$ and that the theorem holds for all exponential groups of lower dimensions. As a first consequence of this assumption we can assume that the kernel of π is discrete, hence the center Z of G either is trivial or $Z \cong \mathbb{R}$ and then the restriction $\pi|_Z$ is not trivial.

We fix a minimal noncentral ideal $\mathfrak{a}$ in $\mathfrak{g}$. Let $\mathfrak{z}$ denote the center of $\mathfrak{g}$, thus $\dim \mathfrak{z} \leq 1$. There are four possibilities for $\mathfrak{a}$:

(a_1) $\mathfrak{z} \cap \mathfrak{a} = (0)\,, \dim \mathfrak{a} = 1\,,$

(a_2) $\mathfrak{z} \cap \mathfrak{a} = (0)\,, \dim \mathfrak{a} = 2\,,$

(b_1) $\mathfrak{z} \subset \mathfrak{a}\,, \quad \dim \mathfrak{a} = 2\,,$

(b_2) $\mathfrak{z} \subset \mathfrak{a}\,, \quad \dim \mathfrak{a} = 3\,.$

In all cases $\mathfrak{a}$ is abelian. This is clear for (a_1), (a_2) and (b_1). In case (b_2) the quotient $\mathfrak{a}/\mathfrak{z}$ is a minimal, hence abelian ideal in $\mathfrak{g}/\mathfrak{z}$. Because $\mathfrak{a}/\mathfrak{z}$ is minimal and $\dim \mathfrak{a}/\mathfrak{z} = 2$, the algebra $\mathfrak{g}$ acts nontrivially on $\mathfrak{a}/\mathfrak{z}$, thus the annihilator $\mathfrak{h}$ of $\mathfrak{a}/\mathfrak{z}$ is an ideal of codimension ≤ 2. But $\dim \mathfrak{g}/\mathfrak{h} = 2$ is excluded, because G is exponential. So there exists a basis $\{u, v, z\}$ for $\mathfrak{a}$ with $\mathfrak{z} = \mathbb{R}z$ and $a \in \mathfrak{g}$ with $[a, u] = u - \lambda v$, $[a, v] = \lambda u + v$, $\lambda \neq 0$, $[u, v] \in \mathfrak{z}$. Then $0 = [a, [u, v]] = [[a, u], v] + [u, [a, v]] = 2[u, v]$, hence $[u, v] = 0$.

The normal subgroup $A = \exp \mathfrak{a}$ is isomorphic with $\mathbb{R}^j$, $j = 1, 2$ or 3, and actually we may consider the exponential mapping exp as an identification of $\mathfrak{a}$ with A. Then the restriction of the inner automorphism, defined by an element $g \in G$, on A is the same as the restriction of Ad g on $\mathfrak{a}$, i.e. $\exp(\mathrm{Ad}\ g(x)) = g \cdot \exp x \cdot g^{-1}$ or briefly

$\mathrm{Ad}\ g(x) = gxg^{-1}$. In this picture the dual group $\widehat{A}$ is the linear dual $\mathfrak{a}^*$ of $\mathfrak{a}$ and the action of G is the coadjoint action on $\mathfrak{a}^*$.

Let $\mathfrak{c}$ be the centralizer of $\mathfrak{a}$ in $\mathfrak{g}$ and $C = \exp \mathfrak{c}$ the corresponding normal subgroup of G. Then the dimension of the quotient $F := G/C$ is 1 in the cases (a_1), (a_2), 1 or 2 in case (b_1) and 3 in case (b_2). We will study the action of G, resp. F on $\widehat{A} = \mathfrak{a}^*$ in these cases.

(a_1): Let $\mathfrak{a} = \mathbb{R}u$. Then $\mathfrak{g} = \mathbb{R}a \oplus \mathfrak{c}$ where we can choose $a \in \mathfrak{g}$ so that $[a, u] = u$, hence $F \cong \exp \mathbb{R}a$ and $\mathrm{Ad}\ \exp ta(u) = e^t u$. Let $\mathfrak{a}^* = \mathbb{R}\varphi$ with $\varphi(u) = 1$. Then the coadjoint orbits are $\omega_0 = \{0\}$, $\omega_+ = \{t\varphi\,; t > 0\}$ and $\omega_- = \{t\varphi\,; t < 0\}$. The orbit ω_0 corresponds to the trivial representation of A. As π is nontrivial on A it follows that – with notation as in (26) – the hull ω_π equals $\overline{\omega_+} = \{t\varphi\,; t \geq 0\}$ or $\overline{\omega_-}$ and contains ω_+, resp. ω_- as a dense, open subset. We have $\mathrm{Ad}^* \exp ta(\varphi) = e^{-t}\varphi$, so C is the stabilizer of φ in G and $(\exp ta)C \to e^{-t}\varphi$ is a homeomorphism from G/C onto ω_+. Similarly with ω_-. It follows from (26), that π is induced from some $\sigma \in \widehat{C}$. Since $\dim C = n-1$ we know that σ is monomial, hence induction in stages shows that π is monomial.

(a_2): Let $\mathfrak{a} = \mathbb{R}u \oplus \mathbb{R}v$. Then again $\dim F = 1$, $\mathfrak{g} = \mathbb{R}a \oplus \mathfrak{c}$. In this case we can identify $\mathfrak{a}$, resp. A with $\mathbb{C}$: $\alpha u \oplus \beta v \to \alpha + i\beta \in \mathbb{C}$, moreover there exists $q = 1 + i\lambda \in \mathbb{C}$, $\mathbb{R} \ni \lambda \neq 0$, such that $\mathrm{Ad}\ \exp ta(z) = e^{qt}z = e^t e^{i\lambda t}z$. We identify also $\mathfrak{a}^*$ with $\mathbb{C}$: $\zeta \in \mathbb{C}$ corresponds to the linear form $z \to \mathrm{Re}\,(z\zeta)$. Then $\mathrm{Ad}^* \exp ta(\zeta) = e^{-qt}\zeta$. For $\zeta = 1$ the orbit

$$\omega_1 = \{e^{(1+i\lambda)t}\,;\, t \in \mathbb{R}\}$$

is a spiral, which meets $\mathbb{T} = \{z \in \mathbb{C}; |z| = 1\}$ in 1. For general $\zeta \neq 0$ there exists $c \in \mathbb{T}$ such that the orbit of ζ equals $\omega_c := c\,\omega_1 = \{cz; z \in \omega_1\}$, hence ω_c is a spiral, which meets $\mathbb{T}$ in c. Thus the orbit space $\widehat{A}/G$ can be identified with $\{0\} \cup \mathbb{T}$. The open subsets of $\widehat{A}/G$ are the ordinary open subsets of $\mathbb{T}$ and the whole orbit space. Every nonempty closed set contains the zero–orbit $\omega_0 = \{0\}$ and $\overline{\omega_c} = \omega_c \cup \{0\}$. The same arguments as in case (a_1) show that again π is monomial.

(b_1): If $\dim F = 1$ we have $\mathfrak{g} = \mathbb{R}a \oplus \mathfrak{c}$ and there exist a basis $\{u, z\}$ of $\mathfrak{a}$ with $[a, u] = z \in \mathfrak{z}$, which means that $\mathbb{R}a \oplus \mathfrak{a}$ is the Heisenberg algebra und $(\exp \mathbb{R}a) \ltimes A$ the corresponding Heisenberg group. If $\{u^*, z^*\}$ is the basis of $\mathfrak{a}^* = \widehat{A}$, dual to $\{u, z\}$, then the orbits in $\mathfrak{a}^*$ are the points ξu^*, $\xi \in \mathbb{R}$, and the lines $\omega_s = \mathbb{R}u^* \oplus sz^*$, $0 \neq s \in \mathbb{R}$. The point orbits correspond to representations trivial on Z. As $\pi|_Z$ is non trivial, $\omega_\pi = \omega_s$ for some $s \neq 0$. Again our claim follows as above.

Now let $\dim F = 2$. Then F acts nontrivially on $\mathfrak{a}/\mathfrak{z}$, hence there exists $b \in \mathfrak{g}$ with $[b, u] \equiv u(\bmod \mathfrak{z})$. It is an easy exercise to show that one can find elements $a, b \in \mathfrak{g}$ and a basis $\{u, z\}$ of $\mathfrak{a}$ such that $\mathfrak{g} = \mathbb{R}a \oplus \mathbb{R}b \oplus \mathfrak{c}$ and

$$[a, u] = z\,,\ [b, u] = u\,,\ [a, b] \equiv -b (\bmod \mathfrak{c})\,.$$

This implies $\mathrm{Ad}\ \exp ta(u) = u + tz$, $\mathrm{Ad}\ \exp sb(u) = e^s u$, hence if we think of the elements of $\mathfrak{a}$ as column vectors, in particular if we set $u = \binom{1}{0}$, $z = \binom{0}{1}$, then F acts as the group of all matrices $\begin{pmatrix} e^x & 0 \\ y & 1 \end{pmatrix}$. As in this model $\mathfrak{a}^*$ is the space of vectors (ξ, η), we see that we have the following orbits in $\mathfrak{a}^*$:

$$0; \quad \omega_\pm = \{(\pm t, 0)\,;\, t > 0\}\,; \quad \omega_s = \{(t, s)\,;\, t \in \mathbb{R}\}\,,\ s \neq 0\,.$$

The ω_s are closed and as before we have $\omega_\pi = \omega_s$ for some $s \neq 0$. The matrices stabilizing the element $(0, s) \in \omega_s = \omega_\pi$ are those of the form $\begin{pmatrix} e^x & 0 \\ 0 & 1 \end{pmatrix}$. So the stabilizer of the character χ_0 of A, given by $(0, s) \in \omega_s$, is the closed subgroup $\mathrm{H} = (\exp \mathbb{R}a) \ltimes C$ of G. Again π is monomial.

(b_2): In this case $\mathfrak{a}$ defines a two dimensional ideal $\mathfrak{a}^c = \mathbb{C}w \oplus \mathbb{C}z$ in the complexified algebra $\mathfrak{g}^{\mathbb{C}}$. If $w = u + iv$ with $u, v \in \mathfrak{g}$, then $\mathfrak{a} = \mathbb{R}u \oplus \mathbb{R}v \oplus \mathbb{R}z$. Let φ and ψ be the $\mathbb{C}$–valued linear forms on $\mathfrak{g}$ with

$$[x, w] = \varphi(x)w + \psi(x)z \ .$$

As $\mathfrak{g}$ is exponential and $\varphi(x)$ an eigenvalue of ad x it follows that $\varphi(x) = q\varphi_0(x)$ with $\varphi_0 \in \mathfrak{g}^*$ and $q = 1 + i\lambda$, $\lambda \in \mathbb{R}$. Moreover $\lambda \neq 0$, because $\mathfrak{a}$ is minimal noncentral. Let $a \in \mathfrak{g}$ with $\varphi_0(a) = 1$. We can assume that $\psi(a) = 0$, because otherwise we could substitute $w' = w + q^{-1}\psi(a)z$ for w. Thus $[a, w] = qw$. Let $\psi = \psi_1 + i\psi_2$ with $\psi_1, \psi_2 \in \mathfrak{g}^*$. We claim that $\varphi_0, \psi_1, \psi_2$ are linearly independent.

Suppose that $\alpha_0\varphi_0 + \alpha_1\psi_1 + \alpha_2\psi_2 = 0$ for some $\alpha_j \in \mathbb{R}$. Then $\varphi_0(a) = 1$, $\psi(a) = 0$ yields $\alpha_0 = 0$, hence $\alpha_1\psi_1 + \alpha_2\psi_2 = 0$. Now $\alpha_1 \neq 0$ or $\alpha_2 \neq 0$ would imply $\psi = c\psi_0$ with $\psi_0 \in \mathfrak{g}^*$ and $0 \neq c \in \mathbb{C}$. Let $c = 1$ (otherwise take $c^{-1}w$ instead of w). By minimality of $\mathfrak{a}$ the functional ψ_0 is independent of φ_0, so take $b \in \mathfrak{g}$ with $\varphi_0(b) = 0$, $\psi_0(b) = 1$. But then the Jacobi identity yields

$$[[a, b], w] = [[a, w], b] + 0 = q[w, b] = -qz = \psi_0([a, b])z \ ,$$

hence $\psi_0([a, b]) = -q$, a contradiction, because $\psi_0 \in \mathfrak{g}^*$. It follows that there exist elements $a, b, c \in \mathfrak{g}$ with $\mathfrak{g} = \mathbb{R}a \oplus \mathbb{R}b \oplus \mathbb{R}c \oplus \mathfrak{c}$ and

$$\begin{aligned} [a, u] &= u - \lambda v \,, & [b, u] &= z \,, & [c, u] &= 0 \ , \\ [a, v] &= \lambda u + v \,, & [b, v] &= 0 \,, & [c, v] &= z \ . \end{aligned}$$

If we represent the elements $\xi_1 u + \xi_2 v + \xi_3 z \in \mathfrak{a}$ by column vectors $x = \begin{pmatrix} \xi_1 \\ \xi_2 \\ \xi_3 \end{pmatrix}$ then the matrices of the restrictions of ad a, ad b, ad c on $\mathfrak{a}$ are

$$\begin{pmatrix} 1 & -\lambda & 0 \\ \lambda & 1 & 0 \\ 0 & 0 & 0 \end{pmatrix} , \quad \begin{pmatrix} 0 & 0 & 0 \\ 0 & 0 & 0 \\ 1 & 0 & 0 \end{pmatrix} , \quad \begin{pmatrix} 0 & 0 & 0 \\ 0 & 0 & 0 \\ 0 & 1 & 0 \end{pmatrix} .$$

It follows that the quotient group $F = G/C$ acts on $\mathfrak{a}$ as the group of all matrices

$$\begin{pmatrix} e^r \cos \lambda r & -e^r \sin \lambda r & 0 \\ e^r \sin \lambda r & e^r \cos \lambda r & 0 \\ s & t & 1 \end{pmatrix} =: g(r, s, t) \ , \qquad r, s, t \in \mathbb{R} \ .$$

To obtain the coadjoint action on $\mathfrak{a}^*$ one has to identify $\mathfrak{a}^*$ with $\mathbb{R}^3$, i.e. the $\chi \in \mathfrak{a}^*$ are vectors $(\chi_1, \chi_2, \chi_3) \in \mathbb{R}^3$, and to let the $g(r, s, t)$ act as usually from the right. In particular the orbit of $\chi_0 = (\mu, \nu, \zeta)$ with $\zeta \neq 0$ is the plane

$$\omega_\zeta := \{(\xi_1, \xi_2, \zeta) \, ; \, \xi_1, \xi_2 \in \mathbb{R}\}$$

and the stabilizer of $\chi_0 = (0, 0, \zeta)$ is

$$H := H_{\chi_0} = (\exp \mathbb{R}a) \ltimes C .$$

Again, because $\pi|_Z$ is nontrivial, we have $\omega_\pi = \omega_\zeta$ with $\zeta \neq 0$, hence the situation is exactly the same as in case (b_1), $\dim F = 2$. It follows that also in this last case π is induced from some $\sigma \in \widehat{H}$, hence π is monomial and (29) is proved.

Hidden in the arguments of the proof of (29) is the following useful observation:

(30) *Let M be a non trivial irreducible $\mathfrak{g}$ module and $\mathfrak{n}$ its annihilator in $\mathfrak{g}$. Then* $\dim \mathfrak{g}/\mathfrak{n} = 1$. *If $\mathfrak{h}$ is a subalgebra of $\mathfrak{g}$, either $\mathfrak{h}$ annihilates M or $\mathfrak{h}$ acts irreducibly on M, in particular, if $\mathfrak{m}$ is a noncentral minimal ideal in $\mathfrak{g}$, then either* $[\mathfrak{h}, \mathfrak{m}] = 0$ *or* $[\mathfrak{h}, w] = \mathfrak{m}$ *for every $w \in \mathfrak{m}$, $w \neq 0$.*

Proof. $\mathfrak{g}/\mathfrak{n}$ acts faithfully as a solvable Lie algebra of linear transformations on M, hence $\dim \mathfrak{g}/\mathfrak{n} \leq 2$. If $\dim \mathfrak{g}/\mathfrak{n} = 2$, then some of these transformations would have nonzero purely imaginary eigenvalues, which is impossible, because $\mathfrak{g}$ is exponential. If the subalgebra $\mathfrak{h}$ acts nontrivially, i.e. $\mathfrak{h} \not\subset \mathfrak{n}$, then $\mathfrak{h} + \mathfrak{n} = \mathfrak{g}$, hence $\mathfrak{h}$ acts irreducibly.

Proposition (29) has a fundamental consequence: Let $\pi \in \widehat{G}$ be induced from the character χ of the connected closed subgroup H, so χ is trivial on the commutator subgroup H', of H. If $\mathfrak{h}$ is the subalgebra of $\mathfrak{g}$, corresponding to H, then the exponential mapping E defines a group isomorphism from $\mathfrak{h}/\mathfrak{h}'$ onto H/H', hence $\chi(\exp x) = e^{if_0(x)}$ for some linear form f_0 on $\mathfrak{h}/\mathfrak{h}'$, or, which is essentially the same, for some $f_0 \in \mathfrak{h}^*$ with $f_0(\mathfrak{h}') = 0$. Clearly this f_0 extends in many ways to a linear form f on $\mathfrak{g}$, i.e. there exists $f \in \mathfrak{g}^*$ with $f_0 = f|_{\mathfrak{h}}$. For an easier formulation of this result we define:

Definition. For $f \in \mathfrak{g}^*$ let B_f be the symplectic form on $\mathfrak{g}$ defined by

$$B_f(x, y) = f([x, y]), \qquad x, y \in \mathfrak{g} .$$

A subalgebra $\mathfrak{h} \subset \mathfrak{g}$ is called *subordinate* to f, if $\mathfrak{h}$ is isotropic with respect to B_f, i.e. if $f(\mathfrak{h}') = 0$. If $\mathfrak{h}$ is subordinate to f, then the formula

$$\chi(f, \mathfrak{h})(E(x)) = e^{if(x)}, \quad x \in \mathfrak{h} ,$$

defines an abelian character $\chi(f, \mathfrak{h})$ of the subgroup $H = E(\mathfrak{h})$. We write

$$\pi(f, \mathfrak{h}, G) = \operatorname{ind}_H^G \chi(f, \mathfrak{h})$$

for the representation, induced by $\chi(f, \mathfrak{h})$ on G.

The result of our above discussion can now be formulated as follows:

For every $\pi \in \widehat{G}$ there exist $f \in \mathfrak{g}^*$ and subalgebras $\mathfrak{h} \subset \mathfrak{g}$, subordinate to f, such that π is equivalent to $\pi(f, \mathfrak{h}, G)$.

It is now clear how to proceed: The first step is to determine, when, for given $f \in \mathfrak{g}^*$ and subordinate $\mathfrak{h}$, $\pi(f, \mathfrak{h})$ is irreducible. An immediate necessary condition follows from (28):

(31) *If $\pi(f, \mathfrak{h})$ is irreducible, then no proper superalgebra of $\mathfrak{h}$ is subordinate to f.*

Proof. Let $\mathfrak{j}$ be a subordinate superalgebra of $\mathfrak{h}$ in $\mathfrak{g}$ and $J = E(\mathfrak{j})$ the corresponding closed subgroup. Then $\chi(f, \mathfrak{j}) \in \widehat{J}$ extends $\chi(f, \mathfrak{h})$, hence $\sigma :=$ ind ${}_H^J \chi(f, \mathfrak{h})$ is equivalent to $\lambda_{J,H} \otimes \chi(f, \mathfrak{h})$ with the left regular representation $\lambda_{J,H}$ of J in $L^2(J/H)$. Thus σ and a fortiori $\pi(f, \mathfrak{h}) =$ ind ${}_J^G \sigma$ are reducible, if $J \neq H$, i.e. $\mathfrak{j} \neq \mathfrak{h}$.

A sufficient condition for a subalgebra $\mathfrak{h}$ to be maximal subordinate to f is of course that $\mathfrak{h}$ is a maximal isotropic subspace with respect to B_f. For a subset $\mathfrak{x} \in \mathfrak{g}$ let

$$\mathfrak{x}^f := \{y \in \mathfrak{g}\,;\ B_f(x, y) = 0, \ \text{all } x \in \mathfrak{x}\}$$

i.e. $\mathfrak{x}^f = \{y \in \mathfrak{g}\,;\ f([y, \mathfrak{x}]) = 0\}$, in particular $\mathfrak{g}^f$ is the *radical* of B_f, i.e. the largest subspace of $\mathfrak{g}$, such that B_f is well-defined on the quotient $\mathfrak{g}/\mathfrak{g}^f$ by $B_f(\dot{x}, \dot{y}) = B_f(x, y)$ for $\dot{x} = x + \mathfrak{g}^f$, $\dot{y} = y + \mathfrak{g}^f$. On this quotient B_f is necessarily nondegenerate. We set

$$i(f, \mathfrak{g}) = \frac{1}{2} \dim(\mathfrak{g}/\mathfrak{g}^f)\ .$$

Then a subspace $\mathfrak{m} \subset \mathfrak{g}$ is maximal isotropic with respect to B_f, resp. to f, if and only if $B_f|_{\mathfrak{m}\times\mathfrak{m}} = 0$, i.e. $f([\mathfrak{m}, \mathfrak{m}]) = 0$, and $\dim(\mathfrak{g}/\mathfrak{m}) = i(f, \mathfrak{g})$.

Definition. A *polarization* for $f \in \mathfrak{g}^*$ in $\mathfrak{g}$ is a subalgebra $\mathfrak{p}$ in $\mathfrak{g}$, subordinate to f and with $\dim(\mathfrak{g}/\mathfrak{p}) = i(f, \mathfrak{g})$. The set of all polarizations for f in $\mathfrak{g}$ will be denoted by $P(f, \mathfrak{g})$.

Thus $\mathfrak{p} \in P(f, \mathfrak{g})$ if and only if $\mathfrak{p}' \subset \mathfrak{p}$, $f(\mathfrak{p}') = 0$ and $\dim(\mathfrak{g}/\mathfrak{p}) = i(f, \mathfrak{g})$. Furthermore every polarization $\mathfrak{p}$ for f contains $\mathfrak{g}^f$: If $\mathfrak{m}$ is an isotropic subspace of $\mathfrak{g}$, then also $\mathfrak{m} + \mathfrak{g}^f$ is isotropic, hence maximality implies $\mathfrak{m} \supset \mathfrak{g}^f$.

Using induction on $\dim \mathfrak{g} = n$ we will show:

(32) *If $\pi(f, \mathfrak{h})$ is irreducible, then $\mathfrak{h}$ is a polarization for f.*

Proof. Let us assume that $\mathfrak{h}$ contains an ideal $\mathfrak{a} \neq 0$ with $f(\mathfrak{a}) = 0$ and let $A = E(\mathfrak{a})$ be the corresponding normal subgroup. Then f defines uniquely an element $\dot{f} \in \dot{\mathfrak{g}} := \mathfrak{g}/\mathfrak{a}$ and $\dot{\mathfrak{h}} = \mathfrak{h}/\mathfrak{a}$ is subordinate to $\dot{f}$. If κ is the canonical homomorphism from G onto $\dot{G} = G/A$, then obviously $\pi(f, \mathfrak{h}) = \pi(\dot{f}, \dot{\mathfrak{h}}) \circ \kappa$. It follows that $\pi(\dot{f}, \dot{\mathfrak{h}})$ is irreducible and consequently $\dot{\mathfrak{h}} \in P(\dot{f}, \dot{\mathfrak{g}})$. As $i(f, \mathfrak{g}) = i(\dot{f}, \dot{\mathfrak{g}})$ and $\dim \mathfrak{g}/f = \dim \dot{\mathfrak{g}}/\dot{\mathfrak{h}}$ we conclude that $\mathfrak{h} \in P(f, \mathfrak{g})$.

(33) *Let* $\mathfrak{m}$ *be an ideal in* $\mathfrak{g}$. *If one of the following conditions holds:*
(i) $f(\mathfrak{h} \cap \mathfrak{m}) = 0$, (ii) $f(\mathfrak{m}' + [\mathfrak{h}, \mathfrak{m}]) = 0$, (iii) $\mathfrak{m}' = 0$, $\mathfrak{h} \subset \mathfrak{m}^f$,
then $\mathfrak{m}$ *is contained in* $\mathfrak{h}$.

This follows in the cases (i) and (ii) from the fact that the restriction $f_1 = f|_{\mathfrak{h}}$ extends to a homomorphism $\tilde{f}$ of the subalgebra $\tilde{\mathfrak{h}} = \mathfrak{h} + \mathfrak{m}$: In case (i) we set $\tilde{f}(x+y) = f_1(x) = f(x)$ for $x \in \mathfrak{h}$, $y \in \mathfrak{m}$. As $\tilde{\mathfrak{h}}' \subset \mathfrak{h}' + \mathfrak{m}$ we see that $\tilde{f}(\tilde{\mathfrak{h}}') = 0$. (31) implies $\tilde{\mathfrak{h}} = \mathfrak{h}$, i.e. $\mathfrak{m} \subset \mathfrak{h}$. The condition (ii) means that f is a homomorphism on $\tilde{\mathfrak{h}} = \mathfrak{h} + \mathfrak{m}$, so again $\tilde{\mathfrak{h}} = \mathfrak{h}$, $\mathfrak{m} \subset \mathfrak{h}$. If $\mathfrak{m}' = 0$, then (ii) reduces to $f([\mathfrak{h}, \mathfrak{m}]) = 0$, i.e. $\mathfrak{h} \subset \mathfrak{m}^f$, hence also (iii) implies $\mathfrak{m} \subset \mathfrak{h}$.

(34) *If* $\mathfrak{m}$ *is a nontrivial ideal in* $\mathfrak{g}$ *then* $\mathfrak{h} \cap \mathfrak{m} \neq 0$.

Proof. If $\mathfrak{h} \cap \mathfrak{m} = 0$, then $\mathfrak{l} = \mathfrak{h} \oplus \mathfrak{m}$ is a semidirect extension of $\mathfrak{m}$ by $\mathfrak{h}$, hence $L := E(\mathfrak{l}) = H \ltimes M$ is the semidirect product of $H = E(\mathfrak{h})$ and $M = E(\mathfrak{m})$. Obviously $\chi(f, \mathfrak{h})$ extends to a character of L. But this contradicts the irreducibility of $\pi(f, \mathfrak{h}, G)$, see (28).

Now let $\mathfrak{a}$ be a minimal ideal in $\mathfrak{g}$, $\mathfrak{a} \not\subset \mathfrak{z}$. Then (30) gives $\mathfrak{a} = [\mathfrak{g}, \mathfrak{a}]$, hence either $f(\mathfrak{a}) = 0$ or $\mathfrak{a}^f \neq \mathfrak{g}$. We saw already that $f(\mathfrak{a}) = 0$ implies $\mathfrak{h} \in P(f, \mathfrak{g})$. So let us assume that $\mathfrak{a}^f \neq \mathfrak{g}$. We claim that $\mathfrak{h}$ contains $\mathfrak{a}$. This follows from (33), (ii), if $[\mathfrak{h}, \mathfrak{a}] = 0$ and from (30) together with (34), if $[\mathfrak{h}, \mathfrak{a}] \neq 0$. From $\mathfrak{a} \subset \mathfrak{h}$ we get $\mathfrak{h} \subset \mathfrak{h}^f \subset \mathfrak{a}^f$, so we can induce $\chi(f, \mathfrak{h})$ to G in stages via $A^f := E(\mathfrak{a}^f)$. As $\pi(f_1, \mathfrak{h}, A^f)$, $f_1 = f|_{\mathfrak{a}^f}$, is irreducible and $\dim \mathfrak{a}^f < n$ we conclude that $\mathfrak{h}$ is a polarization for f_1 in $\mathfrak{a}^f$, which means that $\mathfrak{h} = \mathfrak{h}^{f_1}$. Finally $\mathfrak{h}^f \subset \mathfrak{a}^f$ and $\mathfrak{h}^{f_1} = \mathfrak{a}^f \cap \mathfrak{h}^f$ give $\mathfrak{h}^f = \mathfrak{h}^{f_1} = \mathfrak{h}$, which proves $\mathfrak{h} \in P(f, \mathfrak{g})$.

Our results show that in order to prove (32) we can now suppose that the center $\mathfrak{z}$ of $\mathfrak{g}$ is one dimensional, i.e. $\mathfrak{z} = \mathbb{R}z$, and contained in every nonzero ideal, moreover $\mathfrak{z} \subset \mathfrak{h}$ and $f(z) = 1$. If $\mathfrak{a}$ is a noncentral ideal, then $[\mathfrak{g}, \mathfrak{a}]$ is a nonzero ideal, hence $z \in [\mathfrak{g}, \mathfrak{a}]$ and consequently $\mathfrak{a}^f \neq \mathfrak{g}$. If $\mathfrak{a} \subset \mathfrak{h}$, then the same proof as above shows that $\mathfrak{h} \in P(f, \mathfrak{g})$. Thus we can also suppose that $\mathfrak{z}$ is the only nontrivial ideal contained in $\mathfrak{h}$.

Under these assumptions we choose a minimal noncentral ideal $\mathfrak{a}$ in $\mathfrak{g}$, hence $2 \leq \dim \mathfrak{a} \leq 3$, $\mathfrak{h} \cap \mathfrak{a} = \mathfrak{z}$ and $\mathfrak{k} := \mathfrak{h} + \mathfrak{a}$ is a proper superalgebra of $\mathfrak{h}$. Let $\mathfrak{h}_0 = \{x \in \mathfrak{h} ; f(x) = 0\}$. As f vanishes on $\mathfrak{h}'$ we see that $\mathfrak{h}_0$ is an ideal in $\mathfrak{h}$ and $\mathfrak{h} = \mathfrak{h}_0 \oplus \mathfrak{z}$, thus

$$\mathfrak{k} = \mathfrak{h}_0 \oplus \mathfrak{a} , \qquad K = H_0 \ltimes A ,$$

where K, H_0 and A are the subgroups corresponding to $\mathfrak{k}$, $\mathfrak{h}_0$ and $\mathfrak{a}$. The representation $\pi_1 = \operatorname{ind}_H^K \chi(f, \mathfrak{h})$ is irreducible and $\pi = \operatorname{ind}_K^G \pi_1$. We claim that

$$\{x \in \mathfrak{a};\ [x, \mathfrak{h}] = 0\} = \mathfrak{z}\ .$$

The left hand side is a subalgebra $\mathfrak{a}_0$ of $\mathfrak{a}$, containing $\mathfrak{z}$. Hence $\mathfrak{a}_0 = \mathfrak{z} \oplus \mathfrak{b}$ with $\mathfrak{b} \cap \mathfrak{h} = 0$. Then $\tilde{\mathfrak{h}} := \mathfrak{h} \oplus \mathfrak{b}$ is a subalgebra of $\mathfrak{k}$ with $\tilde{\mathfrak{h}}' = \mathfrak{h}'$, so $f|_{\mathfrak{h}}$ extends to a homomorphism of $\tilde{\mathfrak{h}}$ into $\mathbb{R}$. As π_1 is irreducible this implies $\mathfrak{b} = 0$.

Next we see that $\mathfrak{h}_1 := \{x \in \mathfrak{h}_0;\ [x, \mathfrak{a}] = 0\}$ in an ideal in $\mathfrak{k}$, on which f vanishes. Hence π_1 is trivial on $H_1 = E(\mathfrak{h}_1)$. This allows us (as long as we stay in $\mathfrak{k}$!) to assume that $\mathfrak{h}_1 = 0$, which means that $\mathfrak{h}_0$ acts faithfully on $\mathfrak{a}$. Let

$$\mathfrak{x} = \{x \in \mathfrak{h}_0;\ [x, \mathfrak{a}] \subset \mathfrak{z}\}\ .$$

Then

$$[x, y] = \beta(x, y)z$$

defines a bilinear form β on $\mathfrak{x} \times \mathfrak{a}/\mathfrak{z}$. We claim that β is nondegenerate. We distinguish two cases:

1. $\mathfrak{k} \neq \mathfrak{g}$, i.e. $\dim \mathfrak{k} < n$. Then, by induction, $\mathfrak{h}$ is a polarization for $f_1 = f|_{\mathfrak{k}}$ in $\mathfrak{k}$, thus $\mathfrak{k}^{f_1} \subset \mathfrak{h}$ and B_{f_1} is nondegenerate on $\mathfrak{k}/\mathfrak{k}^{f_1}$.

Let $\dim \mathfrak{k}/\mathfrak{k}^{f_1} = 2d$, hence $d = 1$ or $d = 2$. In case $d = 2$ there exist $x_1, x_2 \in \mathfrak{h}_0$, $y_1, y_2 \in \mathfrak{a}$ with $f([x_i, y_k]) = \delta_{ik}$. For $x \in \mathfrak{x}$, $x \neq 0$, we have $[x, \mathfrak{a}] \neq 0$, because $\mathfrak{x}$ acts faithfully on $\mathfrak{a}$. This implies $\dim \mathfrak{x} \leq 2$ and the nondegeneracy of β, if $\dim \mathfrak{x} = 2$. If $\dim \mathfrak{x} = 1$, then we can choose x_2 in $\mathfrak{x}$, hence $\mathfrak{h}_0 = \mathbb{R}x_1 \oplus \mathbb{R}x_2$, $x_1 \notin \mathfrak{x}$, i.e. ad x_1 acts irreducibly on $\mathfrak{a}/\mathfrak{z}$. Clearly we can take the $y_j \in \mathfrak{a}$ so that $f(y_j) = 0$. Then $[x_1, y_1] = \alpha_1 y_1 + \alpha_2 y_2 + z$ and $\alpha_2 \neq 0$, because the eigenvalue of the restriction of ad x_1 onto $\mathfrak{a}/\mathfrak{z}$ has nonzero imaginary part. As $\mathfrak{x}$ is an ideal in $\mathfrak{h}_0$ we have $[x_1, x_2] = \xi x_2$, hence $0 = [[x_1, x_2]y_1] = [[x_1, y_1], x_2] + [x_1, [x_2, y_1]] = -\alpha_2 z + 0$, i.e. $\alpha_2 = 0$, a contradiction.

If $d = 1$ and $\mathfrak{x} \neq 0$, then obviously β is nondegenerate, more precisely in this case $\mathfrak{a} = \mathbb{R}y \oplus \mathbb{R}z$ and either $\mathfrak{h}_0 = \mathfrak{x} = \mathbb{R}x$ with $[x, y] = z$ or $\mathfrak{h} = \mathbb{R}a \oplus \mathbb{R}x$ with $[a, y] = y$, $[a, x] = -x$, $[x, y] = z$. If $\mathfrak{x} = 0$, then $\mathfrak{h}_0 = \mathbb{R}a$ with $[a, y] = y$ for properly chosen $y \in \mathfrak{a}$, hence $\mathfrak{k}$ is the semidirect sum of $\mathfrak{h} = \mathbb{R}a \oplus \mathbb{R}z$ and the ideal $\mathbb{R}y$. But then as before (29) shows that π_1 would be reducible, hence $\mathfrak{x} \neq 0$.

2. $\mathfrak{k} = \mathfrak{g}$, i.e. $\mathfrak{g} = \mathfrak{h}_0 \oplus \mathfrak{a}$. In this case $[\mathfrak{h}_0, \mathfrak{a}] = [\mathfrak{g}, \mathfrak{a}]$ is a nonzero ideal in $\mathfrak{g}$, hence $[\mathfrak{h}_0, \mathfrak{a}] \supset \mathfrak{z}$. Furthermore, the annihilator $\mathfrak{d}$ of $\mathfrak{a}/\mathfrak{z}$ is also an ideal. If $\mathfrak{d} \supset \mathfrak{h}$, i.e. $\mathfrak{d} = \mathfrak{g}$, we have $\mathfrak{h}_0 = \mathfrak{x}$ and β is nondegenerate. Moreover, in this case the minimality of $\mathfrak{a}$ implies $\dim \mathfrak{a} = 2$. If $\mathfrak{d} \not\supset \mathfrak{h}$, then $\mathfrak{d} \cap \mathfrak{h} = \mathfrak{z}$, $\mathfrak{x} = 0$, $\mathfrak{h}_0 = \mathbb{R}a$. But then again $\mathfrak{a} = \mathfrak{a}_0 \oplus \mathfrak{z}$ with $[a, \mathfrak{a}_0] = \mathfrak{a}_0$, hence $\mathfrak{a}$ is not minimal noncentral. As this was excluded we see that β is never degenerate.

The nondegeneracy of β has an important consequence. Let

$$\mathfrak{y} = \{y \in \mathfrak{a};\ f(y) = 0\}$$

hence $\dim \mathfrak{y} = d$, $\mathfrak{a} = \mathfrak{y} \oplus \mathbb{R}z$, $\mathfrak{k} = \mathfrak{h} \oplus \mathfrak{y}$. If $\mathfrak{h}_0 \neq \mathfrak{x}$, i.e. if $\mathfrak{h}_0 = \mathbb{R}a \oplus \mathfrak{x}$, then $\mathfrak{y}$ is not necessarily ad a invariant: There exists a linear transformation A of $\mathfrak{y}$ and a linear functional $\zeta : \mathfrak{a} \to \mathbb{R}$ with $[a, y] = Ay + \zeta(y) \cdot z$ for all $y \in \mathfrak{y}$. Since β is nondegenerate, there exists $x_0 \in \mathfrak{x}$ with $\zeta(y) = \beta(x_0, y)$, which means that $\mathfrak{y}$ is stable under ad $(a - x_0)$. Since also $\mathfrak{h}_0 = \mathbb{R}(a - x_0) \oplus \mathfrak{x}$ we can assume that already $[a, \mathfrak{y}] = \mathfrak{y}$. So finally we find the following list of possible cases for $\mathfrak{k}$:

$$(\mathrm{a_d}) \qquad \mathfrak{k} = \mathfrak{x} \oplus \mathfrak{y} \oplus \mathbb{R}z = \mathfrak{x} \oplus \mathfrak{a}$$

with $\dim \mathfrak{k} = 2d+1$ and bases $\{x_i\}$, resp. $\{y_j\}$ for $\mathfrak{x}$, resp. $\mathfrak{y}$ such that $[x_i, y_j] = \delta_{ij}z$, i.e. $\mathfrak{k}$ is the Heisenberg algebra $\mathfrak{h}_d$, $d = 1$ or 2, and $f(\mathfrak{x}) = f(\mathfrak{y}) = 0$, $f(z) = 1$.

$$(\mathrm{b_d}) \qquad \mathfrak{k} = \mathbb{R}a \oplus \mathfrak{x} \oplus \mathfrak{y} \oplus \mathbb{R}z = \mathbb{R}a \oplus \mathfrak{e}$$

with $\mathfrak{e} = \mathfrak{x} \oplus \mathfrak{y} \oplus \mathbb{R}z = \mathfrak{h}_d$, $[a, \mathfrak{x}] = \mathfrak{x}$, $[a, \mathfrak{y}] = \mathfrak{y}$ and $f(a) = f(\mathfrak{x}) = f(\mathfrak{y}) = 0$, $f(z) = 1$.

In both cases we see that the subalgebra

$$\hat{\mathfrak{h}} := \begin{cases} \mathfrak{y} \oplus \mathbb{R}z & \text{in case } (\mathrm{a_d}) \\ \mathbb{R}a \oplus \mathfrak{y} \oplus \mathbb{R}z & \text{in case } (\mathrm{b_d}) \end{cases}$$

is subordinate to f and has the same dimension as $\mathfrak{h}$, moreover, $\hat{\mathfrak{h}}$ contains the ideal $\mathfrak{a}$, hence $\pi(f, \hat{\mathfrak{h}}, G) \in \widehat{G}$ implies $\hat{\mathfrak{h}} \in P(f, \mathfrak{g})$ and in turn also $\mathfrak{h} \in P(f, \mathfrak{g})$. Therefore (32) will follow if we can show that $\pi(f, \hat{\mathfrak{h}}, G)$ and $\pi(f, \mathfrak{h}, G)$ are equivalent. This is certainly a consequence of

(35) *Let $f_1 = f|_{\mathfrak{k}} \in \mathfrak{k}^*$. The representations $\pi(f_1, \mathfrak{h}, K)$ and $\pi(f_1, \hat{\mathfrak{h}}, K)$ of K are equivalent.*

Proof. We treat only the case $(\mathrm{b_2})$, the others are simpler and the Heisenberg case is classic anyhow. We can realize K as $\mathbb{R} \times \mathbb{C} \times \mathbb{C} \times \mathbb{R}$ with product given by

$$(t, x, y, z)(t', x', y', z') = (t + t', e^{qt'}x + x', e^{-\bar{q}t'}y + y', z + z' + (e^{t'q}x, y'))$$

with $q = 1 + i\lambda$, $0 \neq \lambda \in \mathbb{R}$ and $(x, y) = \operatorname{Re} x\bar{y}$ for $x, y \in \mathbb{C}$. Furthermore, $H = E(\mathfrak{h}) = \mathbb{R} \times \mathbb{C} \times \{0\} \times \mathbb{R}$ and $\chi(f, \mathfrak{h}) = \chi$ is given by

$$\chi(t, x, 0, z) = e^{iz} .$$

The mapping $q : y \to q(y) = (0, 0, y, 0)H$ identifies $\mathbb{C}$ with the quotient space K/H and defines the smooth section $s : q(y) \to (0, 0, y, 0) \in K$.

The formula

$$(t, x, y, z)^{-1}(0, 0, w, 0) = (0, 0, e^{-\bar{q}t}w - y, 0)(-t, -e^{-qt}x, 0, -z - (x, e^{-\bar{q}t}w - y))$$

for $(t, x, y, z) \in K$, $(0, 0, w, 0) \in Y = E(\mathfrak{y})$ shows that the action of K on K/H in terms of the section q, i.e. expressed as an action on $\mathbb{C}$, is given by

$$(t, x, y, z)^{-1}(w) = e^{-\overline{q}t}w - y \ ,$$

in particular the Lebesgue measure on $\mathbb{C}$ is relatively K invariant with modulus $\Delta(t, x, y, z) = e^{2t}$. Moreover it is now clear that $\pi_1 := \pi(f, \mathfrak{h}, K)$ can be realized on $\mathfrak{H}(\pi_1) = L^2(\mathbb{C})$ by

$$(\pi_1(t, x, y, z)g)(w) = e^{-t}e^{i(z+(x,e^{-\overline{q}t}w-y))}g(e^{-\overline{q}t}w - y) \ .$$

For $\hat{\mathfrak{h}}$ and $\widehat{H} = E(\hat{\mathfrak{h}}) = \mathbb{R}\times\{0\}\times\mathbb{C}\times\mathbb{R}$ the subgroup $X = E(\mathfrak{x}) = \{0\}\times\mathbb{C}\times\{0\}\times\{0\}$ parametrizes $K/\widehat{H}$ and we get

$$(t, x, y, z)^{-1}(0, v, 0, 0) = (0, e^{qt}v - x, 0, 0)(-t, 0, -e^{\overline{q}t}y, -z + (e^{qt}v, y)) \ ,$$

hence $\pi_2 := \pi(f, \hat{\mathfrak{h}}, K)$ is given by

$$(\pi_2(t, x, y, z)g)(v) = e^t e^{i(z-(e^{qt}v,y))}g(e^{qt}v - x)$$

We claim that π_1 and π_2 are unitarily equivalent. We set $a = (t, x, y, z) \in K$. If $\mathcal{F}$ denotes the Fourier transformation on $L^2(\mathbb{C})$, then a straightforward computation shows that for $g \in L^2(\mathbb{C})$ we have

$$\begin{aligned}(\mathcal{F}\pi_1(a)g)(v) &= \int (\pi_1(a)g)(w)e^{-i(w,v)}dw = \\ &= e^t e^{i(z-(e^{qt}v,y))}(\mathcal{F}g)(e^{qt}v - x)\end{aligned}$$

which means that $\mathcal{F}\pi_1(a)g = \pi_2(a)\mathcal{F}g$ and proves our claim.

The condition $\mathfrak{h} \in P(f, \mathfrak{g})$ is necessary for the irreducibility of $\pi(f, \mathfrak{h}, G)$, but not sufficient: Let $\mathfrak{g} = \mathbb{R}a \oplus \mathbb{R}b$ with $[a, b] = b$, $f \in \mathfrak{g}^*$ with $f(a) = 0$, $f(b) = 1$. Then $\mathfrak{h} = \mathbb{R}a \in P(f, \mathfrak{g})$, but $\pi(f, \mathfrak{h}) = \lambda_H$, $H = E(\mathfrak{h})$, is not irreducible. In general $\pi(f, \mathfrak{h}, G)$ depends only on the restriction $f|_{\mathfrak{h}} = f_1 \in \mathfrak{h}^*$. Now it is plausible that such extensions of f_1 which are conjugate to f under some Ad^*x with $\mathrm{Ad}\, x(\mathfrak{h}) = \mathfrak{h}$, will define equivalent representations. The following definition takes care of this fact.

Definition. A polarization $\mathfrak{p} \in P(f, \mathfrak{g})$ satisfies *Pukanszky's condition* or is a *Pukanszky polarization*, if

$$f + \mathfrak{p}^{\perp} = \mathrm{Ad}^*P(f)$$

with $P = E(\mathfrak{p})$, in other words if for every $g \in \mathfrak{g}^*$ with $g|_{\mathfrak{p}} = f|_{\mathfrak{p}}$ there exists $x \in P$ with $g = \mathrm{Ad}^*x(f)$. The set of Pukanszky polarizations for f is denoted by $\mathrm{Puk}(f, \mathfrak{g})$.

The next proposition characterizes the Pukanszky polarizations:

Proposition 4. *For a polarization* $\mathfrak{p}$ *for* f *the following properties are equivalent:*
(i) $\mathfrak{p} \in \operatorname{Puk}(f, \mathfrak{g})$, *i.e.* $f + \mathfrak{p}^\perp = \operatorname{Ad}^* P(f)$
(ii) $f + \mathfrak{p}^\perp \subset \operatorname{Ad}^* G(f)$
(iii) $\mathfrak{p} \in P(f + h, \mathfrak{g})$ *for all* $h \in \mathfrak{p}^\perp$

Proof. The implication (i)$\Longrightarrow$(ii) is trivial. If $f + h = \operatorname{Ad}^* a(f)$ for some $a \in G$, then clearly $i(f + h, \mathfrak{g}) = i(f, \mathfrak{g})$, moreover, $h \in \mathfrak{p}^\perp$ implies that $\mathfrak{p}$ is subordinate to $f + h$. But $\dim \mathfrak{g}/\mathfrak{p} = i(f + h, \mathfrak{g})$, hence $\mathfrak{p} \in P(f + h, \mathfrak{g})$. This proves (ii) $\Longrightarrow$ (iii).

To show (iii) $\Longrightarrow$ (i) we observe that for any $l \in \mathfrak{g}^*$ with $l(\mathfrak{p}') = 0$ we have $\operatorname{Ad}^* P(l) - l \subset \mathfrak{p}^\perp$ i.e. $\operatorname{Ad}^* P(l) \subset l + \mathfrak{p}^\perp$, in particular $\operatorname{Ad}^* P(f + h) \subset f + \mathfrak{p}^\perp$ if (iii) holds. Furthermore (iii) implies $\dim \operatorname{Ad}^* P(f + h) = \dim(f + \mathfrak{p}^\perp)$, hence $\operatorname{Ad}^* P(f + h)$ is open in $f + \mathfrak{p}^\perp$. As the flat subspace $f + \mathfrak{p}^\perp$ is the disjoint union of the $\operatorname{Ad}^* P$–orbits there can be only one orbit: $f + \mathfrak{p}^\perp = \operatorname{Ad}^* P(f)$.

It is trivial that there exist maximal isotropic subspaces for B_f for any $f \in \mathfrak{g}^*$, however, it is not clear that they can be choosen to be subalgebras of $\mathfrak{g}$ and everything but trivial that $\operatorname{Puk}(f, \mathfrak{g})$ for a given $f \in \mathfrak{g}^*$ is non empty. Fortunately there is a very elegant and useful construction of Pukanszky polarizations in exponential Lie algebras. In § 1 we had defined Jordan–Hölder bases by means of certain composition series of $\mathfrak{g}$. These series were sequences $\{\mathfrak{g}_k\}_k$ of subalgebras $\mathfrak{g}_k$ of $\mathfrak{g}$ such that $\mathfrak{g}_{k+1}$ is an ideal in $\mathfrak{g}_k$ for $k = 1, 2, \ldots, n = \dim \mathfrak{g}$, $\mathfrak{g}_{n+1} = 0$, and which contain a composition series, thus either $\mathfrak{g}_k$ is an ideal in $\mathfrak{g}$ or $\mathfrak{g}_{k-1}$ and $\mathfrak{g}_{k+1}$ are ideals in $\mathfrak{g}$ and $\mathfrak{g}_{k-1}/\mathfrak{g}_{k+1}$ is a two dimensional irreducible $\mathfrak{g}$ module. We will call these series *Jordan–Hölder sequences* or JH *sequences.*

(36) *Let* $\mathfrak{h}$ *be a subalgebra in the exponential Lie algebra* $\mathfrak{g}$. *If* $\mathcal{S} = \{\mathfrak{g}_j\}_j$ *is a JH sequence of* $\mathfrak{g}$, *then* $\mathcal{S}_0 = \{\mathfrak{g}_j \cap \mathfrak{h}\}_j$ *is a JH sequence of* $\mathfrak{h}$.

Proof. Let $\mathfrak{h}_j = \mathfrak{g}_j \cap \mathfrak{h}$. Of course $\mathcal{S}_0$ is to be understood as the chain of *different* $\mathfrak{h}_{j_k}$ with $\dim \mathfrak{h}_{j_{k-1}} = \dim \mathfrak{h}_{j_k} + 1$. If $\mathfrak{h}_j$ is no ideal in $\mathfrak{h}$, then $\mathfrak{g}_j$ is no ideal in $\mathfrak{g}$, hence $\mathfrak{g}_{j-1}$ and $\mathfrak{g}_{j+1}$ are ideals in $\mathfrak{g}$ and $\mathfrak{h}_{j-1}$ and $\mathfrak{h}_{j+1}$ are ideals in $\mathfrak{h}$, moreover

$$\mathfrak{h}_{j-1}/\mathfrak{h}_{j+1} \cong (\mathfrak{h}_{j-1} + \mathfrak{g}_{j+1})/\mathfrak{g}_{j+1} = \mathfrak{g}_{j-1}/\mathfrak{g}_{j+1} \ ,$$

hence $\mathfrak{h}_{j-1}/\mathfrak{h}_{j+1}$ is an irreducible $\mathfrak{g}$ module. Since $\mathfrak{h}_j$ is no ideal in $\mathfrak{h}$, it is also a non trivial $\mathfrak{h}$ module. Then (30) implies that $\mathfrak{h}_{j-1}/\mathfrak{h}_{j+1}$ is $\mathfrak{h}$ irreducible, which proves our claim.

We use this result for

Proposition 5. *Let* $\mathcal{S} = \{\mathfrak{g}_k\}_k$ *be a Jordan–Hölder sequence of the exponential Lie algebra* $\mathfrak{g}$. *For* $f \in \mathfrak{g}^*$ *let* $f_k = f|_{\mathfrak{g}_k}$ *be the restriction of* f *to* $\mathfrak{g}_k$, *thus* $f_k \in \mathfrak{g}_k^*$. *If* *we write* $\mathfrak{r}_k = \mathfrak{g}_k^{f_k}$ *for the radical of the symplectic form* B_{f_k} *in* $\mathfrak{g}_k$, *then*

$$\mathfrak{v} := \mathfrak{v}(f, \mathcal{S}) := \sum_{k=1}^{n} \mathfrak{r}_k$$

is a Pukanszky polarization for f *in* $\mathfrak{g}$.

Proof. We use induction on $\dim \mathfrak{g} = n$. Obviously $\mathcal{S}_2 = \{\mathfrak{g}_k\}_{k=2,\ldots,n}$ is a JH sequence in $\mathfrak{g}_2$, hence $\mathfrak{v}_2 := \sum_{k=2}^{n} \mathfrak{r}_k = \mathfrak{v}(f_2, \mathcal{S}_2) \in$ Puk $(f_2, \mathfrak{g}_2)$, in particular $[\mathfrak{v}_2, \mathfrak{v}_2] \subset \mathfrak{v}_2$. We show first that $\mathfrak{v}$ is a subalgebra, i.e. that $a \in \mathfrak{r}_1$, $u \in \mathfrak{r}_i$, $i > 1$, implies $[a, u] \in \mathfrak{v}$. This is clear if $\mathfrak{g}_i$ is an ideal in $\mathfrak{g}$: If in this case $x \in \mathfrak{g}_i$, then $[[a, u]x] = [[a, x], u] + [a, [u, x]]$. But $[a, x] \in \mathfrak{g}_i$ and $u \in \mathfrak{r}_i$, hence $f_i([[a, x], u]) = 0$, similarly, $f_i([a, [u, x]]) = 0$, because $a \in \mathfrak{r}_1$. It follows that $f_i([[a, u], x]) = 0$, which means $[a, u] \in \mathfrak{r}_i \subset \mathfrak{v}$.

Let us now assume that $\mathfrak{g}_i$ is not an ideal. Then $\mathfrak{g}_{i-1}$ and $\mathfrak{g}_{i+1}$ are ideals and $\mathfrak{g}_{i-1}/\mathfrak{g}_{i+1}$ is $\mathfrak{g}$ irreducible. We set $v = [a, u]$ and claim that

$$f([v, \mathfrak{g}_{i+1}]) = 0 \,. \tag{37}$$

Indeed, for $x \in \mathfrak{g}_{i+1}$ we have $[v, x] = [[a, x], u]+[a, [u, x]]$, hence $[a, x] \in \mathfrak{g}_{i+1} \subset \mathfrak{g}_i$, $a \in \mathfrak{r}_1$ and $u \in \mathfrak{r}_i$ imply $f([v, x]) = 0$. If $v \in \mathfrak{g}_{i+1}$, then (37) implies $v \in \mathfrak{r}_{i+1}$. If $v \notin \mathfrak{g}_{i+1}$, then also $u \notin \mathfrak{g}_{i+1}$, hence $\mathfrak{g}_i = \mathbb{R}u \oplus \mathfrak{g}_{i+1}$ and since ad a acts irreducibly on $\mathfrak{g}_{i-1}/\mathfrak{g}_{i+1}$ we get

$$\mathfrak{g}_{i-1} = \mathbb{R}v \oplus \mathfrak{g}_i = \mathbb{R}v \oplus \mathbb{R}u \oplus \mathfrak{g}_{i+1} \,. \tag{38}$$

Next suppose that $f([u, v]) = 0$. From (37) and (38) we obtain $f([v, \mathfrak{g}_{i-1}]) = 0$, i.e. $v \in \mathfrak{r}_{i-1}$. Thus to prove $[\mathfrak{v}, \mathfrak{v}] \subset \mathfrak{v}$ we have to show that $f([u, v]) \neq 0$ is impossible.

Let $[a, v] = \alpha v+\beta u+y$ with $y \in \mathfrak{g}_{i+1}$. As $[[a, v], u] = [[a, u], v]+[a, [v, u]] = [a, [v, u]]$ we have $f([[a, v], u]) = 0$. On the other hand $[[a, v], u] = \alpha[v, u] + [y, u]$, hence $0 = \alpha f([v, u])$. But $\alpha \neq 0$, because otherwise $[a, u] \equiv v(\bmod\ \mathfrak{g}_{i+1})$ and $[a, v] \equiv \beta u\ (\bmod\ \mathfrak{g}_{i+1})$ would imply that ad$a$ would have purely imaginary eigenvalues. Thus $f([v, u]) = 0$ and $\mathfrak{v}' \subset \mathfrak{v}$ is proved.

It is evident that f vanishes on $\mathfrak{v}'$, so left over is to show that $\dim \mathfrak{g}/\mathfrak{v} = i(f, \mathfrak{g})$ and $\mathfrak{v}$ satisfies Pukanszky's condition. This will follow immediately from

(39) *Let* $\mathfrak{h}$ *be an ideal in* $\mathfrak{g}$ *with* $\dim \mathfrak{g}/\mathfrak{h} = 1$, $f \in \mathfrak{g}^*$, $f_1 = f|_{\mathfrak{h}} \in \mathfrak{h}^*$ *and* $\mathfrak{p}$ *a subalgebra of* $\mathfrak{g}$ *such that* $\mathfrak{p} = \mathfrak{g}^f + \mathfrak{p} \cap \mathfrak{h}$. *Then* $\mathfrak{p} \in$ Puk $(f, \mathfrak{g})$ *if and only if* $\mathfrak{p}_0 := \mathfrak{p} \cap \mathfrak{h} \in$ Puk $(f_1, \mathfrak{h})$.

Proof. Assume first that $\mathfrak{g}^f \subset \mathfrak{h}$, hence $\mathfrak{p} = \mathfrak{p}_0 \subset \mathfrak{h}$. Since $i(f, \mathfrak{g}) = i(f_1, \mathfrak{h})+1$ it is clear that $\mathfrak{p} \in P(f, \mathfrak{g})$ if and only if $\mathfrak{p} \in P(f_1, \mathfrak{h})$. Let $\mathfrak{p} \in$ Puk $(f, \mathfrak{g})$ and $h_1 \in \mathfrak{h}^*$

with $h_1(\mathfrak{p}) = 0$. h_1 extends to some $h \in \mathfrak{p}^\perp$ in $\mathfrak{g}^*$. Then proposition 4, (3) shows that $\mathfrak{p} \in P(f+h, \mathfrak{g})$, hence $\mathfrak{p} \in P(f_1 + h_1, \mathfrak{h})$ and then also $\mathfrak{p} \in \operatorname{Puk}(f_1, \mathfrak{h})$. With $\mathfrak{p} \in \operatorname{Puk}(f_1, \mathfrak{h})$, given $h \in \mathfrak{p}^\perp$ and $h_1 = h|_{\mathfrak{h}}$ the same argument works also in the opposite direction, thus (39) is proved if $\mathfrak{g}^f \subset \mathfrak{h}$. If $\mathfrak{g}^f \not\subset \mathfrak{h}$, then $\mathfrak{g}^f$ contains an element a with $\mathfrak{g} = \mathbb{R}a \oplus \mathfrak{h}$. Then clearly also $\mathfrak{g}^f = \mathbb{R}a \oplus \mathfrak{h}^{f_1}$ and $\mathfrak{p} = \mathbb{R}a \oplus \mathfrak{p}_0$, moreover $i(f, \mathfrak{g}) = i(f_1, \mathfrak{h})$, consequently $\mathfrak{p} \in P(f, \mathfrak{g})$ if and only if $\mathfrak{p}_0 \in P(f_1, \mathfrak{h})$. Suppose that $\mathfrak{p} \in \operatorname{Puk}(f, \mathfrak{g})$, i.e. $f + \mathfrak{p}^\perp = \operatorname{Ad}^*P(f)$, $P = E(\mathfrak{p})$. With $R = E(\mathbb{R}a)$ and $P_0 = E(\mathfrak{p}_0)$ evidently $P = R \ltimes P_0 = P_0R$ and because $a \in \mathfrak{g}^f$ we have also $\operatorname{Ad}^*R(f) = f$, hence $\operatorname{Ad}^*P(f) = \operatorname{Ad}^*P_0(f)$. Restriction to $\mathfrak{h}$ yields $f_1 + \mathfrak{p}_0^\perp = \operatorname{Ad}^*P_0(f_1)$ in $\mathfrak{h}^*$, which proves that $\mathfrak{p}_0 \in \operatorname{Puk}(f_1, \mathfrak{h})$. On the other hand, if $\mathfrak{p}_0 \in \operatorname{Puk}(f_1, \mathfrak{h})$ then $f_1 + \mathfrak{p}_0^\perp = \operatorname{Ad}^*P_0(f_1)$, which implies $f + \mathfrak{p}^\perp \subset \operatorname{Ad}^*P_0(f) + \mathfrak{h}^\perp$ in $\mathfrak{g}^*$. Let $\mathfrak{h}^\perp = \mathbb{R}k$ with $k(a) = 1$. Then for $f + l \in f + \mathfrak{p}^\perp$ there exists an element $p \in P_0$ and $\xi \in \mathbb{R}$ with $f + l = \operatorname{Ad}^*pf + \xi k$. Now clearly $\operatorname{Ad}^*p'f - f \in \mathfrak{p}^\perp$ for all $p' \in P$, hence $a \in \mathfrak{p}$ implies $(f + l)(a) = f(a) = (\operatorname{Ad}^*pf)(a) + \xi = f(a) + \xi$, i.e. $\xi = 0$ and consequently $f + l = \operatorname{Ad}^*pf$. Thus we have proved $f + \mathfrak{p}^\perp \subset \operatorname{Ad}^*P_0(f) = \operatorname{Ad}^*P(f)$. This finishes the proof of (39).

To apply (39) to the proof of proposition 5 we set $\mathfrak{h} = \mathfrak{g}_2$ and $\mathfrak{p} = \mathfrak{v}$. Since $\mathfrak{v} = \mathfrak{g}^f + \mathfrak{v}_2$ and $\mathfrak{v}_2 \in \operatorname{Puk}(f|_{\mathfrak{h}}, \mathfrak{h})$ by induction, we see that $\mathfrak{p} = \mathfrak{g}^f + \mathfrak{p} \cap \mathfrak{h}$ and $\mathfrak{p} \cap \mathfrak{h} = \mathfrak{p}_0 = \mathfrak{v}_2$. It follows that $\mathfrak{v} \in \operatorname{Puk}(\mathfrak{g}, f)$, and proposition 5 is proved. As an important corollary we note:

(40) *a) Let* $\{\mathfrak{a}_j\}_j$ *be a decreasing sequence of ideals in* $\mathfrak{g}$ *and* $f \in \mathfrak{g}^*$. *There exists a Pukanszky polarization* $\mathfrak{p}$ *for* f *in* $\mathfrak{g}$ *with* $\mathfrak{p} \cap \mathfrak{a}_j \in \operatorname{Puk}(f|_{\mathfrak{a}_j}, \mathfrak{a}_j)$ *for all* j.
b) Let $\mathfrak{a}$ *be a commutative ideal in* $\mathfrak{g}$. *For every* $f \in \mathfrak{g}^*$ *there exists a Pukanszky polarization* $\mathfrak{p}$ *for* f *containing* $\mathfrak{a}$.

Proof. a) Refine $\{\mathfrak{a}_j\}$ to a JH sequence $\mathcal{S}$ of $\mathfrak{g}$ and set $\mathfrak{p} = \mathfrak{v}(f, \mathcal{S})$. Then $\mathfrak{a}_j = \mathfrak{g}_i$ for some i, hence $\mathfrak{p} \cap \mathfrak{a}_j \supset \mathfrak{v}_i = \sum_{k=i}^{n} \mathfrak{r}_k = \mathfrak{v}(f_i, \{\mathfrak{g}_k\}_{k \geq i}) \in \operatorname{Puk}(f_i, \mathfrak{g}_i)$. Since $\mathfrak{p} \cap \mathfrak{a}_j$ is subordinate to f_i we have $\mathfrak{p} \cap \mathfrak{a}_j = \mathfrak{v}_i \in \operatorname{Puk}(f|_{\mathfrak{a}_j}, \mathfrak{a}_j)$. b) follows from a) by taking $\{\mathfrak{a}_j\} = \{\mathfrak{a}\}$.

Polarizations of the kind described in proposition 5 we will call *Vergne polarizations.*

Theorem 7. *Let* $f \in \mathfrak{g}^*$, $\mathfrak{p} \in P(f, \mathfrak{g})$ *and* $\pi = \pi(f, \mathfrak{p}, G)$. *Then* π *is irreducible if and only if* $\mathfrak{p}$ *satisfies Pukanszky's condition, i.e.* $\mathfrak{p} \in \operatorname{Puk}(f, \mathfrak{g})$. *Moreover, in this case* π *is independent of the choice of* $\mathfrak{p}$ *in* $\operatorname{Puk}(f, \mathfrak{g})$.

Proof by induction on $n = \dim G$. If f is zero on a nontrivial ideal $\mathfrak{a}$, then $\mathfrak{a}$ is contained in all $\mathfrak{p} \in P(f, \mathfrak{g})$ and π is trivial on $A = E(\mathfrak{a})$, so the situation is essentially reduced to the quotient group G/A and we can apply induction. We leave the details to the reader and assume for the rest of the proof that $f(\mathfrak{a}) \neq 0$

for every ideal $\mathfrak{a} \neq 0$ in $\mathfrak{g}$; in particular the dimension of the center $\mathfrak{z}$ of $\mathfrak{g}$ is 0 or 1.

Let $\pi = \pi(f, \mathfrak{p}, G)$ be irreducible and let us fix a minimal noncentral ideal $\mathfrak{a}$ in $\mathfrak{g}$. We distinguish two cases with respect to $f \in \mathfrak{g}^*$, $\mathfrak{p} \in P(f, \mathfrak{g})$ and $\mathfrak{a}$:
A. $\mathfrak{p}$ contains $\mathfrak{a}$. Then $\mathfrak{a}^f$ is a proper subalgebra of $\mathfrak{g}$ and $\mathfrak{p} \subset \mathfrak{a}^f$. In the cases $(\mathrm{a_d})$ – see the proof of (29) – $\mathfrak{a}^f$ coincides with the centralizer $\mathfrak{c}$ of $\mathfrak{a}$ in $\mathfrak{g}$ and we can write $\mathfrak{g} = \mathbb{R}x \oplus \mathfrak{c} = \mathbb{R}x \oplus \mathfrak{a}^f$. Then $[x, \mathfrak{a}] = \mathfrak{a}$ and as $f(\mathfrak{a}) \neq 0$ there is an $y \in \mathfrak{a}$ with $f([x, y]) = 1$, i.e. $(\mathrm{ad}^* yf)(x) = 1$. Evidently $\mathrm{ad}^* yf$ vanishes on $\mathfrak{c}$ and $(\mathrm{ad}^* y)^2 f = 0$, hence $(\mathrm{Ad}^* \exp ty) f = f + th$, where $h = \mathrm{ad}^* yf$ is a generator of $(\mathfrak{a}^f)^\perp$ in $\mathfrak{a}^*$. Thus, with $A = E(\mathfrak{a})$, we have shown

$$f + (\mathfrak{a}^f)^\perp \subset \mathrm{Ad}^* A(f) \ . \tag{41}$$

In cases $(\mathrm{b_d})$ we have $\mathfrak{a} = \mathfrak{a}_0 \oplus \mathbb{R}z$ with $\mathfrak{a}_0 = \{y \in \mathfrak{a};\ f(y) = 0\}$. For $d = 1$ we set $\mathfrak{a}_0 = \mathbb{R}y$. Then there are $\varphi, \psi \in \mathfrak{g}^*$ such that

$$[x, y] = \varphi(x) y + \psi(x) z \quad \text{for all} \quad x \in \mathfrak{g} \ .$$

From $f(y) = 0$, $f(z) = 1$ and the last equation we derive $\mathrm{ad}^* yf = \psi$, hence $\mathfrak{a}^f = \ker(\mathrm{ad}^* yf) = \ker \psi$ and $(\mathfrak{a}^f)^\perp = \mathbb{R}\psi$ which again implies (41).

For $d = 2$ there exist $u, v \in \mathfrak{a}_0$, $q = 1 + i\lambda \in \mathbb{C}$, $\varphi, \psi_1, \psi_2 \in \mathfrak{g}^*$ such that for $w = u + iv \in \mathfrak{g}^{\mathbb{C}}$ and $\psi = \psi_1 + i\psi_2$ we have $[x, w] = q\varphi(x) w + \psi(x) z$, hence $f([x, u]) = \psi_1(x)$, $f([x, v]) = \psi_2(x)$, which means

$$\psi_1 = \mathrm{ad}^* uf \quad , \quad \psi_2 = \mathrm{ad}^* vf \ .$$

Since $\mathfrak{a}^f = \ker \psi_1 \cap \ker \psi_2$ we get

$$(\mathfrak{a}^f)^\perp = \mathrm{ad}^* \mathfrak{a}(f) \ ,$$

so also in case $(\mathrm{b_2})$ be inclusion (41) is true.

Now let us assume that $\pi = \pi(f, \mathfrak{p}, G)$ is irreducible. We induce in stages via $A^f := E(\mathfrak{a}^f)$. Then also $\pi_1 := \pi(f|_{\mathfrak{a}^f}, \mathfrak{p}, A^f)$ is irreducible, which implies that $\mathfrak{p} \in \mathrm{Puk}(f|_{\mathfrak{a}^f}, \mathfrak{a}^f)$ and that π_1 and $\pi = \mathrm{ind}_{A^f}^G \pi_1$ are independent of polarizations containing $\mathfrak{a}$. The fact that $\mathfrak{p}$ as a polarization for π_1 has Pukanszky's property means that in $\mathfrak{g}^*$ the inclusion

$$f + \mathfrak{p}^\perp \subset \mathrm{Ad}^* P f + (\mathfrak{a}^f)^\perp$$

holds. We conclude from (41) that

$$f + \mathfrak{p}^\perp \subset \mathrm{Ad}^* P(f + (\mathfrak{a}^f)^\perp) \subset \mathrm{Ad}^* P \mathrm{Ad}^* A(f) = \mathrm{Ad}^* P(f) \ ,$$

i.e. $f + \mathfrak{p}^\perp = \mathrm{Ad}^* P(f)$. Thus $\mathfrak{p} \in \mathrm{Puk}(f, \mathfrak{g})$.
B. $\mathfrak{p}$ does not contain $\mathfrak{a}$. Then $\mathfrak{k} := \mathfrak{p} + \mathfrak{a}$ is a proper superalgebra of $\mathfrak{p}$. As we have seen the irreducibility of π excludes now the cases $(\mathrm{a_d})$. In the course of

proving (32) we have shown that in this case $\mathfrak{k}$ contains another polarization $\hat{\mathfrak{p}}$ with $\mathfrak{a} \subset \hat{\mathfrak{p}}$ such that, with $f_1 = f|_{\mathfrak{k}}$, the representations $\pi(f_1, \mathfrak{p}, K)$ and $\pi(f_1, \hat{\mathfrak{p}}, K)$ are equivalent, hence $\pi(f, \mathfrak{p}, G)$ and $\pi(f, \hat{\mathfrak{p}}, G)$ are equivalent and $\hat{\mathfrak{p}} \in \text{Puk}\,(f, \mathfrak{g})$, because $\mathfrak{a} \subset \hat{\mathfrak{p}}$. But also $\mathfrak{p} \in \text{Puk}\,(f, \mathfrak{g})$. This follows directly from the explicit description ($\mathrm{a_d}$), resp. ($\mathrm{b_d}$) of $\mathfrak{k}$ in case $\mathfrak{k} = \mathfrak{g}$, we omit the easy computations. If $\mathfrak{k} \neq \mathfrak{g}$, then $\mathfrak{p} \in \text{Puk}\,(f_1, \mathfrak{k})$ by induction, hence $f + \mathfrak{p}^\perp \subset \text{Ad}^*P(f) + \mathfrak{k}^\perp = \text{Ad}^*P(f + \mathfrak{k}^\perp)$. On the other hand $\hat{\mathfrak{p}} \in \text{Puk}\,(f, \mathfrak{g})$ and $\hat{\mathfrak{p}} \subset \mathfrak{k}$ implies $f + \mathfrak{k}^\perp \subset f + \hat{\mathfrak{p}}^\perp \subset \text{Ad}^*G(f)$, hence

$$f + \mathfrak{p}^\perp \subset \text{Ad}^*P(f + \mathfrak{k}^\perp) \subset \text{Ad}^*P\text{Ad}^*G(f) = \text{Ad}^*G(f)\ ,$$

which proves our claim.

To prove that Pukanszky's condition is sufficient for irreducibility we only have to show that every Puk $(f, \mathfrak{g})$ contains at least one polarization $\mathfrak{p}$, for which $\pi = \pi(f, \mathfrak{p}, G)$ is irreducible. Since we know already that an irreducible π is independent of $\mathfrak{p}$, it will follow that *every* Pukanszky polarization defines an irreducible representation.

Let $\mathfrak{a}$ be a fixed minimal noncentral ideal and $\mathfrak{e}$ its centralizer. In cases ($\mathrm{a_1}$) and ($\mathrm{a_2}$) $\mathfrak{e}$ is an ideal of codimension 1. We take a Vergne polarization $\mathfrak{p}$ in Puk $(f, \mathfrak{g})$, defined by a JH sequence containing $\mathfrak{a}$ and $\mathfrak{e}$ *and assume that* $\mathfrak{g}^f \subset \mathfrak{e}$, hence $\mathfrak{a} \subset \mathfrak{p} \subset \mathfrak{e}$. Let $A = E(\mathfrak{a})$, $P = E(\mathfrak{p})$ and $H = E(\mathfrak{e})$ be the corresponding subgroups, so A is normal in G and central in H, moreover $A \subset P$. Let $\mathfrak{g} = \mathbb{R}a \oplus \mathfrak{e}$, hence ad a acts irreducibly on $\mathfrak{a}$. Again we identify A with $\mathbb{R}$ in case ($\mathrm{a_1}$) and with $\mathbb{C}$ in case ($\mathrm{a_2}$). Then

$$\exp ta \cdot x \cdot \exp(-ta) = e^{qt} x \tag{42}$$

for $t \in \mathbb{R}$, $x \in A$ and $q = 1$ in case ($\mathrm{a_1}$), $q = 1 + i\lambda$, $\lambda \neq 0$, in case ($\mathrm{a_2}$).

If $f_1 = f|_{\mathfrak{e}}$, then $\mathfrak{p} \in \text{Puk}\,(f_1, \mathfrak{e})$, hence $\pi_1 := \pi(f_1, \mathfrak{p}, H) \in \widehat{H}$ by induction, and $\pi := \pi(f, \mathfrak{p}, G) = \text{ind}_H^G \pi_1$. The character $\chi_1 = \chi(f_1, \mathfrak{p})$ is nontrivial on A, so we can assume that $\chi_1(x) = e^{i\,Re\,x}$ for $x \in A$ (i.e. $\chi_1(x) = e^{ix}$ in the real case $A = \mathbb{R}$). Since A is central in H it follows that $\pi_1(x) = e^{i\,Re\,x}\mathbf{1}$ for $x \in A$. Now $G = R \ltimes H$ with $R := E(\mathbb{R}a)$, hence $t \to (\exp ta)H \in \overline{G} = G/H$ parametrizes $\overline{G}$. Let $\pi_1^t \in \widehat{H}$ be the representation $y \to \pi_1(\exp ta \cdot y \cdot \exp(-ta))$. Then (42) yields $\pi_1^t(x) = e^{i\,Re\,e^{qt}x} \cdot \mathbf{1}$. We extend the π_1^t to representations of $L^1(H)^b = \mathcal{M}(H)$, the algebra of all bounded Radon measures on H. Then the restriction of π_1^t to $L^1(A) \subset \mathcal{M}(H)$ is just given by $\pi_1^t(\varphi) = \hat{\varphi}(e^{\overline{q}t}) \cdot \mathbf{1}$, which implies immediately that the kernels of the π_1^t are all different in $L^1(A)$, hence also in $L^1(H)$: For $s, t \in \mathbb{R}$, $s \neq t$, we take $\varphi \in L^1(A)$ with $\hat{\varphi}(e^{\overline{q}t}) = 0$, $\hat{\varphi}(e^{\overline{q}s}) = 1$. Then $\varphi \star L^1(H) \subset \ker_{L^1(H)} \pi_1^t$, but $\pi_1^s(\varphi \star L^1(H)) = \pi_1^s(L^1(H)) \neq 0$. From theorem 6 we derive that $\pi = \text{ind}\ \pi_1$ is irreducible.

Now let us assume that $\mathfrak{z} = \mathbb{R}z$ is the center of $\mathfrak{g}$ and the only minimal nonzero ideal in $\mathfrak{g}$. Then we are in case ($\mathrm{b_1}$) or ($\mathrm{b_2}$) and $\mathfrak{w} = \mathfrak{a}/\mathfrak{z}$ is a minimal $\mathfrak{g}$ module. Again let $\mathfrak{e}$ the centralizer of $\mathfrak{a}$ in $\mathfrak{g}$ and $\mathfrak{d}$ the annihilator of $\mathfrak{w}$. If $\mathfrak{d} = \mathfrak{g}$, then $\mathfrak{w}$ is a trivial $\mathfrak{g}$ module, thus $\dim \mathfrak{w} = 1$, $\mathfrak{a} = \mathbb{R}y \oplus \mathbb{R}z$, $f(y) = 0$ and there exists $x \in \mathfrak{g}$ with $\mathfrak{g} = \mathbb{R}x \oplus \mathfrak{e}$, $[x, y] = z$. Then as before, $A = E(\mathfrak{a})$ is central in$H = E(\mathfrak{e})$ and $\mathfrak{g}^f$ is contained in $\mathfrak{e}$. This implies the existence of a Vergne polarization $\mathfrak{p}$ with $\mathfrak{a} \subset \mathfrak{p} \subset \mathfrak{e}$. For $a = \exp(\xi_1 y + \xi_2 z) \in A$ we have now $\chi_1(a) = e^{i\xi_2}$, hence $\pi_1(a) = e^{i\xi_2} \cdot \mathbf{1}$ and – notations analogous to our first case – $\pi_1^t(a) = e^{i(\xi_2 + t\xi_1)} \cdot \mathbf{1}$. Theorem 6 shows that $\pi = \operatorname{ind}_H^G \pi_1 \in \widehat{G}$ in the same way as before.

Now suppose that $\mathfrak{g} \neq \mathfrak{d}$. Then $\mathfrak{g} = \mathbb{R}a \oplus \mathfrak{d}$ and ad a acts irreducibly on $\mathfrak{w}$. We take a Vergne polarization $\mathfrak{p}$ for a JH sequence through $\mathfrak{a}$, $\mathfrak{e}$ and $\mathfrak{d}$ *and assume again that* $\mathfrak{g}^f \subset \mathfrak{d}$, hence also $\mathfrak{p} \subset \mathfrak{d}$. We claim that even $\mathfrak{p} \subset \mathfrak{e}$: Let $f_1 = f|_{\mathfrak{d}}$ and $x \in \mathfrak{d}^{f_1}$. On p 40 we had seen that the bilinear form β defined on $(\mathfrak{d}/\mathfrak{e}) \times (\mathfrak{a}/\mathfrak{z})$ by $[x, y] = \beta(x, y)z$ is nondegenerate. Since $x \in \mathfrak{d}^{f_1}$ implies $\beta(x, y) = f([x, y]) = 0$ for all $y \in \mathfrak{a}$ we get $x \in \mathfrak{e}$. The same argument works (in case (b$_2$)) for any algebra $\mathfrak{j}$ properly between $\mathfrak{e}$ and $\mathfrak{d}$.

The subalgebra $\mathfrak{h} := \mathbb{R}a \oplus \mathfrak{e}$ is not an ideal in $\mathfrak{g}$, however $\mathfrak{p}$ is also a Pukanszky polarization in $\mathfrak{h}$ for $f_1 = f|_{\mathfrak{h}}$, so $\pi_1 := \pi(f_1, \mathfrak{p}, H)$, $H = E(\mathfrak{h})$, is irreducible. We will now apply theorem 5 to show that then also $\pi = \operatorname{ind}_H^G \pi_1$ is irreducible. To this end it is sufficient to prove that the subalgebra $\mathfrak{Q} \subset \mathcal{L}(\operatorname{ind} \mathfrak{H}(\pi_1))$ of multiplication operators with bounded measurable functions on G/H is contained in the von Neumann algebra $\pi(G)''$.

It is easy to see that we can choose the element a in $\mathfrak{h}$ so that $\mathfrak{a}_0 = \{x \in \mathfrak{a}; f(x) = 0\}$ is invariant under ad a. This implies that we can identify A with $A_0 \times \mathbb{R}$, where $A_0 = E(\mathfrak{a}_0) = \mathbb{R}$ in case (b$_1$) and $A_0 = \mathbb{C}$ in case (b$_2$) and such that (42) holds for $x \in A_0$. Thus if φ is a function in the representation space $\mathfrak{h}(\pi_1)$, $x = \exp ta \cdot c \in H$ with $c \in E(\mathfrak{e})$, and $(y, s) \in A_0 \times \mathbb{R}$, then

$$(\pi_1(y, s)\varphi)(x) = \varphi((-y, -s) \exp ta \cdot c) = \varphi(\exp ta \cdot c \cdot (-e^{-qt} y, s)) = e^{is}\varphi(x)$$

hence $\pi_1(y, s) = e^{is}\mathbf{1}$.

For the next steps in our proof we consider only case (b$_2$), the arguments for (b$_1$) are the same, only the computations are simpler. So now $\mathfrak{a}_0 = \mathbb{R}u_1 \oplus \mathbb{R}u_2$, $\mathfrak{d} = \mathbb{R}b_1 \oplus \mathbb{R}b_2 \oplus \mathfrak{e}$, where b_i and u_j are so choosen that $[b_i, u_j] = \delta_{ij}z$, $\exp(\xi_1 u_1 + \xi_2 u_2) = \xi_1 + i\xi_2 \in A_0 = \mathbb{C}$. Observe that $\{b_1, b_2\}$ is a coexponential basis for $\mathfrak{h}$ in $\mathfrak{g}$ with $\exp s_1 b_1 \exp s_2 b_2 H = \exp(s_1 b_1 + s_2 b_2)H$, the mapping $(s_1, s_2) \to e(s_1, s_2)H \in G/H$ is a diffeomorphism from $\mathbb{R}^2$ onto G/H. Furthermore,

$$e(s_1, s_2)(t_1 + it_2, r)e(s_1, s_2)^{-1} = (t_1 + it_2, r + s_1 t_1 + s_2 t_2) \ .$$

Now let ξ be an element in the representation space of $\pi = \operatorname{ind}_H^G \pi_1$. Then we have for $y = t_1 + it_2 \in A_0$ and $\tilde{y} = (y, 0) \in A$:

$$\begin{aligned}(\pi(\tilde{y})\xi)(e(s_1, s_2) \cdot \exp ra \cdot c) &= \xi((-y, 0)e(s_1, s_2) \cdot \exp ra \cdot c) = \\ &= \xi(e(s_1, s_2) \cdot \exp ra \cdot c \cdot (-e^{-qr} y, s_1 t_1 + s_2 t_2)) = \\ &= e^{i(s_1 t_1 + s_2 t_2)} \xi(e(s_1, s_2) \exp ra \cdot c) \ .\end{aligned}$$

This yields for any $g \in L^1(A_0) = L^1(\mathbb{R}^2) \subset \mathcal{M}(G)$ with $x = \exp ra \cdot c \in H$, $t = (t_1, t_2)$, $s = (s_1, s_2) \in \mathbb{R}^2$ and $(t, s) = t_1 s_1 + t_2 s_2$:

$$\pi(g)\xi(e(s)x) = \int_{\mathbb{R}^2} g(t)e^{i(t,s)}dt \cdot \xi(e(s)x) = \hat{g}(s)\xi(e(s)x)$$

hence $\pi(g)$ is the multiplication operator with $\hat{g}$, thus $\hat{g} \in \pi(G)''$. As these $\hat{g}$ form a dense subalgebra of $\mathcal{C}_\infty(G/H)$ and consequently are weakly dense in $\mathfrak{Q}$, it follows that $\mathcal{G} \subset \pi(G)''$. From Theorem 5 we derive that $\pi(G)' = \pi(G)' \cap \mathfrak{Q}' \doteq \pi_1(H)' \doteq \mathbb{C}$, hence $\pi \in \widehat{G}$.

Finally we have to discuss the following situation: $\mathfrak{g}$ contains an ideal $\mathfrak{h}$ of codimension 1, which does not contain $\mathfrak{g}^f$, i.e. $\mathfrak{g} = \mathfrak{h} + \mathfrak{g}^f$.

Our next proposition will handle this situation, but first we need the following fact:

(43) *Let the exponential group G act on the finite dimensional real vectorspace V, so that the transformations $T_x : \xi \to x\xi \in V$, $x \in G$ have no eigenvalues $\zeta \in \mathbb{C}$ with $|\zeta| = 1$, except for $\zeta = 1$. Then also the Lie algebra $\mathfrak{g}$ acts on V and has no purely imaginary eigenvalues different from 0. For $\xi \in V$ let $\mathfrak{g}^\xi = \{X \in \mathfrak{g}; X\xi = 0\}$ be the stabilizer of ξ in $\mathfrak{g}$. Then $E(\mathfrak{g}^\xi) = \{\exp X ; X \in \mathfrak{g}^\xi\}$ is the stabilizer $G^\xi := \{x \in G; x\xi = \xi\}$ of ξ in G.*

Proof. Obviously $E(\mathfrak{g}^\xi) \subset G^\xi$, thus it suffices to show that $X \in \mathfrak{g}$, $\exp X\xi = \xi$ implies $X\xi = 0$. We decompose V into the Jordan components of X: $V = \sum^\oplus V_j$, $XV_j \subset V_j$ and $X|_{V_j} = X_j$ has exactly one eigenvalue λ_j on V_j. Then $x = \exp X$ restricts to $x_j = \exp X_j$ on V_j and has the eigenvalue e^{λ_j} on V_j. Now $e^{\lambda_j} = 1$ if and only if $\lambda_j = 0$, i.e. X_j is nilpotent. Let $\xi = \sum \xi_j$ with $\xi_j \in V_j$, hence $x\xi = \sum x_j\xi_j = \sum \xi_j$ if and only if $x_j\xi_j = \xi_j$ for all j. This implies $\xi_j = 0$ for $\lambda_j \neq 0$ and consequently $X^{m+1}\xi = 0$ for some $m \in \mathbb{N}$. Thus $x\xi = e^X\xi = (\mathbf{1} + X + \cdots + \frac{1}{m!}X^m)\xi = \xi$ implies $(\mathbf{1} + \frac{1}{2}X + \cdots + \frac{1}{m!}X^{m-1})X\xi = 0$ and finally $X\xi = 0$.

Corollary: *The stabilizer G^f of $f \in \mathfrak{g}^*$ in G with respect to the coadjoint representation of G in $\mathfrak{g}^*$ is equal to $E(\mathfrak{g}^f)$, in particular G^f is always connected.*

Proposition 6. *Let $\mathfrak{h}$ be an ideal of codimension 1 in $\mathfrak{g}$, $f \in \mathfrak{g}^*$, $f_1 = f|_\mathfrak{h} \in \mathfrak{h}^*$ the restriction,*

$$\Omega_f = \{\mathrm{Ad}^* x f \in \mathfrak{g}^* ; x \in G\} \subset \mathfrak{g}^*$$

the orbit of f in $\mathfrak{g}^*$ and $\Omega_{f_1} \subset \mathfrak{h}^*$ the orbit of f_1 in $\mathfrak{h}^*$. Let $\mathfrak{p}$ be a polarization for f and $\mathfrak{p}_0 = \mathfrak{p} \cap \mathfrak{h} \in P(f_1, \mathfrak{h})$. If $\mathfrak{g}^f \not\subset \mathfrak{h}$ and $f_1 \neq 0$, then

(i) *the restriction mapping $l \longrightarrow l_1 = l|_\mathfrak{h} \in \mathfrak{h}^*$ from $\mathfrak{g}^*$ to $\mathfrak{h}^*$ maps Ω_f bijectively onto Ω_{f_1},*

(ii) *the restriction* $\pi' = \pi(f, \mathfrak{p}, G)|_H$ *of the representation* $\pi = \pi(f, \mathfrak{p}, G)$ *to* H *is equivalent to* $\pi_1 = \pi(f_1, \mathfrak{p}_0, H)$

(iii) π *is irreducible if and only if* π_1 *is irreducible.*

Proof. As $\mathfrak{g}^f \not\subset \mathfrak{h}$ and $f_1 \neq 0$ we can choose $a \in \mathfrak{g}^f$ with $f(a) = 0$ and $\mathfrak{g} = \mathbb{R}a \oplus \mathfrak{h}$. We set $R = E(\mathbb{R}a)$, hence $G = R \ltimes H$. Since R is contained in the stabilizer G^f of f, we conclude that $\Omega_f = \{\mathrm{Ad}^*xf\,;\, x \in H\}$ and consequently that the restriction $l \longrightarrow l_1$ maps Ω_f onto Ω_{f_1}. If Ad^*xf, $x \in H$, also restricts to f_1, i.e. if $(\mathrm{Ad}^*xf)_1 = \mathrm{Ad}^*x \cdot f_1 = f_1$, then $x \in H^{f_1}$. From $\mathfrak{h}^{f_1} = \mathfrak{g}^f \cap \mathfrak{h}$, see (39), and the corollary of (43) it follows that $H^{f_1} = E(\mathfrak{g}^f \cap \mathfrak{h}) = G^f \cap H$, hence $x \in G^f$ and $\mathrm{Ad}^*xf = f$. This proves (i).

To prove (ii) we denote the representation spaces of π, π' and π_1 by $\mathfrak{H}$ and $\mathfrak{H}_1$ and remark that $\mathfrak{p} = \mathbb{R}a \oplus \mathfrak{p}_0$ and $P = R \ltimes P_0 = E(\mathfrak{p})$. As $f(a) = 0$ the character $\chi := \chi(f, \mathfrak{p})$ is trivial on R, hence $\chi(r, y) = \chi_1(y)$ with $\chi_1 = \chi(f_1, \mathfrak{h})$ and $\chi_1(y^s) = \chi_1(y)$ with our usual notation $(0, y^s) = (-s, e)(0, y)(s, e)$.

Recall that $\mathfrak{H}$ is the space of all measurable $\mathbb{C}$–valued functions φ satisfying

$$\varphi((s, x)(t, y)) = \overline{\chi_1(y)}\varphi(s, x) \tag{44}$$

for all $(s, x) \in G$ and $(t, y) \in P$ and $\int\limits_{G/P} |\varphi|(\overline{x})^2 d\overline{x} < \infty$. Similarly $\mathfrak{H}_1$ consists of all ψ with

$$\psi(xy) = \overline{\chi_1(y)}\psi(x) \ , \quad y \in P_0 \ , \tag{45}$$

and $\int\limits_{H/P_0} |\psi|(\overline{x})^2 d\overline{x} < \infty$. For $y = e$, $t = -s$ we obtain from (44):

$$\varphi(s, x) = \varphi(0, x^{-s})$$

This suggests defining an operator U on $\mathfrak{H}_1$ by

$$U\psi(s, x) = \psi(x^{-s})$$

for $(s, x) \in G$. Clearly $U\psi$ is measurable and for $(t, y) \in P$ we have

$$U\psi((s, x)(t, y)) = \psi(x^{-s}y^{-(s+t)}) = \overline{\chi_1(y)}\psi(x^{-s}) = \overline{\chi_1(y)}U\psi(s, x) \ ,$$

hence $U\psi$ satisfies (45).

To compute the norms in $\mathfrak{H}$ and $\mathfrak{H}_1$ we take a coexponential basis $\mathcal{B} = \{b_j\}_j$ for $\mathfrak{p}$ in $\mathfrak{g}$. As $a \in \mathfrak{p}$ we can take all b_j in $\mathfrak{h}$ and since a $\mathfrak{p}$ module is a fortiori a $\mathfrak{p}_0$ module we are allowed to assume that $\mathcal{B}$ is also a coexponential basis for $\mathfrak{p}_0$ in $\mathfrak{h}$. Next, let c be an extension of χ_P^G onto G, see p 10. Then the restriction $c|_H$ is also an extension of $\chi_{P_0}^H$, because $\mathfrak{g}/\mathfrak{p}$ and $\mathfrak{h}/\mathfrak{p}_0$ are evidently isomorphic $\mathfrak{p}_0$ modules. It follows that the same formula in theorem 3 defines both integrals on G/P and

H/P_0, with $\overline{\xi} = E(\xi)P \in G/P$ or $\overline{\xi} = E(\xi)P_0 \in H/P_0$, where

$$E(\xi) := E(\xi_1 b_1)E(\xi_2 b_2)\ldots E(\xi_d b_d)\ .$$

Now let $\psi \in \mathfrak{H}_1$. Then $U\psi(0, E(\xi)) = \psi(E(\xi))$,

$$|U\psi|(\overline{\xi}) = |U\psi(0, E(\xi))| = |\psi(E(\xi))| = |\psi|(\overline{\xi})$$

and consequently the norm of $U\psi$ in $\mathfrak{H}$ is the same as the norm of ψ in $\mathfrak{H}_1$.

If $\varphi \in \mathfrak{H}$ is continuous, then $\psi : y \longrightarrow \varphi(0, y)$ is in $\mathfrak{H}_1$ and $U\varphi = \psi$. It follows that U maps $\mathfrak{H}_1$ unitarily onto $\mathfrak{H}$. Finally, for $x \in H$ and the real character c, defined above, we have

$$\begin{aligned}(U\pi_1(x)\psi)(s, y) &= (\pi_1(x)\psi)(y^{-s}) = c(x)^{\frac{1}{2}}\psi(x^{-1}y^{-s}) = \\ &= c(x)^{\frac{1}{2}}\psi(((x^s)^{-1}y)^{-s}) = c(x)^{\frac{1}{2}}(U\psi)(s, (x^s)^{-1}y) = \\ &= c(x)^{\frac{1}{2}}U\psi((0, x)^{-1}(s, y) = (\pi(0, x)U\psi)(s, y)\end{aligned}$$

which implies $\pi_1 = U^*\pi' U$ and proves proposition 6 (ii), hence the proposition, because (iii) is a trivial consequence of (ii).

Now let $\mathfrak{p} \in$ Puk $(f, \mathfrak{g})$. Then $\mathfrak{p}_0 \in$ Puk $(f_1, \mathfrak{h})$ by (39), hence $\pi_1 \in \widehat{H}$ by induction and so finally $\pi \in \widehat{G}$ by proposition 6 (iii). This finishes also the proof of theorem 7.

Again let us take an f in $\mathfrak{g}^*$. Then proposition 5 guaranties the existence of a Pukanszky polarization $\mathfrak{p}$ for f and from theorem 7 we see that the representation $\pi(f, \mathfrak{p}, G)$ of G is irreducible and independent of $\mathfrak{p}$, i.e. we obtain a mapping $f \to \pi(f) = \pi(f, \mathfrak{p}, G)$ from $\mathfrak{g}^*$ into the unitary dual $\widehat{G}$ of G. Moreover, after (29) this mapping is onto $\widehat{G}$. For $a \in G$ let ι_a be the inner automorphism of G, defined by a. The composed representation $\pi(f) \circ \iota_a$ obviously is equivalent to $\pi(f)$ and it is easily verified that $\pi(f, \mathfrak{p}) \circ \iota_a = \pi(\mathrm{Ad}^*a^{-1}f, \mathrm{Ad}\, a(\mathfrak{p}))$, which shows, that the equivalence class of $\pi(f)$ depends only on the orbit

$$\Omega_f := \{\mathrm{Ad}^*xf\,;\ x \in G\} \subset \mathfrak{g}^*$$

of the linear form $f \in \mathfrak{g}^*$. Thus

$$K : \Omega_f \to \pi(f) \in \widehat{G} \tag{46}$$

defines a mapping K from the *orbit space*

$$\Omega(\mathfrak{g}^*) := \{\Omega_f\,;\ f \in \mathfrak{g}^*\}$$

of all Ad^*G orbits in $\mathfrak{g}^*$ onto the unitary dual $\widehat{G}$ of G.

Definition. The mapping K from $\Omega(\mathfrak{g}^*)$ onto $\widehat{G}$ is called the *Kirillov map*.

Our aim is to prove

Theorem 8. *The Kirillov map is a bijection from $\Omega(\mathfrak{g}^*)$ onto $\widehat{G}$.*

We know already that K is well defined and onto, so all we have to do is to prove the injectivity of K. For this we apply again induction.

Let f_1 and f_2 be linear forms in $\mathfrak{g}^*$ such that $\pi_1 = K(\Omega_{f_1})$ and $\pi_2 = K(\Omega_{f_2})$ are equivalent. If $\mathfrak{a}$ is a nontrivial ideal in $\mathfrak{g}$ with $f_1(\mathfrak{a}) = 0$, then π_1 is trivial on $A = E(\mathfrak{a})$, hence also π_2 is trivial on A, which implies that $f_2(\mathfrak{a}) = 0$. But then f_1 and f_2 define $\overline{f}_1$ and $\overline{f}_2$ on $\overline{\mathfrak{g}} := \mathfrak{g}/\mathfrak{a}$ and equivalent representations $K(\Omega_{\overline{f}_1})$ and $K(\Omega_{\overline{f}_2})$ on $\overline{G} = G/A$. Hence $\overline{f}_2 = \mathrm{Ad}^*\overline{x}\overline{f}_1$ for some $\overline{x} \in \overline{G}$. This means $\overline{f}_2 = (\mathrm{Ad}^* x f_1)^-$ for $x \in G$ and consequently $f_2 = \mathrm{Ad}^* x f_1$, i.e. $\Omega_{f_2} = \Omega_{f_1}$.

Now suppose that f_1 and f_2 are nontrivial on all nonzero ideals in $\mathfrak{g}$. Then the dimension of the center $\mathfrak{z}$ of $\mathfrak{g}$ is at most 1 and if $\mathfrak{z} \neq 0$ we can choose $z \in \mathfrak{z}$ so that $f_1(z) = f_2(z) = 1$, because $\pi_j(\exp tz) = e^{itf_j(z)}, t \in \mathbb{R}, j = 1, 2$, and clearly π_1 and π_2 coincide on $E(\mathfrak{z})$. If $\mathfrak{a}$ is a minimal noncentral ideal and $\mathfrak{c}$ its centralizer in $\mathfrak{g}$, then in particular $f_j([\mathfrak{g}, \mathfrak{a}]) \neq 0$, hence $\mathfrak{g} \neq \mathfrak{a}^{f_j} \supset \mathfrak{c}$. If $\dim \mathfrak{g}/\mathfrak{c} = 1$, then necessarily $\mathfrak{a}^{f_1} = \mathfrak{a}^{f_2} = \mathfrak{c}$. If $\dim \mathfrak{g}/\mathfrak{c} \geq 2$, then $0 \neq \mathfrak{z} \subset \mathfrak{a}$ and $\mathfrak{g}$ acts irreducibly and non trivially on $\mathfrak{a}/\mathfrak{z}$. In § 4, p 34 f, we had computed, how the adjoint representation of G in the two possible cases acts on $\mathfrak{a}$, and consequently how the coadjoint representation acts on $\mathfrak{a}^*$. It follows from these results in particular that all $l \in \mathfrak{a}^*$ with $l(z) = 1$ belong to the same orbit. In our present case this implies that there exists $x \in G$ so that f_1 and $\mathrm{Ad}^* x f_2$ coincide on $\mathfrak{a}$, hence we can suppose that already f_1 and f_2 coincide on $\mathfrak{a}$. But then also $\mathfrak{a}^{f_1} = \mathfrak{a}^{f_2}$.

Now we take $\mathfrak{p}_i \in \mathrm{Puk}\,(f_i, \mathfrak{g})$ with $\mathfrak{a} \subset \mathfrak{p}_i$. Then we have $\mathfrak{a} \subset \mathfrak{p}_i = \mathfrak{p}_i^{f_i} \subset \mathfrak{a}^{f_i} = \mathfrak{b}$ for $i = 1, 2$. Let $B = E(\mathfrak{b})$ and set $\sigma_i = \pi(f_i, \mathfrak{p}_i, B)$; hence $\pi_i = \mathrm{ind}_B^G \sigma_i$. It follows that σ_1 and σ_2 are irreducible and from the corollary of theorem 5 we conclude that they are equivalent. Since $\mathfrak{b} \neq \mathfrak{g}$ our induction hypothesis shows that the restrictions $f_i|_{\mathfrak{b}}$ lie in the same B orbit in $\mathfrak{b}^*$. So we can assume that already f_1 and f_2 coincide on $\mathfrak{b}$, i.e. $f_2 - f_1 \in \mathfrak{b}^\perp$. As $\mathfrak{p}_1 \in \mathrm{Puk}\,(f_1, \mathfrak{g})$, hence $f_1 + \mathfrak{p}_1^\perp \subset \Omega_{f_1}$, we obtain finally

$$f_2 \in f_1 + \mathfrak{b}^\perp \subset f_1 + \mathfrak{p}_1^\perp \subset \Omega_{f_1} ,$$

and consequently $\Omega_{f_2} = \Omega_{f_1}$. This finishes the proof of theorem 8.

§ 5 Kernels of restricted and induced representations

The Kirillov correspondence allows us to replace or identify the rather abstract set $\widehat{G}$ of equivalence classes of irreducible unitary representations of G by resp. with the relatively concrete set $\Omega(\mathfrak{g}^*)$ of G orbits of $\mathfrak{g}^*$. Now it is well known that both $\widehat{G}$ and $\Omega(\mathfrak{g}^*)$ are also topological spaces: The topology of $\widehat{G}$ is defined by the natural projection from $\widehat{G}$ onto the space Prim $C^*(G)$ of primitive ideals of $C^*(G)$ with its Jacobson topology, while $\Omega(\mathfrak{g}^*)$ is endowed with the quotient topology, defined by the projection from the euclidean space $\mathfrak{g}^*$ onto $\Omega(\mathfrak{g}^*)$. The *Kirillov conjecture*, which we will prove in part 3 of this book, states that K is indeed a homeomorphism.

But the description of $\widehat{G}$ via $\Omega(\mathfrak{g}^*)$ is already useful in situations not involving the topologies. As an example – and because we will need the results later – we treat the following problems: Let H be a closed connected subgroup of G, $\pi \in \widehat{G}$ and $\sigma \in \widehat{H}$. How can one describe the kernel of the restriction $\pi|_H$ in $C^*(H)$ and the kernel of the induced representation $\mathrm{ind}_H^G \sigma$ in $C^*(G)$? To treat these questions we remind the reader of the following facts (see e.g. [7]). Let $\mathcal{A}$ be a C^* algebra and Prim $\mathcal{A}$ its primitive ideal space. The *hull* $h(\mathcal{J})$ of a closed ideal $\mathcal{J} \subset \mathcal{A}$ is the set of all $\mathcal{P} \in$ Prim $\mathcal{A}$ with $\mathcal{J} \subset \mathcal{P}$. These hulls are exactly the closed sets of Jacobson's T_0–topology. A subset $S \subset h(\mathcal{J})$ is dense in $h(\mathcal{J})$, i.e. $h(\mathcal{J}) = \overline{S} =$ closed hull of S in Prim $\mathcal{A}$, if and only if $\mathcal{J} = \bigcap_{\mathcal{P} \in S} \mathcal{P}$.

In this context the following proposition is most useful. It is a special case of a more general theorem by Fell on the "continuity of induction", see [9], in particular theorem 4.2.

Proposition 7. *Let K be a closed subgroup of the locally compact group G and let σ and τ be unitary representations of K. If* $\ker_* \sigma \subset \ker_* \tau$ *in $C^*(K)$, then* $\ker_* \mathrm{ind}_K^G \sigma \subset \ker_* \mathrm{ind}_K^G \tau$ *in $C^*(G)$.*

Proof. We denote the space of $\sigma, \tau, \pi_\sigma := \mathrm{ind}_K^G \sigma$ and $\pi_\tau := \mathrm{ind}_K^G \tau$ by $\mathfrak{H}_\sigma, \mathfrak{H}_\tau, \mathfrak{L}_\sigma$ and $\mathfrak{L}_\tau$. The condition $\ker_* \sigma \subset \ker_* \tau$ is equivalent to the fact that there exists a total subset $T \subset \mathfrak{H}_\sigma$ such that all coefficients $\Phi_\xi(x) = (\sigma(x)\xi|\xi)$ for $\xi \in T$ are uniform limits on compacta in K of sums of positive definite functions on K, associated to τ, see [7] § 18. In this case *every* positive definite function associated to σ is such a limit. The same is true for π_σ and π_τ, moreover, if T is a subspace, then more generally the coefficients $\Phi_{\xi,\vartheta}(x) = (\sigma(x)\xi|\vartheta)$, $\xi, \vartheta \in T$, are limits uniformly on compacta of corresponding coefficients associated to τ if and only if $\ker_* \sigma \subset \ker_* \tau$ and similarly for π_σ and π_τ.

In (9), § 3, we had defined $u\#\alpha$ for $u \in \mathcal{C}_0(G)$ and $\alpha \in \mathfrak{H}_\sigma$ by $(u\#\alpha)(x) = \int_K u(xy)\sigma(y)^*\alpha\, dy$ and we saw that these functions span a dense subspace of $\mathfrak{L}_\sigma$. If χ is the modulus of the relatively invariant measure on G/K, then

$$\Phi_{u,\alpha}(x) = \chi(x)^{\frac{1}{2}} \int_{G/K} (u\#\alpha(x^{-1}y)|u\#\alpha(y))d\overline{y}$$

defines a positive definite function on G, associated to π_σ. For $y \in K$ and $x \in G$ let

$$W_u(x, y) = \chi(x)^{\frac{1}{2}} \int_K u(x^{-1}ky)\overline{u(k)}dk \ .$$

A simple computation shows that

$$\Phi_{u,\alpha}(x) = \int_K W_u(x, y)\Phi_\alpha(y)dy$$

with $\Phi_\alpha = \Phi_{\alpha,\alpha}$. In the same way we get of course positive definite functions $\Phi_{u,\xi}$, associated to π_τ, for any $\xi \in \mathfrak{H}_\tau$. Let C be compact in G and D the compact support of u. Then $W_u(x, y) = 0$ for $x \in C$ and $y \notin F := D^{-1}CD \cap K$, hence $\Phi_{u,\alpha}(x) = \int_F W_u(x, y)\Phi_\alpha(y)dy$ for $x \in C$ and similarly for the $\Phi_{u,\xi}$, $\xi \in \mathfrak{H}_\tau$. Now suppose that $\ker_* \sigma \subset \ker_* \tau$. Then Φ_α is the uniform limit of sums of the Φ_ξ, $\xi \in \mathfrak{H}_\tau$, uniformly on F. Since W_u is continuous and bounded on $C \times F$ it follows that also $\Phi_{u,\alpha}$ is the uniform limit on C of the corresponding sums of the $\Phi_{\alpha,\xi}$. This proves proposition 7. As a consequence we note that $\ker_* \sigma = \ker_* \tau$ implies $\ker_* \mathrm{ind}_K^G \sigma = \ker_* \mathrm{ind}_K^H \tau$, more generally we have the following

Corollary. *Let* $\{\tau_i\}_{i \in I}$ *be a family of unitary representations of* K. *If* $\ker_* \sigma = \bigcap_{i \in I} \ker_* \tau_i$, *then* $\ker_* \mathrm{ind}_K^G \sigma = \bigcap_{i \in I} \ker_* \mathrm{ind}_K^G \tau_i$.

This follows from applying the result for the special case $\ker_* \sigma = \ker_* \tau$ for just one τ to the representation $\tau = \sum_{i \in I}^{\oplus} \tau_i$ and the observations that $\mathrm{ind}_K^G \tau = \sum_{i \in I}^{\oplus} \mathrm{ind}_K^G \tau_i$ and the kernel of a direct sum–representation is the intersection of the kernels of the summands.

Before applying this result to our group algebras we fix some notations: In (46) we had already defined $\pi(f)$ for $f \in \mathfrak{g}^*$ as $\pi(f) = K(\Omega_f)$. Since $\pi(f)$ depends only on the whole orbit $\Omega = \Omega_f$, we will also use the notation $\pi(\Omega)$. In the other direction we will write $\Omega(\pi)$ for the orbit, corresponding to $\pi \in \widehat{G}$, i.e. we set $\Omega(\pi) = K^{-1}(\pi)$ with the inverse $K^{-1} : \widehat{G} \to \Omega(\mathfrak{g}^*)$ of the Kirillov map.

Now we fix $\pi \in \widehat{G}$. Clearly the restriction $\Omega(\pi)|_\mathfrak{h}$ of $\Omega(\pi)$ onto $\mathfrak{h}$ is $\mathrm{Ad}^* H$ invariant, hence we can define

$$\Omega_\mathfrak{h}(\pi) := \{\Theta \in \Omega(\mathfrak{h}^*) \, ; \, \Theta \subset \Omega(\pi)|_\mathfrak{h}\}$$

i.e. $\Omega_\mathfrak{h}(\pi)$ is the subset in $\Omega(\mathfrak{h}^*)$ of all $\mathrm{Ad}^* H$ orbits in $\mathfrak{h}^*$, contained in the restriction $\Omega(\pi)|_\mathfrak{h}$. We claim

Theorem 9. *The set* $\{\pi(\Theta) \, ; \, \Theta \in \Omega_\mathfrak{h}(\pi)\}$ *is dense in the hull of the kernel* $\ker_* \pi|_H$ *of the restriction* $\pi|_H$ *in* $C^*(H)$, *i.e.:*

$$\ker_* \pi|_H = \bigcap_{\Theta \in \Omega_\mathfrak{h}(\pi)} \ker_* \pi(\Theta) \ .$$

Proof by induction on $\dim G + \dim G/H =: d(G, H)$. We need the following observation: Let $\mathcal{A}$ be a C^*–algebra and $\mathcal{B}$ a C^*–subalgebra of the adjoint algebra $\mathcal{A}^b$, see [18]. If π is a representation of $\mathcal{A}$, then π extends uniquely to $\mathcal{A}^b$. We denote the restriction of this extension to $\mathcal{B}$ by $\pi|_{\mathcal{B}}$. For $\mathcal{A} = C^*(G)$, $\mathcal{B} = C^*(H)$ we have $\pi|_{C^*(H)} = \pi|_H$. Now suppose that $\ker_*\pi = \bigcap_{\tau\in T} \ker_*\tau$ for some set T of representations of $\mathcal{A}$. Then $\ker_*\pi|_{\mathcal{B}} = \bigcap_{\tau\in T} \ker_*\tau|_{\mathcal{B}}$. To prove this fact it suffices to show that for $c \in \mathcal{B}$ with $\tau|_{\mathcal{B}}(c) = 0$ for all $\tau \in T$ also $\pi|_{\mathcal{B}}(c) = 0$ holds. But this is clear, because for any $a \in \mathcal{A}$ and $\tau \in T$ we have $ca \in \mathcal{A}$, $\tau(ca) = \tau|_{\mathcal{B}}(c)\tau(a) = 0$, hence $\pi|_{\mathcal{B}}(c)\pi(a) = \pi(ca) = 0$ and, as π is nondegenerate, $\pi|_{\mathcal{B}}(c) = 0$.

Now assume that there exists a closed connected subgroup K strictly between G and H. Then $d(G, K) < d(G, H)$ and $d(K, H) < d(G, H)$, consequently by induction $\ker_*\pi|_K = \bigcap_{\Theta\in\Omega_{\mathfrak{k}}(\pi)} \ker_*\pi(\Theta)$ and $\ker_*\pi(\Theta)|_H = \bigcap_{\Psi\in\Omega_{\mathfrak{h}}(\pi(\Theta))} \ker_*\pi(\Psi)$ for all $\Theta \in \Omega_{\mathfrak{k}}(\pi)$. Thus

$$\ker_*\pi|_H = \bigcap_{\Theta\in\Omega_{\mathfrak{k}}(\pi)} \ker_*\pi(\Theta)|_H = \bigcap_{\Theta\in\Omega_{\mathfrak{k}}(\pi)} \bigcap_{\Psi\in\Omega_{\mathfrak{h}}(\pi(\Theta))} \ker_*\pi(\Psi) =$$

$$= \bigcap_{\Psi\in\Omega_{\mathfrak{h}}(\pi)} \ker_*\pi(\Psi) .$$

So the proof is reduced to the case of a maximal connected subgroup H.

Let us assume that H is normal, hence $G = \mathbb{R} \ltimes H$ is a semidirect product. If we are in the situation of proposition 6, then $\pi|_H = \pi(\Omega|_{\mathfrak{h}})$ is irreducible and $\Omega_{\mathfrak{h}}(\pi) = \{\Omega|_{\mathfrak{h}}\}$, so we are done. In the other case we can choose $f \in \Omega(\pi)$ and a Pukanszky polarization $\mathfrak{p}$ for f, contained in $\mathfrak{h}$, hence $\mathfrak{p} \in \mathrm{Puk}\,(f|_{\mathfrak{h}}, \mathfrak{h})$. Then $\pi = \mathrm{ind}_H^G\pi_1$ with $\pi_1 = \pi(f_1)$, $f_1 = f|_{\mathfrak{h}}$. The representation space $\mathfrak{H}(\pi)$ is $L^2(\mathbb{R}, \mathfrak{H}(\pi_1))$ and for $\varphi \in \mathfrak{H}(\pi)$, $y \in H$ we have $(\pi(y)\varphi)(s) = \pi_1(y^s)\varphi(s)$, hence $(\pi(c)\varphi(s) = \pi_1(c^s)\varphi(s)$ for $c \in C^*(H)$ and where $c \to c^s$ is the automorphism of $C^*(H)$, extending the automorphism on $L^1(H)$, which is induced by the automorphism $y \to y^s$ on H (recall that $(-s, e)(0, y)(s, e) = (0, y^s)$ on $G = \mathbb{R} \ltimes H$). It follows that $\ker_*\pi$ is the intersection of the kernels of the representations ${}^s\pi_1 : e \to \pi_1(c^s)$ of $C^*(H)$. But clearly ${}^s\pi_1 = \pi((\mathrm{Ad}^*sf)|_{\mathfrak{h}})$ and since $\Omega(\pi)|_{\mathfrak{h}} = \bigcap_{s\in\mathbb{R}} \Omega(\mathrm{Ad}^*sf|_{\mathfrak{h}})$ also in this case the theorem follows. We remark that from what we have shown it follows more generally, that the theorem holds for all normal subgroups in G.

Now suppose that H is maximal, but not normal. We claim:

(47) *There exists an ideal* $\mathfrak{a} \subset \mathfrak{g}$ *with* $\dim \mathfrak{g}/\mathfrak{a} = 1$ *such that* $\mathfrak{g} = \mathfrak{h} + \mathfrak{a}$ *and* $\mathfrak{h} \cap \mathfrak{a} = \mathfrak{b}$ *is an ideal in* $\mathfrak{g}$.

Proof by induction on $\dim \mathfrak{g} = n$. (47) is certainly true if $n \leq 2$. If $\mathfrak{h}$ contains an ideal $\mathfrak{i} \neq 0$, then (47) holds for $\mathfrak{g}/\mathfrak{i}$ and $\mathfrak{h}/\mathfrak{i}$, thus evidently also for $\mathfrak{g}$ and $\mathfrak{h}$.

Now assume that $\mathfrak{h}$ contains no nonzero ideal. Let $\mathfrak{m}$ be a minimal ideal in $\mathfrak{g}$ and $\mathfrak{a}$ its centralizer. Then $\mathfrak{h}+\mathfrak{m}=\mathfrak{g}$ and $[\mathfrak{h},\mathfrak{m}]\not\subset\mathfrak{h}$ because $\mathfrak{h}$ is no ideal. This implies $\mathfrak{a}\neq\mathfrak{g}$, hence $\dim(\mathfrak{g}/\mathfrak{a})=1$ and $\mathfrak{g}=\mathfrak{h}+\mathfrak{a}$. Finally, $[\mathfrak{h},\mathfrak{h}\cap\mathfrak{a}]\subset\mathfrak{h}\cap\mathfrak{a}$ and $[\mathfrak{m},\mathfrak{h}\cap\mathfrak{a}]=0$ imply that $\mathfrak{b}:=\mathfrak{h}\cap\mathfrak{a}$ is an ideal in $\mathfrak{g}$.

Let A, resp. B be the connected subgroups of G corresponding to $\mathfrak{a}$ and $\mathfrak{b}$, and choose $a\in\mathfrak{h}$, such that $\mathfrak{g}=\mathbb{R}a\oplus\mathfrak{a}$, $\mathfrak{h}=\mathbb{R}a\oplus\mathfrak{b}$. Then we have

$$G=\mathbb{R}\ltimes A\supset H=\mathbb{R}\ltimes B\ .$$

Suppose that π is induced from A, i.e. $\pi=\operatorname{ind}_A^G\pi_1$ for some $\pi_1\in\widehat{A}$. This is the case if and only if $\mathfrak{g}^l\subset\mathfrak{a}$ for $l\in\Omega(\pi)$. If $\mathfrak{H}_1$ is the space of π_1 then $\operatorname{ind}\mathfrak{H}_1=L^2(\mathbb{R},\mathfrak{H}_1)$ is the induced space and π is given by

$$(\pi(s,x)\xi)(t)=\pi_1(x^{t-s})\xi(t-s)$$

for $\xi\in\operatorname{ind}\mathfrak{H}_1$ and $(s,x)\in\mathbb{R}\ltimes A$. This yields immediately that the restriction $\pi|_H$ is induced from the restriction $\sigma:=\pi_1|_B$ of π_1 onto $B:\pi|_H=\operatorname{ind}_B^H\sigma$.

Since $\dim A<\dim G$ we can apply induction and see that $\ker_*\sigma=\ker_*\pi_1|_B=\bigcap_{\Theta\in\Omega_{\mathfrak{b}}(\pi_1)}\ker_*(\Theta)$, hence by the corollary of proposition 7:

$$\ker_*\pi|_H=\bigcap_{\Theta\in\Omega_{\mathfrak{b}}(\pi_1)}\ker_*\operatorname{ind}_B^H\pi(\Theta)\ . \tag{48}$$

To determine the representations $\operatorname{ind}_B^H\pi(\Theta)$ and their kernels we prove a

(49) **Lemma.** *Let $\mathfrak{h}$ be an ideal of codimension 1 in $\mathfrak{g}$ and H the corresponding normal subgroup. For a fixed orbit $\Theta\in\Omega(\mathfrak{h}^*)$ we set*

$$A_{\mathfrak{g}}(\Theta)=\{\Omega\in\Omega(\mathfrak{g}^*)\,;\,\Theta\subset\Omega|_{\mathfrak{h}}\}$$

and $\pi=\operatorname{ind}_H^G\pi(\Theta)$. The following conditions are equivalent:
(α) π is irreducible, i.e. $\pi=\pi(\Omega)$ for some $\Omega\in\Omega(\mathfrak{g}^*)$.
(β) $A_{\mathfrak{g}}(\Theta)$ contains exactly one orbit Ω.
(γ) For all $f\in\mathfrak{g}^*$ with $f|_{\mathfrak{h}}\in\Theta$ we have $\mathfrak{g}^f\subset\mathfrak{h}$.

Moreover, π is reducible if and only if $\pi(\Theta)$ extends to a representation $\pi(\Omega)$ of G on $\mathfrak{H}(\pi(\Theta))$, i.e. if $\pi(\Theta)=\pi(\Omega)|_H$ for some $\Omega=\Omega_f\in\Omega(\mathfrak{g}^*)$ and π is equivalent to the tensor product of $\pi'=\pi(\Omega)$ and the regular representation $\lambda_{G,H}:\pi\cong\pi'\otimes\lambda_{G,H}$. If $q\in\mathfrak{g}^*$ generates $\mathfrak{h}^\perp$, i.e. if $\mathfrak{h}^\perp$ is the kernel of q in $\mathfrak{g}$ and if $f_t=f+tq^*$ for $t\in\mathbb{R}$, then $A_{\mathfrak{g}}(\Theta)=\{\Omega_{f_t};t\in\mathbb{R}\}$. Finally, in both cases

$$\ker_*\ \operatorname{ind}_H^G\pi(\Theta)=\bigcap_{\Omega\in A_{\mathfrak{g}}(\Theta)}\ker_*\pi(\Omega)\ . \tag{50}$$

Proof. Suppose that $\pi = \operatorname{ind}_H^G \pi(\Theta)$ is irreducible, hence $\pi = \pi(\Omega) = K(\Omega)$, $\Omega \in \Omega(\mathfrak{g}^*)$, and let $g \in \mathfrak{g}^*$ with $g_0 = f|_{\mathfrak{h}} \in \Theta$. Then there exists $\mathfrak{p} \in \operatorname{Puk}(g_0, \mathfrak{h})$ with $\pi(\Theta) = \operatorname{ind}_P^H \chi(g_0, \mathfrak{p})$ and consequently $\pi = \operatorname{ind}_P^G \chi(g_0, \mathfrak{p})$, $P = E(\mathfrak{p})$. This shows that $\mathfrak{p}$ is a Pukanszky polarisation for g and implies $\Omega = \Omega_g$ and, after proposition 6, also $\mathfrak{g}^g \subset \mathfrak{h}$. Thus (α) implies (β) and (γ).

Now we take $q^* \neq 0$ in $\mathfrak{g}^*$ with $q^*(\mathfrak{h}) = 0$ and $q \in \mathfrak{g}$ with $q^*(q) = 1$. Then the kernel of q in $\mathfrak{g}^*$, i.e. $\mathbb{R}q^{\perp}$, can be identified with the dual $\mathfrak{h}^*$ of $\mathfrak{h}$, so we can write $\mathfrak{g}^* = \mathbb{R}q^* \oplus \mathfrak{h}^*$ and consider Θ as part of $\mathfrak{h}^*$. (Attention: This identification is not compatible with the coadjoint actions of G or H on $\mathfrak{g}^*$ and $\mathfrak{h}^*$!). Now condition (β) means that for any $g_0 \in \Theta$ all extensions g onto $\mathfrak{g}$ are conjugate under $\operatorname{Ad}^* G$, hence $\Omega = \mathbb{R}q^* \oplus \Theta$ is "the" orbit in $A_{\mathfrak{g}}(\Theta)$. Since the restriction of Ω onto $\mathfrak{h}$, i.e. the projection into $\mathfrak{h}^*$, is not injective, proposition 6, (i) guarantees that $\mathfrak{g}^f \subset \mathfrak{h}$ for all $f \in \Omega$. Thus (γ) follows from (β).

The implication $(\gamma) \Rightarrow (\alpha)$ is trivial, because $\mathfrak{g}^g \subset \mathfrak{h}$ for $g \in \mathfrak{g}^*$ with $g_0 = f|_{\mathfrak{h}}$ implies that $\pi = \pi(\Omega_g)$ is irreducible and proves the equivalence of (α), (β) and (γ).

If π is reducible, then (γ) implies the existence of $f \in \mathfrak{g}^*$ with $f_0 = f|_{\mathfrak{h}} \in \Theta$ with $\mathfrak{g}^f \not\subset \mathfrak{h}$. From proposition 6 we see that in this case $\pi(\Omega_f)$ extends $\pi(\Omega_{f_0}) = \pi(\Theta)$. If, on the other hand, $\pi(\Theta)$ is the restriction of $\pi(\Omega)$ for $\Omega \in \Omega(\mathfrak{g}^*)$, then we conclude from (27) and (28) that π is equivalent to $\pi(\Omega) \otimes \lambda_{G,H}$ and reducible.

We will study this case, so let $\pi' := \pi(\Omega)$ be an extension of $\pi(\Theta)$ onto $G = \mathbb{R} \ltimes H$ and $\pi = \pi' \otimes \lambda_{G,H} = \operatorname{ind}_H^G \pi(\Theta)$. The representation $\lambda_{G,H}$ is basically the regular representation of $G/H \cong \mathbb{R}$, hence by Fourier transformation it is equivalent to the representation $\hat{\lambda}$ of $G = \mathbb{R} \ltimes H$ on $L^2(\mathbb{R})$, defined by $(\hat{\lambda}(s, y)\varphi)(t) = e^{ist}\varphi(t)$ for $\varphi \in L^2(\mathbb{R})$, $(s, y) \in \mathbb{R} \ltimes H$. Consequently $\operatorname{ind}_H^G \pi(\Theta)$ is equivalent to $\pi' \otimes \hat{\lambda}$. This last tensor product acts in $\mathfrak{H}(\pi') \otimes L^2(\mathbb{R}) = L^2(\mathbb{R}, \mathfrak{H}(\pi'))$ and can explicitly be described by

$$((\pi' \otimes \hat{\lambda})(c)\varphi)(t) = \pi'_t(c)(\varphi(t)) \tag{51}$$

for $\varphi \in L^2(\mathbb{R}, \mathfrak{H}(\pi'))$ and π'_t the irreducible representation $\pi'_t(s, y) = e^{its}\pi'(s, y)$ for $(s, y) \in G$. It follows that $\ker_*(\operatorname{ind} \pi(\Theta)) = \bigcap_t \ker_* \pi'_t$.

We have to determine the orbits $\Omega(\pi'_t)$. We fix an element f in Ω and a polarization $\mathfrak{p} \in \operatorname{Puk}(f, \mathfrak{g})$. As $\mathfrak{p} \not\subset \mathfrak{h}$ and f is nontrivial on $\mathfrak{p}_0 := \mathfrak{p} \cap \mathfrak{h}$, we can choose an element $a \in \mathfrak{p}$ with $\mathfrak{p} = \mathbb{R}a \oplus \mathfrak{p}_0$, $f(a) = 0$. Then also $\mathfrak{g} = \mathbb{R}a \oplus \mathfrak{h}$, $E(\mathfrak{p}) = P = R \ltimes P$ and $G = R \ltimes H$ with $R = E(\mathbb{R}a)$. Let $q \in \mathfrak{g}^*$ be the form with $q(a) = 1$, $q(\mathfrak{h}) = 0$. For every real t we define $f_t = f + tq$, hence the f_t are exactly the extensions of $f|_{\mathfrak{h}} \in \Theta$, moreover, all f_t define the same symplectic form B_f on $\mathfrak{g}$. It follows that $\mathfrak{p} \in P(f_t, \mathfrak{g})$ for all $t \in \mathbb{R}$. We set $\chi_t = \chi(f_t, \mathfrak{p})$, in particular $\chi_0 = \chi(f, \mathfrak{p})$, and define a character γ_t of G by $\gamma_t(\exp sa \cdot y) = e^{its}$ for $y \in H$.

Now clearly χ_t coincides on P_0 with χ_0 and on R with γ_t, while $\gamma_t(P_0) = \chi_0(R) = \{1\}$. Thus $\chi_t = \gamma_t|_R \otimes \chi_0$ and $\mathrm{ind}_P^G \chi_t = \gamma_t \otimes \mathrm{ind}_P^G \chi_0 = \pi'_t$, by (27). This proves that $\Omega(\pi'_t) = \Omega_{f_t}$ and that $A_{\mathfrak{g}}(\Theta) = \{\Omega_{f_t}; t \in \mathbb{R}\}$.

Formula (49) is trivial if $\pi' = \mathrm{ind}_H^G \pi(\Theta)$ is irreducible and follows easily from (51) if π' is reducible. So lemma (49) is proved.

We go back to the kernels of $\mathrm{ind}_B^H \pi(\Theta)$ for the $\Theta \in \Omega_{\mathfrak{b}}(\pi_1)$. These were the orbits of the restrictions of elements f in $\Omega(\pi)$. If we decompose $\mathfrak{g}$ into the sum $\mathfrak{g}^* = \mathbb{R}a \oplus \mathfrak{c} \oplus \mathfrak{b}$ with $\mathfrak{a} = \mathfrak{c} \oplus \mathfrak{b}$ and $\mathfrak{h} = \mathbb{R}a \oplus \mathfrak{b}$, then we can write $\mathfrak{g}^* = \mathbb{R}q \oplus \mathfrak{c}^* \oplus \mathfrak{b}^*$ with $\mathbb{R}q = \mathfrak{a}^\perp$, $\mathfrak{c}^* = \mathfrak{h}^\perp$ and $\mathfrak{b}^* = (\mathbb{R}a \oplus \mathfrak{c})^\perp$. In this decomposition the corresponding projections of $\mathfrak{g}^*$ identify $\mathfrak{b}^*$ with the dual of $\mathfrak{b}$ and $\mathbb{R}q \oplus \mathfrak{b}$ with the dual $\mathfrak{h}^*$ of $\mathfrak{h}$. Now let Θ be an orbit in $\Omega_{\mathfrak{b}}(\pi)$. This means that there exists $f \in \Omega$ such that $f|_{\mathfrak{b}} \in \Theta$. If $f = \alpha q + \ell + f_0$ with $\ell \in \mathfrak{c}^*$, $f_0 \in \mathfrak{b}^*$, then, with the identifications described above, $f_0 = f|_{\mathfrak{b}} \in \mathfrak{b}^*$ and $f' = \alpha q + f_0 = f|_{\mathfrak{h}} \in \mathfrak{h}^*$, furthermore with the notation of the lemma, $f'_t = tq + f' = (\alpha + t)q + f_0$. Now we observe that Ω is saturated with respect to $\mathfrak{a}^\perp$, because $\mathfrak{g}^f \subset \mathfrak{a}$. This means that $f + tq \in \Omega$ for all $t \in \mathbb{R}$. Hence $f'_t = (f + tq)|_{\mathfrak{h}}$ and $\Omega_{f'_t} \in \Omega_{\mathfrak{b}}(\pi)$. Since $\Omega_{f'_t} \in A_{\mathfrak{h}}(\Theta)$, lemma (49) and formula (48) give the wanted result.

Now assume that $\pi = \pi(f)$ is not induced from A. Then $\mathfrak{g}^f \not\subset \mathfrak{a}$ and $\pi|_A = \pi_1$ is irreducible and corresponds to the orbit of $f_1 = f|_{\mathfrak{a}}$ in $\mathfrak{a}^*$. Let $\mathfrak{p} \in \mathrm{Puk}\,(f, \mathfrak{g})$ be a Vergne polarization passing through $\mathfrak{a}$ and $\mathfrak{b}$, hence $\mathfrak{p}_1 = \mathfrak{p} \cap \mathfrak{a} \in \mathrm{Puk}\,(f_1, \mathfrak{a})$ and $\mathfrak{p}_1$ and $\mathfrak{p}_2 = \mathfrak{p} \cap \mathfrak{b}$ are ad x invariant for all $x \in \mathfrak{g}^f$. Since $[\mathfrak{a}, \mathfrak{a}] \subset \mathfrak{b}$ and $\mathfrak{g} = \mathfrak{g}^f + \mathfrak{a}$ this implies that $(\mathfrak{p}_1 + \mathfrak{b})/\mathfrak{b}$ is an invariant submodule of the irreducible ad $\mathfrak{g}$ module $\mathfrak{a}/\mathfrak{b}$. It follows that either

(α) $$\mathfrak{p}_1 + \mathfrak{b} = \mathfrak{a} \quad \text{or}$$

(β) $$\mathfrak{p}_1 \subset \mathfrak{b}\ .$$

We claim that in case (α) also the restrictions $\pi|_B = \pi_1|_B$ are irreducible.

For $i = 1, 2$ let $P_i = E(\mathfrak{p}_i)$, $\chi_i = \chi(f, \mathfrak{p})|_{P_i} \in \widehat{P}_i$, $\mathfrak{H}_1 = \mathfrak{H}(\mathrm{ind}_{P_1}^A \chi_1)$ and $\mathfrak{H}_2 = \mathfrak{H}(\mathrm{ind}_{P_2}^B \chi_2)$. Condition (α) implies that $A = P_1 B = B P_1$, hence $\gamma : xP_2 \to xP_1$, $x \in B$, defines a homeomorphism from B/P_2 onto A/P_1 which commutes with the action of B on these two quotient spaces. The modulus for the relatively invariant measure $d_2\overline{x}$ on B/P_2 is the restriction of the modulus for $d_1\overline{x}$ on A/P_1, hence $d_2\overline{x}$ coincides with the induced measure $\gamma^*(d_1\overline{x})$.

Now let ξ be a continuous function in $\mathfrak{H}_1$ and let $\beta(\xi) = \xi|_B$ its restriction onto B. Then $\beta(\xi)(xy) = \overline{\chi_2(y)}\beta(\xi)(x)$ for $x \in B$ and $y \in P_2$, moreover for the norms of $\beta(\xi)$ and ξ we get

$$|\beta(\xi)|^2 = \int_{B/P_2} |\xi|(\overline{x})^2 d_2\overline{x} = \int_{A/P_1} |\xi|(\overline{x})^2 d_1\overline{x} = |\xi|^2$$

which shows that $\beta : \xi \to \beta(\xi)$ defines an isometry from $\mathfrak{H}_1$ onto $\mathfrak{H}_2$. A direct calculation shows that β commutes with the left translations $\Lambda(y)$ for $y \in B$, hence β intertwines the restriction of $\mathrm{ind}_{P_1}^A \chi_1 = \pi_1$ onto B with $\mathrm{ind}_{P_2}^B \chi_2 = \pi_2$. But $\mathfrak{p}_2$ is a Pukanszky polarization for $f|_{\mathfrak{b}}$ in $\mathfrak{b}$, consequently $\pi_2 = \pi(f|_{\mathfrak{b}})$ is irreducible. It follows that then also $\pi|_B$ and $\pi|_H$ are irreducible.

To finish case (α) we have to show that $\pi|_H = \pi'$ corresponds to the orbit of $f' = f|_{\mathfrak{h}}$ in $\mathfrak{h}^*$. From (α) and $\mathfrak{p} \not\subset \mathfrak{a}$ we derive $\mathfrak{p} + \mathfrak{b} = \mathfrak{g}$, hence $\mathfrak{p} \cap \mathfrak{h} + \mathfrak{b} = \mathfrak{h}$ and consequently $\mathfrak{p} \cap \mathfrak{h} \not\subset \mathfrak{a}$. So we can take $a \in \mathfrak{p} \cap \mathfrak{h}$, clearly with $f(a) = 0$, so that $\mathfrak{g} = \mathbb{R}a \oplus \mathfrak{a}$, $\mathfrak{h} = \mathbb{R}a \oplus \mathfrak{b}$, $\mathfrak{p} = \mathbb{R}a \oplus \mathfrak{p}_1$ and $\mathfrak{q} := \mathfrak{p} \cap \mathfrak{h} = \mathbb{R}a \oplus \mathfrak{p}_2$. For the corresponding groups we get $G = R \ltimes A$, $H = R \ltimes B$, $P = R \ltimes P_1$ and $Q := E(\mathfrak{p} \cap \mathfrak{h}) = R \ltimes P_2$, with $R = E(\mathbb{R}a)$.

In proposition 6, we have shown how π can be realized on the space $\mathfrak{H}(\pi_1)$ and the same arguments show that $\mathfrak{H}(\pi_2)$ can be taken as the space of $\rho := \mathrm{ind}_Q^H \chi(f', \mathfrak{q})$. Furthermore, we had defined the isomorphism β from $\mathfrak{H}(\pi_1)$ onto $\mathfrak{H}(\pi_2)$, which in–tertwined $\pi_1|_B$ and π_2, hence $\pi|_B$ and π_2 are intertwined. Evidently also $\rho|_B$ is transfered into π_2. On the other hand, we leave it to the reader to verify that also $\pi|_R$ and $\rho|_R$ are transfered to the same representation τ of R on $\mathfrak{H}(\pi_2)$, given by $(\tau(r)\xi)(x) = c(r)^{\frac{1}{2}}\xi(r^{-1}xr)$ for $r \in R$, $x \in B$ and with a modular function c as described in proposition 6. It follows that ρ and π' are equivalent, hence $\mathfrak{q} \in \mathrm{Puk}\,(f', \mathfrak{h})$ and $\pi' = \pi(\Omega_{f'})$.

Finally we have to treat the case (β), i.e. $\mathfrak{p}_1 \subset \mathfrak{b}$. We assume first, that $\mathfrak{g}^f \cap \mathfrak{h} \not\subset \mathfrak{a}$, so that we can choose $a \in \mathfrak{g}^f$ with $\mathfrak{g} = \mathbb{R}a \oplus \mathfrak{a}$, $\mathfrak{h} = \mathbb{R}a \oplus \mathfrak{b}$ and $\mathfrak{p} = \mathbb{R}a \oplus \mathfrak{q}$ with $\mathfrak{q} = \mathfrak{p} \cap \mathfrak{a} = \mathfrak{p} \cap \mathfrak{b} \subset \mathfrak{b}$. Then $\mathfrak{p}$ lies in $\mathfrak{h}$ and consequently π is induced from H : $\pi = \mathrm{ind}_H^G \sigma$ with $\sigma = \mathrm{ind}_P^H \chi(f, \mathfrak{p}) \in \widehat{H}$, $P = E(\mathfrak{p})$. With $R = E(\mathbb{R}a)$ and $Q = E(\mathfrak{q})$ we can write $G = R \ltimes A$, $H = R \ltimes B$, $P = R \ltimes Q$. As ad a acts irreducibly on $\mathfrak{a}/\mathfrak{b}$ we can assume – as before – that $\mathfrak{a}$ contains an element $u \notin \mathfrak{b}$ such that $[a, u] \equiv u (\mathrm{mod}\ \mathfrak{b})$ in case $\dim \mathfrak{a}/\mathfrak{b} = 1$ or elements u and v with $\mathfrak{a} = \mathbb{R}u \oplus \mathbb{R}v \oplus \mathfrak{b}$ and $[a, u] \equiv u - \lambda v (\mathrm{mod}\ \mathfrak{b})$, $[a, v] \equiv \lambda u + v (\mathrm{mod}\ \mathfrak{b})$ if $\dim \mathfrak{a}/\mathfrak{b} = 2$. For the rest of our proof we will discuss only the second case.

Let $\mathfrak{c} = \mathbb{R}u \oplus \mathbb{R}v$, hence $\mathfrak{a} = \mathfrak{c} \oplus \mathfrak{b}$ and $A = CB$, where $C = E(\mathfrak{c})$. We parametrize C with the complex numbers $\mathbb{C}$ by setting $e(x + iy) = \exp(xu + yv)$. Then it is easy to see that with $q = 1 + i\lambda$

$$\exp t\, a\, e(z) \exp(-ta) \equiv e(e^{qt} z) (\mathrm{mod}\ B)$$

or in brief $E(ta)e(z)E(-ta)B = e(e^{qt}z)B$.
As $H = RB = BR$, this implies $E(ta)e(z)H = e(e^{qt}z)H$. Let $\mathbb{T} = \{z \in \mathbb{C}; |z| = 1\}$. Then $(t, \zeta) \to e^{qt}\zeta$ maps $\mathbb{R} \times \mathbb{T}$ diffeomorphically onto $\mathbb{C}^\times = \mathbb{C} \setminus \{0\}$ and

$$p : (t, \zeta) \to p(t, \zeta) = E(ta)e(\zeta)H$$

maps $\mathbb{R} \times \mathbb{T}$ diffeomorphically onto the open subset $\overline{G}_H = \{xH; x \notin H\}$ of the quotient space $\overline{G} = G/H$.

Since the coset H as a point in $\overline{G}$ has measure 0 with respect to the relatively invariant measure on $\overline{G}$ we see that $\mathrm{ind}_H^G \mathfrak{H}(\sigma)$, the space of $\mathrm{ind}_H^G \sigma = \pi$, can be defined as the space of functions from $G_H := \complement_G H$ into $\mathfrak{H}(\sigma)$, satisfying the relevant conditions, in particular the identity $\xi(xy) = \sigma(y)^*\xi(x)$ for $x \in G_H$, $y \in H$. Although this space is not invariant under the left translation $\Lambda(x)$ for arbitrary x in G, it certainly is stable under $\Lambda(y)$ for $y \in H$, hence we can take it as representation space for the restriction $\pi' = \pi|_H$.

If dz denotes the Lebesgue measure on $\mathbb{C}$ and if we set $\tilde{z} = e(z)H \in \overline{G}$, then the relatively invariant measure on $\overline{G}$ is given by $\int_{\mathbb{C}} f(\tilde{z})dz$, with modulus $c(t, x) = e^t$ for $(t, x) \in G = \mathbb{R} \ltimes A$. This follows from the identities

$$E(ta)\tilde{z} = (e^{qt}z)^{\sim}, \; e(w)\tilde{z} = (w+z)^{\sim}, \; y\tilde{z} = \tilde{z}, \quad y \in B .$$

Then the measure $p^*(dz)$ on $\mathbb{R} \times \mathbb{T}$, defined by $p : \mathbb{R} \times \mathbb{T} \to \overline{G}_H = \overline{G} \setminus \{H\}$ is equal to $\frac{1}{2\pi}e^{2t}dt\,dy$, i.e. we have for $f \in \mathscr{C}_0(\overline{G}_H)$

$$(52) \qquad \int_{\overline{G}_H} f(\tilde{z})dz = \frac{1}{2\pi}\int_{\mathbb{R}}\int_0^{2\pi} f(E(ta)e(e^{i\varphi})H)e^{2t}dt\,dy .$$

We use this formula to define a unitary operator Q from $L^2(\mathbb{R}\times\mathbb{T}, \mathfrak{H}(\sigma); e^{2t}dt\,dy) =: \mathfrak{L}$ onto $\mathfrak{H}(\pi')$: Let $\vartheta \in \mathfrak{L}$, i.e. ϑ is a measurable $\mathfrak{H}(\sigma)$ valued function with $|\vartheta|^2 = \frac{1}{2\pi}\int_{\mathbb{R}}\int_0^{2\pi} |\vartheta(t, e^{i\varphi})|^2 e^{2t}dt\,d\varphi < \infty$. Because $(z, y) \to e(z)y$, $y \in H$, is a homeomorphism from $\mathbb{C} \times H$ onto G, the formula

$$(t, \zeta, y) \to E(ta)e(\zeta)y = e(e^{qt}\zeta) \cdot e(ta)y$$

defines a homeomorphism from $\mathbb{R} \times \mathbb{T} \times H$ onto G_H. We define the function $Q\vartheta$ on G_H by

$$Q\vartheta(E(ta)e(\zeta)y) = \sigma(y)^*\vartheta(t, \zeta) .$$

Then $Q\vartheta$ is measurable on G_H and satisfies $Q\vartheta(xy) = \sigma(y)^*Q\vartheta(x)$ for $x \in G_H$, $y \in H$. From (52) it follows that $|Q\vartheta| = |\vartheta|$, hence $Q : \vartheta \to Q\vartheta$ maps $\mathfrak{L}$ unitarily into $\mathfrak{H}(\pi')$. If $\xi \in \mathfrak{H}(\pi')$ is continuous, then $\vartheta : (t, \zeta) \to \xi(E(ta)e(\zeta))$ clearly lies in $\mathfrak{L}$ and $Q\vartheta = \xi$, hence $Q\mathfrak{L} = \mathfrak{H}(\pi')$.

Let $\zeta \in \mathbb{T}$. We define the representation $\sigma_\zeta \in \widehat{B}$ by

$$\sigma_\zeta(x) = \sigma(e(-\zeta)xe(\zeta)), \quad x \in B ,$$

and, with our usual notation $x^t = E(-ta)xE(ta)$, for $(t, x) \in H = \mathbb{R} \ltimes B$ and $\vartheta \in \mathfrak{L}$ we set

$$(\tilde{\pi}(s, x)\vartheta)(t, \zeta) = e^{\frac{s}{2}}\sigma_\zeta(x^{t-s})\vartheta(t-s, \zeta) .$$

This yields

$$\begin{aligned}
e^{-\frac{s}{2}}(Q\tilde{\pi}(s,x)\vartheta)(E(ta)e(\zeta)y) &= \sigma(y)^*\sigma(e(-\zeta)x^{t-s}e(\zeta))\vartheta(t-s,\zeta) = \\
&= \sigma(y^{-1}e(-\zeta)x^{t-s}e(\zeta))(Q\vartheta)(E((t-s)a)e(\zeta)) = \\
&= (Q\vartheta)(E(-sa)(x^{-1})^{-s}E(ta)e(\zeta)y) = \\
&= e^{-\frac{s}{2}}(\pi'(s,x)Q\vartheta)(E(ta)e(\zeta)y)\ .
\end{aligned}$$

It follows that $\tilde{\pi} : (s,x) \to \tilde{\pi}(s,x)$ is a representation of H in $\mathfrak{L}$, equivalent to π'. Now this $\tilde{\pi}$ is thc "direct integral" of the representations π_ζ of H in $L^2(\mathbb{R}, \mathfrak{H}(\sigma))$, defined by $(\pi_\zeta(s,x)\xi)(t) = e^{\frac{s}{2}}\sigma_\zeta(x^{t-s})\xi(t-s)$ for $\xi \in L^2(\mathbb{R}, \mathfrak{H}(\sigma))$, which, by proposition 3, are nothing else but $\mathrm{ind}_B^H \sigma_\zeta$. Clearly we can write $\mathfrak{L}$ as $L^2(\mathbb{T}, L^2(\mathbb{R}, \mathfrak{H}(\sigma))) = L^2(\mathbb{T}, \text{ ind } \mathfrak{H}(\sigma_\zeta))$ and the operators $\tilde{\pi}(A)$, $A \in C^*(H)$, act "fiberwise" on the $\xi \in L^2(\mathbb{T}, \text{ ind } \mathfrak{H}(\sigma_\zeta))$, i.e. $(\tilde{\pi}(A)\xi)(\zeta) = \pi_\zeta(A)(\xi(\zeta))$. It follows that $\ker_* \pi' = \ker_* \tilde{\pi} = \bigcap_{\zeta\in\mathbb{T}} \ker_* \pi_\zeta = \bigcap_{\zeta\in\mathbb{T}} \ker_* \ \mathrm{ind}_B^H \sigma_\zeta$.

The orbit of the representation $\sigma = \mathrm{ind}_P^H \chi(f, \mathfrak{p})$ of H contains the restriction $f' = f|_{\mathfrak{h}}$ of $f \in \Omega(\pi)$. Because $\mathfrak{g}^f \cap \mathfrak{h} \not\subset \mathfrak{a}$ and $\mathfrak{g}^f \cap \mathfrak{h} \subset \mathfrak{h}^{f'}$ imply $\mathfrak{h}^{f'} \not\subset \mathfrak{b}$ we see that $\sigma_0 = \sigma|_B$ is irreducible, hence $\Omega(\sigma_0) = \Omega_{f_0} \subset \mathfrak{b}^*$, with $f_0 = f'|_{\mathfrak{b}} = f|_{\mathfrak{b}}$. If ι_ζ is the inner automorphism $\iota_\zeta(x) = e(-\zeta)xe(\zeta)$, then the representation σ_ζ is just the composition $\sigma_0 \circ \iota_\zeta$, hence the orbit $\Omega_\zeta := \Omega(\sigma_\zeta)$ contains the restriction f_ζ^0 of the functional $f_\zeta := \mathrm{Ad}^*e(\zeta)f \in \mathfrak{g}^*$ onto $\mathfrak{b}$, i.e. we have $\Omega(\sigma_\zeta) = \mathrm{Ad}^*Bf_\zeta^0$. We claim that all π_ζ are irreducible. Otherwise there would exist $\zeta \in \mathbb{T}$ with $\mathfrak{h}^{f'_\zeta} \not\subset \mathfrak{b}$, with $f'_\zeta = f_\zeta|_{\mathfrak{h}}$. This means that there exist b_1 and b_2 in $\mathfrak{b}$ such that

$$f_\zeta([a+b_1, \mathfrak{b}]) = f([\mathrm{Ad}\, e(\zeta)a + b_2, \mathfrak{b}]) = 0\ .$$

If $e(\zeta) = \exp c$ with $c \in \mathfrak{c}$, then

$$\mathrm{Ad}\, e(\zeta)a = e^{\mathrm{ad}\, c}a \equiv a + [c,a] (\mathrm{mod}\ \mathfrak{b})$$

hence $f([a+[c,a]+b_3, \mathfrak{b}]) = 0$. Now $a \in \mathfrak{g}^f$ implies $f([a,\mathfrak{b}]) = 0$, consequently $f([c', \mathfrak{b}]) = 0$ for $c' = [c,a] + b_3 \in \mathfrak{a} \setminus \mathfrak{b}$. But this contradicts the assumption (β), i.e. $\mathfrak{p}_1 \subset \mathfrak{b}$, and proves our claim.

Now theorem 9 for this subcase follows easily: As $G = HC$, the orbits in $\Omega_{\mathfrak{h}}(\pi)$ are exactly the orbits $\mathrm{Ad}^*Hf'_\zeta$ and these, on the other hand, are the unique orbits in $\mathfrak{h}^*$ for which the restrictions to $\mathfrak{b}$ contains $\Omega(\sigma_\zeta)$, which means that the formula for $\ker_* \pi|_H$ in theorem 9 holds.

If $\mathfrak{g}^f \cap \mathfrak{h} \subset \mathfrak{a}$, then, with our previous notation, there exist $a \in \mathfrak{g}^f$ and $c \in \mathfrak{c}$ with $\mathfrak{g} = \mathbb{R}a \oplus \mathfrak{a}$ and $\mathfrak{h} = \mathbb{R}(a+c) \oplus \mathfrak{b}$. Since $\mathrm{ad}\, a$ acts bijectively on $\mathfrak{a}/\mathfrak{b}$ there exists $m \in \mathfrak{a}$ with $[m,a] \equiv c (\mathrm{mod}\ \mathfrak{b})$, hence $\mathrm{Ad}\ \exp m\,(a) \equiv a + c\,(\mathrm{mod}\ \mathfrak{b})$ and

$$\mathfrak{h} = \mathrm{Ad}\ \exp m\,(\mathfrak{h}_0) \quad \text{with} \quad \mathfrak{h}_0 = \mathbb{R}a \oplus \mathfrak{b}$$

Let $f_m = \mathrm{Ad}^* \exp m(f) \in \Omega$. Then $\mathrm{Ad} \exp m(\mathfrak{g}^f) = \mathfrak{g}^{f_m}$ and $\mathfrak{g}^{f_m} \cap \mathfrak{h} = \mathrm{Ad}\ \exp m(\mathfrak{g}^f \cap \mathfrak{h}_0) \not\subset \mathrm{Ad}\ \exp m(\mathfrak{a}) = \mathfrak{a}$, so with f_m we are in the same situation handled above with $f \in \Omega$. Now obviously theorem 9 depends only on the data $\mathfrak{g}$, $\mathfrak{h}$ and π, resp. Ω, and not on the choice of f, resp. f_m in Ω, which implies that we have proved the theorem.

Let H be a connected closed normal subgroup of G, $\pi \in \widehat{G}$ and π' the restriction of π onto H. It is an immediate consequence of proposition 6 and lemma (49) that π' is irreducible if and only if the restriction map from $\mathfrak{g}^*$ to $\mathfrak{h}^*$ maps $\Omega(\pi)$ bijectively onto an Ad^*H orbit Ω' in $\mathfrak{h}^*$ and that in this case $\pi' = K(\Omega')$. We will show that this is also true for arbitrary, i.e. not necessarily normal subgroups H. The usual induction arguments show that we can assume that H is maximal nonnormal. So let $\mathfrak{a}$, $\mathfrak{b}$, $\mathfrak{h}$ etc. have the same meaning as in (47) ff.

Suppose that π is induced from A, i.e. $\pi = \mathrm{ind}_A^G \pi_1$, and let $\sigma = \pi_1|_B$. In this case we had seen that $\pi' = \mathrm{ind}_B^H \sigma$ is irreducible if and only if σ is irreducible. Again we decompose $\mathfrak{g}$: Let $\mathfrak{g} = \mathbb{R}a \oplus \mathfrak{c} \oplus \mathfrak{b}$ with $\mathfrak{a} = \mathfrak{c} \oplus \mathfrak{b}$ and $\mathfrak{h} = \mathbb{R}a \oplus \mathfrak{b}$. Then $\mathfrak{g}^* = \mathfrak{a}^\perp \oplus \mathfrak{h}^\perp \oplus \mathfrak{b}^\bullet$ with $\mathfrak{b}^\bullet = (\mathbb{R}a \oplus \mathfrak{c})^\perp \cong \mathfrak{b}^*$ and the orbit Ω of π can be written as $\Omega = \mathfrak{a}^\perp \oplus \mathfrak{e} \oplus \Omega^\bullet$, where we may assume that $\mathfrak{e} \oplus \Omega^\bullet \subset \mathfrak{h}^\perp \oplus \mathfrak{b}^\bullet \cong \mathfrak{a}^*$ and $\mathfrak{e} \oplus \Omega^\bullet$ is the orbit of π_1 and that $\dim \mathfrak{e} = 0$ if and only if σ is irreducible. On the other hand the restriction from Ω onto $\mathfrak{h}^* \cong \mathfrak{a}^\perp \oplus \mathfrak{b}^\bullet$ is one–to–one if and only if $\dim \mathfrak{e} = 0$, which proves our claim in the present case.

If $\pi|_A =: \pi_1$ is irreducible, then we had two cases (α) and (β). In case (α) the restrictions $\pi_1|_B = \pi|_B$ and consequently also $\pi|_H$ are irreducible, hence the restriction of Ω onto $\mathfrak{b}$ and a fortiori also onto $\mathfrak{h}$ are one–to–one. In case (β) we had $\pi_1 = \mathrm{ind}_B^A \sigma_0$ and we had seen that π' is equivalent to a direct integral representation $\tilde{\pi}$, based on $\mathbb{T}$, hence π' is reducible. Since in this case $\Omega = l \oplus \mathfrak{h}^\perp \oplus \Omega^\bullet \subset \mathfrak{g}^*$ we see that the restriction map onto $\mathfrak{h}^*$ is not one–to–one. We state our result:

(53) *Let H be a connected closed subgroup of G, π an irreducible representation of G with corresponding orbit $\Omega \subset \mathfrak{g}^*$ and $\pi' = \pi|_H$ the restriction of π onto H. Then π' is irreducible if and only if the restriction from $\mathfrak{g}^*$ to $\mathfrak{h}$ maps Ω one–to–one onto an Ad^*H orbit Ω' in $\mathfrak{h}^*$. In this case Ω' is the orbit of π'.*

As an important corollary of (53) we obtain

(54) *Let π and ρ be irreducible representations of G and Ω and Θ the corresponding orbits. If the tensor product representation $\pi \otimes \rho$ of G is irreducible, then the vector space–sum $\Omega + \Theta$ in $\mathfrak{g}^*$ is an orbit and $\pi \otimes \rho = K(\Omega + \Theta)$.*

Proof. Let $\widetilde{G} = G \times G$ be the direct product of G with itself and $\tilde{\mathfrak{g}} = \mathfrak{g} \oplus \mathfrak{g}$ its Lie algebra. Then π and ρ define canonically the irreducible representation $\pi \times \rho$ of $\widetilde{G}$ in $\mathfrak{H}(\pi) \otimes \mathfrak{H}(\rho)$. One checks easily that the orbit of $\pi \times \rho$ in $\tilde{\mathfrak{g}}^* = \mathfrak{g}^* \oplus \mathfrak{g}^*$ is equal to $\Psi := \Omega \oplus \Theta$. If δ is the isomorphism from G onto the diagonal G^Δ

in $\widetilde{G}$, resp. from$\mathfrak{g}$ onto the diagonal $\mathfrak{g}^{\Delta}$ in $\tilde{\mathfrak{g}}$, then $\pi \otimes \rho = (\pi \times \rho)|_{G^{\Delta}} \circ \delta$ and $\Psi|_{\mathfrak{g}^{\Delta}} \circ \delta = \Omega + \Theta$. Now (54) follows directly from (53).

We mention that the converse of (54) is also true. This follows from theorem 9 and the fact that G is a type I group, which means that the irreducible representations of G are uniquely determined by their kernels in $C^*(G)$, or in other words that the canonical surjection from $\widehat{G}$ onto the primitive ideal-space Prim $C^*(G)$ is bijective.

Our next theorem treats the analogous problem to that theorem 9, but in the opposite direction. Again let $\mathfrak{h}$ be an arbitrary subalgebra of $\mathfrak{g}$ and $H = E(\mathfrak{h}) \subset G$. As in (49) we define for an orbit $\Theta \in \Omega(\mathfrak{h}^*)$: $A_{\mathfrak{g}}(\Theta) = \{\Omega \in \Omega(\mathfrak{g}^*); \Theta \subset \Omega|_{\mathfrak{h}}\}$. If $\sigma = K(\Theta) \in \widehat{H}$ is the corresponding representation of H, then we also write $A_{\mathfrak{g}}(\sigma)$ for $A_{\mathfrak{g}}(\theta)$.

Theorem 10. *If σ is an irreducible representation of H, then $\{\pi(\Omega); \Omega \in A_{\mathfrak{g}}(\sigma)\}$ is dense in the hull of* $\ker_*$ $\mathrm{ind}_H^G\sigma$, *i.e.*

$$\ker_* \mathrm{ind}_H^G\sigma = \bigcap_{\Omega \in A_{\mathfrak{g}}(\sigma)} \ker_* \pi(\Omega).$$

The first steps of the proof are the same as before: Assume that there is a closed connected subgroup K between H and G and theorem 10 holds for G, K and for K, H. Let $\sigma = K(\Omega_\sigma)$ be an irreducible representation of H, $\pi = \mathrm{ind}_H^G\sigma$, and $\tau = \mathrm{ind}_H^K\sigma$, thus $\pi = \mathrm{ind}_K^G\tau$. Then $\ker_* \tau = \bigcap_{\Theta \in A_{\mathfrak{k}}(\Omega_\sigma)} \ker_* \pi(\Theta)$ and the corollary of proposition 7 gives $\ker_* \pi = \ker_* \mathrm{ind}_K^G\tau = \bigcap_{\Theta \in A_{\mathfrak{k}}(\Omega_\sigma)} \bigcap_{\Omega \in A_{\mathfrak{g}}(\Theta)} \ker_* \pi(\Omega) = \bigcap_{\Omega \in A_{\mathfrak{g}}(\Omega_\sigma)} \ker_* \pi(\Omega)$, because it is evident that the restriction of $\Omega \in \Omega(\mathfrak{g}^*)$ to $\mathfrak{k}$ contains some $\Theta \in \Omega(\mathfrak{h}^*)$ with $\Theta|_{\mathfrak{h}} \supset \Omega_\sigma$ if and only if $\Omega|_{\mathfrak{h}} \supset \Omega_\sigma$. Our argument together with formula (50) in Lemma (49) shows that theorem 10 is true for all *normal* subgroups H in G, so left over is to prove it in the case of a nonnormal connected closed maximal subgroup, resp. for a maximal subalgebra $\mathfrak{h}$, which is no ideal in $\mathfrak{g}$. This situation has been described in (47), here we use the same notation: $\mathfrak{a}$ and $\mathfrak{b}$ for the ideals with $\mathfrak{g} = \mathfrak{h} + \mathfrak{a}$, $\mathfrak{h} \cap \mathfrak{a} = \mathfrak{b}$ etc., $\pi = \mathrm{ind}_H^G\sigma$ for $\sigma = K(\Omega_\sigma) \in \Omega(\mathfrak{h}^*)$.

Suppose first that σ is induced from $B : \sigma = \mathrm{ind}_B^H\rho$ for $\rho = K(\Omega_\rho) \in \Omega(\mathfrak{b}^*)$. As $\pi = \mathrm{ind}_H^G\sigma = \mathrm{ind}_B^G\rho$ and B is normal, our remark made above applies, i.e. $\ker_* \pi = \bigcap_{\Omega \in A_{\mathfrak{g}}(\Omega_\rho)} \ker_* \pi(\Omega)$. Now $\sigma = \mathrm{ind}_B^H\rho$ is irreducible, so lemma (49) tells us that the orbit Ω_ρ of σ in $\mathfrak{h}^*$ is the only one with $\Omega_\sigma|_{\mathfrak{b}} \supset \Omega_\rho$, hence for $\Omega \in \Omega(\mathfrak{g}^*)$ the conditions $\Omega|_{\mathfrak{h}} \supset \Omega_\sigma$ and $\Omega|_{\mathfrak{b}} \supset \Omega_\rho$ are equivalent, which means that $A_{\mathfrak{g}}(\Omega_\sigma) = A_{\mathfrak{g}}(\Omega_\rho)$ and proves the theorem in the present case.

Now we assume that σ is not induced. Then for $f' \in \Omega_\sigma$ we have $\mathfrak{h}^{f'} \not\subset \mathfrak{b}$, hence $\mathfrak{h} = \mathbb{R}a \oplus \mathfrak{b}$ with $f'([a, \mathfrak{b}]) = 0$. Since ad a is regular on $\mathfrak{a}/\mathfrak{b}$ we can write $\mathfrak{a} = [a, \mathfrak{c}] \oplus \mathfrak{b}$ if $\mathfrak{a} = \mathfrak{c} \oplus \mathfrak{b}$. It follows that $\mathfrak{g} = \mathbb{R}a \oplus [a, \mathfrak{c}] \oplus \mathfrak{b}$ and that we can extend f' to $f \in \mathfrak{g}^*$ by setting $f|_\mathfrak{h} = f'$, $f|_{[a,\mathfrak{c}]} = 0$. Then we have $f([a, \mathfrak{g}]) = 0$, hence $a \in \mathfrak{g}^f$ and the representation $\pi = \pi(f)$ of G is not induced from A. As in the proof of theorem 9 we take a Vergne polarization $\mathfrak{p} \in \text{Puk}\,(f, \mathfrak{g})$ through $\mathfrak{a}$ and $\mathfrak{b}$ and distinguish the cases (α), $\mathfrak{p}_1 + \mathfrak{b} = \mathfrak{a}$, and (β), $\mathfrak{p}_1 \subset \mathfrak{b}$, with $\mathfrak{p}_1 = \mathfrak{p} \cap \mathfrak{a}$. In case (β) we had seen that σ is induced from $P = E(\mathfrak{p})$, i.e. $\sigma = \text{ind}_P^H \chi(f, \mathfrak{p})$, and $\pi = \text{ind}_H^G \sigma = \text{ind}_P^G \chi(f, \mathfrak{p})$, moreover $\Omega_f \supset f + \mathfrak{h}^\perp$, hence $A_\mathfrak{g}(\sigma) = \{\Omega_f\}$ and consequently theorem 10 holds in case (β).

In case (α) the representations $\pi|_A$ and $\pi|_B$ were irreducible, consequently also $\pi|_H$ is irreducible and equivalent to σ. From (27) we conclude that $\text{ind}_H^G \sigma \cong \pi \otimes \lambda_H$ with the regular representation $\lambda_H$in $L^2(G/H)$. We have to decompose λ_H into irreducible representations. To do this, we observe first, that λ_H is trivial on B, hence it is basically a representation of the quotient G/B, which in the case $\dim \mathfrak{g}/\mathfrak{h} = 2$ is isomorphic to the semidirect product $\mathbb{R} \ltimes \mathbb{C}$ with multiplication $(s, u)(t, v) = (s + t, e^{-qt}u + v)$ for $s, t \in \mathbb{R}$, $u, v \in \mathbb{C}$. The space of λ_H is $L^2(\mathbb{C})$ and the action is given by

$$(\lambda_H(s, u)\xi)(z) = e^{-s}\xi(e^{-qs}z - u)\,.$$

At the end of § 3, in the proof of (28), we had decomposed λ_H via Fourier transform on $\mathbb{C}$ into representations λ_ζ for $\zeta \in \mathbb{T}$ on $L^2(\mathbb{R})$, given by

$$(\lambda_\zeta(s, u)\varphi)(t) = e^{i(u, e^{\bar{q}(t+s)}\zeta)}\varphi(t + s)$$

with $(u, v) = \text{Im}\, u\bar{v}$ for $u, v \in \mathbb{C}$. The direct integral $\int^\oplus \lambda_\zeta d\zeta$ acts on $L^2(\mathbb{T}, L^2(\mathbb{R}, e^{2t}dt))$ and the operator, intertwining the representation $\hat{\lambda}_H$ with the direct integral of the λ_ζ is the unitary operator defined by the mapping $z = e^{\bar{q}t}\zeta \to (\zeta, t)$ from $\mathbb{C}^\times$ onto $\mathbb{T} \times \mathbb{R}$.

Let χ_ζ be the character $\chi_\zeta(z) = e^{-i(z,\zeta)}$ on $\mathbb{C}$. Then λ_ζ can also be written as

$$(\lambda_\zeta(s, u)\varphi)(t) = \chi_\zeta(e^{q(t+s)}u)\varphi(t + s)\ ,$$

which proves: If V is the unitary operator $V\varphi(t) = \varphi(-t)$ on $L^2(\mathbb{R})$, then $V^{-1}\lambda_\zeta V =$ ind χ_ζ, in particular λ_ζ is irreducible and if we identify the Lie algebra of $\mathbb{R} \ltimes \mathbb{C}$ with $\mathbb{R} \oplus \mathbb{C}$, so that $[t \oplus 0, 0 \oplus z] = e^{qt}z \in \mathbb{C}$, then the orbit Ω_ζ of λ_ζ contains the functional $f_\zeta(t \oplus z) = \text{Im}\, z\bar{\zeta}$.

Let $\pi_\zeta = \pi \otimes \lambda_\zeta$. We claim that the representations π_ζ are irreducible. We have $\mathfrak{H}(\pi_\zeta) = \mathfrak{H}(\pi) \otimes L^2(\mathbb{R})$. Since the restriction $\pi|_B$ is irreducible on $\mathfrak{H}(\pi)$ and λ_ζ is trivial on B we see that the von Neumann algebra $\pi_\zeta(G)''$ contains $\mathcal{L}(\mathfrak{H}(\pi)) \otimes \mathbf{1}$, thus any bounded operator S of $\mathfrak{H}(\pi_\zeta)$ commuting with π_ζ has the form $\mathbf{1} \otimes T$, where $T \in \mathcal{L}(L^2(\mathbb{R}))$ commutes with λ_ζ. Hence T and consequently also S is scalar, which proves the irreducibility of π_ζ.

Now we consider the λ_ζ again as representations of G, trivial on B. Then f_ζ is in $\mathfrak{g}^*$ and $\lambda_\zeta = K(\Omega_{f_\zeta})$. As before we decompose $\mathfrak{g}$ and write $\mathfrak{g} = \mathbb{R}a \oplus [a, \mathfrak{c}] \oplus \mathfrak{b}$, $\mathfrak{a} = [a, \mathfrak{c}] \oplus \mathfrak{b}$ and $\mathfrak{h} = \mathbb{R}a \oplus \mathfrak{b}$. Then $\mathfrak{g}^* = \mathfrak{a}^\perp \oplus \mathfrak{h}^\perp \oplus \mathfrak{b}^\bullet$ and $\mathfrak{h}^*$ can be identified with $\mathfrak{a}^\perp \oplus \mathfrak{b}^\bullet$. In this notation $f' = l \oplus f^\bullet$ with $l \in \mathfrak{a}^\perp$, $f^\bullet \in \mathfrak{b}^\bullet \cong \mathfrak{b}^*$ and $f = l \oplus 0 \oplus f^\bullet \in \mathfrak{g}^*$, furthermore $f_\zeta \in \mathfrak{h}^\perp$ and from (53) we conclude that $f + f_\zeta = l \oplus f_\zeta \oplus f^\bullet$ lies in the orbit Ω_ζ of π_ζ. We observe that $\mathrm{Ad}^* \exp \mathbb{R}a$ leaves $\mathfrak{h}^\perp$ fixed and every $g \neq 0$ in $\mathfrak{h}^\perp$ belongs to an $\exp \mathbb{R}a$ orbit of some f_ζ, $\zeta \in \mathbb{T}$, hence $l \oplus g \oplus f^\bullet$ is contained in some Ω_ζ. This proves $A_\mathfrak{g}(\sigma) = \{\Omega_\zeta; \zeta \in \mathbb{T}\} \cup \{\Omega_f\}$. Since $\ker_* \pi \otimes \lambda_H = \bigcap_{\zeta \in \mathbb{T}} \ker_* \pi_\zeta$ to finish the proof of theorem 10 it suffices to show that $\ker_* \pi \otimes \lambda_H \subset \ker_* \pi$. This follows easily from the following two facts: Since G is amenable, the constant 1 on G is a uniform limit on compacta of a sequence $\{\psi_k\}_k$ of positive definite functions ψ_k on G, associated to λ_H, see [8] p. 44. If φ is a positive definite function on G, associated to π, then the products $\varphi\psi_k$ are associated to $\pi \otimes \lambda_H$ and converge uniformly on compacta to φ, hence $\ker_* \pi \supset \ker_* \pi \otimes \lambda_H$, see [7], § 18.

§ 6 Smooth functions and kernel operators

An important consequence of (43) is the fact that the orbit $G\xi$ of ξ in the vector space V under the exponential group G is locally closed, i.e. that every element in $G\xi$ has a neighborhood in $G\xi$ which is closed in V. This is equivalent to the fact that $G\xi$ is open in its closure $\overline{G\xi}$ in V and also with the local compactness of $G\xi$ in the relative topology. To prove our assertion it is sufficient to show that there exists a neighborhood of ξ in V, which is diffeomorphic to the product of a neighborhood of ξ in $G\xi$ and an open set in $\mathbb{R}^m$, $m = \dim V - \dim G\xi$. For this purpose and with the same notation as in (43) we take a coexponential basis $\{c_1, c_2, \dots, c_k\}$ for $\mathfrak{g}^\xi$ in $\mathfrak{g}$, hence $k = \dim(\mathfrak{g}/\mathfrak{g}^\xi) = \dim G\xi$ and the $\alpha_j := c_j\xi$ are linearly independent in V. We choose elements $\alpha_{k+1}, \dots, \alpha_n$ with $n = \dim V$, such that $\{\alpha_i\}_1^n$ is a basis in V. For $x \in \mathbb{R}^k$ let $\varepsilon(x) = E(x_1c_1)E(x_2c_2)\dots E(x_kc_k)$. Then

$$F : (x, y) \to F(x, y) = \varepsilon(x)\xi + \sum_{j=1}^{n-k} y_j\alpha_{k+j}$$

is a smooth mapping of $\mathbb{R}^k \times \mathbb{R}^{n-k} = \mathbb{R}^n$ into V. It follows from (43) that $F(\mathbb{R}^k \times \{0\}) = G\xi$, more precisely, $x \to F(x, 0) \in V$ is a diffeomorphism from $\mathbb{R}^k$ onto $G\xi$. Since $\frac{\partial F}{\partial x_i}(0, 0) = \alpha_i$ for $i \le k$ and $\frac{\partial F}{\partial y_j} = \alpha_{k+j}$ for $1 \le j \le n - k$ it follows that the Jacobian $\frac{\partial F}{\partial(x,y)}$ is regular in 0, hence F is a local diffeomorphism in 0. Also equivalent to the local closedness of all orbits $G\xi$ is the T_0 separation property of the quotient space V/G, so we obtain the corollary that the orbit space

$\Omega(\mathfrak{g}^*)$ for exponential $\mathfrak{g}$ is a T_0 space. But then the bicontinuity of the Kirillov map implies that also the dual $\widehat{G}$, i.e. the dual $C^*(G)^\wedge$ is a T_0 space, i.e. that $C^*(G)$ is postliminal, which means that for every $\pi \in \widehat{G}$ the image $\pi(C^*(G))$ contains the compact operators, see [7], in particular théorème 9.1. This result was known in 1957 [34], so its proof via Kirillov's map is certainly not the simplest. Here we will prove results on the group algebra $L^1(G)$ which will also immediately imply the postliminality of $C^*(G)$. The basic problem is very simple to state: Let π be in $\widehat{G}$. Are there functions f in $L^1(G)$ so that $\pi(f)$ is nonzero compact? Of course one can ask this question for general locally compact type I groups and in many cases the answer is positive, in particular it has been shown that in our case, i.e. G exponential, for every $\pi \in \widehat{G}$ there are functions f in $L^1(G)$ such that $\pi(f)$ is a rank 1 projection, see [21], [25].

As we have seen, for exponential groups G the representation spaces $\mathfrak{H}(\pi)$ basically can be realized as spaces $L^2(\mathbb{R}^m)$, where $\mathbb{R}^m$ is diffeomorphic to the homogeneous space G/P with $P = E(\mathfrak{p})$ if $\pi = \pi(f, \mathfrak{g}, G)$ for $f \in \Omega_{(\pi)}$. So in this case it is natural to ask which kernel operators are contained in $\pi(L^1)$, in particular whether there are Schwartz kernel or split kernel operators in $\pi(L^1)$. If P is normal, then it has been shown in [20] that the quotient $L^1(G)/\ker\pi$ is isomorphic to the twisted L^1 algebra $L^1(G/P, \mathcal{A}_\pi; P_\pi)$ with cocycle P_π and a Banach subalgebra $\mathcal{A}_\pi$ of $\mathcal{C}_b(G/P)$, the bounded continuous functions on G/P. The subalgebra $\mathcal{A}_{\pi,\infty} = A_\pi \cap \mathcal{C}_\infty(G/P)$ is a closed invariant ideal in $\mathcal{A}_\pi$ and the subalgebra $L^1(G/P, \mathcal{A}_{\pi,\infty}; P_\pi)$ is the inverse image under π of the compact operators of $\mathfrak{H}(\pi) = L^2(G/P)$. We have $\mathcal{A}_{\pi,\infty} = \mathcal{A}_\pi$ if and only if the orbit of π is closed, which shows that we will encounter conditions of decay at infinity when we will try to determine smooth kernels K on L^2, resp. functions f in L^1 for which the operators $\widetilde{K}$, resp. $\pi(f)$ are compact or have finite rank. It will turn out that polynomial decay, as in the case of nilpotent groups, is no longer sufficient.

Since we will have to handle exponentially bounded vector valued entire functions, we start with the following useful

Definition. For $c > 0$ and $r \in \mathbb{N}$, $r \geq 1$, let $\mathfrak{E}_r(c)$ be the class of all entire functions $f: \mathbb{C}^r \to \mathbb{C}$ of the form

$$f(z) = \sum_\nu \frac{1}{\nu!} a_\nu z^\nu \quad , \qquad \nu = (\nu_1, \nu_2, \ldots, \nu_r) \in \mathbb{N}^r \ ,$$

with the usual notation $\nu! = \prod_1^r \nu_i!$, $z^\nu = \prod_1^r z_i^{\nu_i}$, for which, with $|\nu| = \sum_1^r |\nu_i|$,

$$\sup_\nu (c^{-|\nu|} |a_\nu|) < \infty \ .$$

The elementary properties of the $\mathfrak{E}_r(c)$ are listed in

(55) (i) $\mathfrak{E}(c)$ *is a vector space.* (ii) *If* $f \in \mathfrak{E}(c)$, $g \in \mathfrak{E}_r(d)$, *then* $fg \in \mathfrak{E}_r(c+d)$, *in short* $\mathfrak{E}_r(c)\mathfrak{E}_r(d) \subset \mathfrak{E}_r(c+d)$. (iii) *Let* $D_j = \frac{\partial}{\partial z_j}$ *and* $D_\mu = \prod_1^r D_j^{\mu_j}$ *for* $\mu \in \mathbb{N}^r$. *Then* $D^\mu \mathfrak{E}_r(c) = \mathfrak{E}_r(c)$ *and* $P(z, D)\mathfrak{E}_r(c) \subset \mathfrak{E}_r(c)$ *for every polynomial* $P(z, D)$ *in the* z_j *and* D_j.

The *proof* of (i) and (iii) is obvious. To prove (ii) let a_ν, resp b_ν be the coefficients of f, resp. g. Then the coefficients c_ν of the product fg are given by

$$c_\nu = \nu! \sum_{\alpha+\beta=\nu} \frac{a_\alpha}{\alpha!} \frac{b_\beta}{\beta!},$$

hence with constant A and B:

$$|c_\nu| \leq AB\nu! \sum_{\alpha+\beta=\nu} \frac{c^{|\alpha|}}{\alpha!} \frac{d^{|\beta|}}{\beta!} =$$

$$= AB\prod_{i=1}^{r} \left\{ \sum_{k=0}^{\nu_i} \binom{\nu_i}{k} c^k d^{\nu_i - k} \right\} = AB\prod_{1}^{r}(c+d)^{\nu_i} = AB(c+d)^{|\nu|}.$$

In our applications the coefficients of the power series and also the arguments always will be real, hence the $f \in \mathfrak{E}_r(c)$ will be real valued real analytic functions and the D_j are the ordinary partial derivations. This applies also to the spaces in our next

Definition. For $n \in \mathbb{N}$, $n \geq 1$, let $\mathfrak{E}_{r,n}(c)$ be the family of functions $f: \mathbb{C}^r \to \mathbb{C}^n$, $f(z) = (f_1(z), f_2(z), \ldots, f_n(z)) \in \mathbb{C}^n$, with $f_i \in \mathfrak{E}_r(c)$, $i = 1, 2, \ldots, n$, in particular $\mathfrak{E}_{r,1}(c) = \mathfrak{E}_r(c)$. The subset of all functions in $\mathfrak{E}_{r,n}(c)$ with real coefficients, hence the $f \in \mathfrak{E}_{r,n}(c)$ with $f(\mathbb{R}^n) \subset \mathbb{R}^n$, will be denoted by $\mathfrak{E}^{\mathbb{R}}_{r,n}(c)$.

Now let $\mathfrak{g}$ again be a solvable exponential Lie algebra and let $\mathfrak{n}$ be a nilpotent ideal. Let$\{\mathfrak{n}^{(j)}\}_0^p$ be the descending central series, thus $\mathfrak{n} = \mathfrak{n}^{(0)} \supset [\mathfrak{n}, \mathfrak{n}] = \mathfrak{n}^{(1)} \supset \cdots \supset \mathfrak{n}^{(p)} \neq 0$, $[\mathfrak{n}, \mathfrak{n}^{(p)}] = 0$. We choose subspaces $\mathfrak{v}_j \subset \mathfrak{n}^{(j)}$ with $\mathfrak{n}^{(j)} = \mathfrak{v}^{(j)} \oplus \mathfrak{n}^{(j+1)}$, hence $\mathfrak{n}^{(p)} = \mathfrak{v}_p$ and

$$\mathfrak{n} = \sum_{j=1}^{p}{}^{\oplus} \mathfrak{v}_j .$$

Using the Campbell–Baker–Hausdorff polynomial, denoted by $x \Diamond y$ for $x, y \in \mathfrak{n}$, we can identify $\mathfrak{n}$ with its corresponding simply connected group N, i.e. we have $\exp x = x$ for $x \in \mathfrak{n}$ and $x \Diamond y$ denotes the product in $N = \mathfrak{n}$.

For $x = \sum_1^p{}^{\oplus} x_j$, $y = \sum_1^p{}^{\oplus} y_j$ in $\mathfrak{n}$, with $x_j, y_j \in \mathfrak{v}_j$, we set

$$x\Diamond y = z(x, y) = \sum_1^p{}^{\oplus} z_j(x, y), \qquad z_j(x, y) \in \mathfrak{v}_j \ ,$$

and claim:

(56) *There exist polynomials* $m_j(x, y) = m_j(x_1, \ldots, x_{j-1}, y_1, \ldots, y_{j-1})$, *independent of* x_i, y_i *for* $i \geq j$, *such that*

$$z_j(x, y) = x_j + y_j + m_j(x, y) \ .$$

This is clear for $j = p$ because $\mathfrak{v}_p$ is central, hence does not contribute to the nonlinear part of $x\Diamond y$. The same argument applied to the quotients $\mathfrak{n}/\mathfrak{n}^{j+1}$ shows that (56) holds for all j.

Next we choose a basis $\{b_i\}_i$ of $\mathfrak{n}$ and indices d_j, $2 \leq d_1 < d_2 < \cdots < d_p = \dim \mathfrak{n} = m$ such that $\mathfrak{v}_j = \sum_{i=d_{j-1}+1}^{d_j} {}^{\oplus}\mathbb{R}b_i$, hence

$$m_j(x, y) = \sum_{i=d_{j-1}+1}^{d_j} \mu_i(x, y) b_i$$

with ordinary polynomials $\mu_i(x, y)$. It follows from (56) that in these coordinates the

$$B_j = \frac{\partial}{\partial x_j} + \sum_{k=d_i+1}^{m} \frac{\partial \mu_k}{\partial y_j}(x, 0)\frac{\partial}{\partial x_k} \text{ for } i \text{ with } d_{i-1} < j \leq d_i \tag{57}$$

are the left invariant vector fields on N, corresponding to $b_j \in \mathfrak{n}$.

Let $|x|$ be a norm on $\mathfrak{g}$. By induction on $\dim \mathfrak{g}$ it is easy to see that we can assume that $|[x, y]| \leq |x||y|$ for all $x, y \in \mathfrak{g}$. Let $\{a_j\}$ be a basis of $\mathfrak{g}$ with $|a_j| = 1$ for all j and set $A_j = \operatorname{ad} a_j|_{\mathfrak{n}}$. Since $|[a_j, x]| \leq |x|$ it follows that $\operatorname{ad} a_j$ and a fortiori A_j have operator norm $\|A_j\| \leq 1$.

For $x = \sum x_j a_j \in \mathfrak{g}$ we have

(58) $$\operatorname{Ad}\ \exp x|_{\mathfrak{n}} = e^{\sum x_j A_j} = \\ = \sum_{k=0}^{\infty} \frac{1}{k!} (\sum_j x_j A_j)^k = \sum_0^{\infty} \frac{1}{k!} \sum_{|\mu|=k} B_\mu x^\mu = \sum_{\mu \in \mathbb{N}^n} \frac{1}{|\mu|!} B_\mu x^\mu$$

with $n = \dim \mathfrak{g}$, $\mu = (\mu_1, \mu_2, \dots, \mu_n)$, $|\mu| = \sum \mu_i$ and matrices, resp. linear transformations B_μ of $\mathfrak{n}$ which are sums of products of $|\mu|$ of the A_j. From the polynomial identity $(\sum_1^n X_j)^k = \sum \frac{k!}{\mu!} X^\mu$, $X^\mu = X_1^{\mu_1} X_2^{\mu_2} \dots X_n^{\mu_n}$, and $\|A_j\| \leq 1$ it follows easily that $\|B_\mu\| \leq \frac{k!}{\mu!}$, hence $\frac{1}{k!} B_\mu = \frac{1}{\mu!} C_\mu$ with $\|C_\mu\| \leq 1$. If also $|b_j| = 1$ for the basis $\{b_i\}$ of $\mathfrak{n}$, then $e^{\sum x_j A_j} b_i = \sum_\mu \frac{1}{\mu!} (C_\mu b_i) x^\mu$ and since $|C_\mu b_i| \leq 1$ we see:

(59) *For $x \in \mathfrak{g}$ and $y = \sum y_i b_i \in \mathfrak{n}$ we have*

$$(\operatorname{Ad}\ \exp x)(y) = \sum_{i=1}^m F_i(x) y_i$$

with functions F_i in $\mathfrak{E}_{n,m}^{\mathbb{R}}(1)$.

Our next aim is to prove

Theorem 11. *Let $\mathfrak{n}$ be a nilpotent ideal in the exponential solvable Lie algebra $\mathfrak{g}$. There exist constants $C > 0$ and $k \in \mathbb{N}$, depending only on $\mathfrak{g}$ and $\mathfrak{n}$, and functions $F_\mu \in \mathfrak{E}_{n,m}^{\mathbb{R}}(C)$, $n = \dim \mathfrak{g}$, $m = \dim \mathfrak{n}$ and $\mu \in \mathbb{N}^n$, $|\mu| \leq k$, such that for*

$$F(x, y) = \sum_{|\mu| \leq k} F_\mu(x) y^\mu$$

the identity

$$\exp(x + y) = \exp x \exp F(x, y)$$

holds.

Here of course the vector $x = (x_1, \dots, x_n) \in \mathbb{R}^n$, resp. $y \in \mathbb{R}^m$ is identified with the element $\sum_1^n x_i a_i \in \mathfrak{g}$, resp. $\sum_1^m y_i b_i \in \mathfrak{n}$, with the bases $\{a_i\}$ of $\mathfrak{g}$ and $\{b_i\}$ of $\mathfrak{n}$, used above.

Proof. We fix $x = \sum_1^n x_i a_i$ in $\mathfrak{g}$. Then the one parameter subgroup $\exp \mathbb{R}x$ together with the nilpotent normal subgroup $N = \exp \mathfrak{n}$ generate a closed subgroup $\exp \mathbb{R}x \cdot N$, which is isomorphic with the semidirect product $\mathbb{R} \ltimes N$. As before let A_i be the restriction of ad a_i onto $\mathfrak{n}$, hence $A = \sum x_i A_i = \operatorname{ad} x|_{\mathfrak{n}}$. In our identification of $\exp \mathbb{R}x \cdot N$ with $\mathbb{R} \ltimes N$ and of N and $\mathfrak{n}$, the element $\exp tx \exp y$, $y \in \mathfrak{n}$, corresponds to $(t, y) \in \mathbb{R} \ltimes \mathfrak{n}$ and the product is given by $(s, u)(t, v) = (s + t, (e^{-tA}u) \Diamond v)$.

We wish to compute the one parameter group $\exp \mathbb{R}(x + y)$ in the coordinates of $\mathbb{R} \ltimes \mathfrak{n}$. For this we use the decomposition $\mathfrak{n} = \sum_1^p {}^{\oplus} \mathfrak{v}_j$ with corresponding basis $\{b_i\}$ used above and set $q = \sum_1^n x_i a_i$, hence $\exp tq = (t, 0) \in \mathbb{R} \ltimes \mathfrak{n}$. The vector fields B_i of the $b_i \in \mathfrak{n}$ are given by (57), while $(s, x)(t, 0) = (s + t, e^{-tA}x)$ implies that q is given by $\frac{\partial}{\partial t} - \sum_{k=1}^m (Ax)_k \frac{\partial}{\partial x_k}$. Hence to $x + y$ corresponds the differential operator

$$(60) \qquad \frac{\partial}{\partial t} + \sum_{k=1}^m \left\{ -(Ax)_k + \sum_{i=1}^k y_i \left(\delta_{ik} + \frac{\partial \mu_k}{\partial y_i}(x, 0)\right) \right\} \frac{\partial}{\partial x_k} .$$

As is well known, from the coefficients in (60) one obtains the differential equations of the coordinates of $\exp t(x+y) = (\varphi_0(t), \varphi(t)) \in \mathbb{R} \ltimes \mathfrak{n}$. Clearly $\dot{\varphi}_0(t) = 1$, hence $\varphi_0(t) = t$. To compute the function $\varphi(t)$ we decompose $\varphi(t)$ into summands $\varphi_j(t) \in \mathfrak{v}_j$, hence $\varphi(t) = \sum_1^p {}^{\oplus} \varphi_j(t)$. Since the $\mathfrak{n}^{(j)}$ are A invariant, the transformation A defines a lower triangular matrix (A_{jl}) with

$$(Az)_j = \sum_{l=1}^j A_{jl} z_l$$

for the j component of Az, $z = \sum_1^p {}^{\oplus} z_j \in \mathfrak{n}$. Here the $A_{jl} z_l$ depend linearly on the coordinates x_i of $x = \sum_1^n x_i a_i$, resp. $A = \sum x_i A_i$ and on the transformations A_i, i.e. on the fixed basis $\{a_i\}$ of $\mathfrak{g}$. From this fact and the expression (56) for $x + y$ we see that there are $\mathfrak{v}_j$–valued polynomials $M_j(\xi; x, y) = M_j(\xi_1, \xi_2, \ldots, \xi_{j-1}; x, y)$, linear in x and y, with $\xi = \sum_1^p {}^{\oplus} \xi_j$, such that

$$(61) \qquad \dot{\varphi}_j(t) = -A_{jj} \varphi_j(t) + M_j(\varphi_1(t), \ldots, \varphi_{j-1}(t); x, y)$$

are the equations for the one parameter group $\varphi(t)$. We can solve these equations successively: For $j = 1$ the function M_1 is independent of ξ, hence the solution is

$$\Phi_1(t; x, y) = \widetilde{e}(-tA_{11})tM_1(x, y)$$

with our old function $\widetilde{e}(z) = \frac{e^z - 1}{z}$.

Let us now assume that for $j \in \mathbb{N}$, $j > 1$, there exist constants $c_j > 0$, $m_j \in \mathbb{N}$ and that for $i < j$ and $\alpha, \beta \in \mathbb{N}^n$, $|\alpha|, |\beta| \leq m_j$, $1 \leq \nu \leq m_j$, there are functions $G_i^{\alpha,\beta,\nu} \in \mathfrak{E}_{n,d_i}(c_j)$, $d_i = \dim \mathfrak{v}_i$, such that

$$\Phi_i(t; x, y) = \sum_{\nu, |\alpha|, |\beta| \leq m_i} x^\alpha y^\beta t^\nu G_i^{\alpha,\beta,\nu}(tx) \tag{62}$$

for all $i < j$ solves the equations (61). This is clearly the case for $j = 2$, with $m_2 = 1$, $c_2 = 1$.

The polynomial M_j depends only on the A_i and the $\{b_j\}$, i.e. on the bases $\{a_i\}$ and $\{b_j\}$, hence the function

$$F(t; x, y) = M_j(\Phi_1(t; x, y), \ldots, \Phi_{j-1}(t; x, y); x, y)$$

is a sum of terms of the form

$$x^\alpha y^\beta t^\nu H(tx)$$

with $|\alpha|, |\beta| \leq m_{j+1}$, $1 \leq \nu \leq m_{j+1}$ and $H \in \mathfrak{E}_{n,d_j}(c'_{j+1})$, with new constants m_{j+1} and c'_{j+1}, determined by m_j, c_j and the polynomial M_j. The solution of $\dot{\varphi} = -A_{jj}\varphi + F$, i.e. of (61) for the index j is explicitly given by

$$\Phi_j(t; x, y) = e^{-tA_{jj}} \int_0^t e^{sA_{jj}} F(s; x, y) ds$$

Now (55) implies $e^{A_{jj}} H \in \mathfrak{E}_{n,d_j}(c'_{j+1} + 1)$, hence we have an expansion

$$e^{sA_{jj}} H(sx) = \sum_\mu \frac{1}{\mu!} h_\mu x^\mu s^{|\mu|},$$

thus

$$\int_0^t e^{sA_{jj}} s^\nu H(sx) ds = \sum_\mu \frac{1}{\mu!} \frac{h_\mu}{|\mu| + \nu + 1} x^\mu t^{|\mu|+\nu+1} = t^{\nu+1} G(tx)$$

with $G(x) = \sum_\mu \frac{1}{\mu!} \frac{h_\mu}{|\mu|+\nu+1} x^\mu \in \mathfrak{E}_{n,d_j}(c_{j+1})$ for $c_{j+1} = c'_{j+1} + 1$. It follows that also Φ_j is of the form (62). Proceeding this way we obtain eventually a function $t \to \Phi(t; x, y) = \sum_1^p {}^\oplus \Phi_j(t; x, y)$, such that $\varphi(t) = \Phi(t; x, y)$ solves all equations

(61), i.e. for which $\{(t, \varphi(t))\}_t$ is the group $\exp \mathbb{R}(x+y)$. In particular we have

$$\exp(-x) \cdot \exp(x+y) = (0, \varphi(1)) = (0, \Phi(1; x, y)) ,$$

i.e.

$$F(x, y) = \Phi(1; x, y) = \sum_{\alpha,\beta,\nu,i} x^\alpha y^\beta G_i^{\alpha,\beta,\nu}(x) = \sum_{|\beta|\le k} y^\beta F_\beta(x)$$

with $F_\beta(x) = \sum x^\alpha G_i^{\alpha,\beta,\nu}(x)$, hence $F_\beta \in \mathfrak{E}^{\mathbb{R}}_{n,m}(C)$ for $C = c_{p+1}$. This proves theorem 11.

We will now introduce special coordinates in the coset spaces G/P of exponential groups G. As we have seen (proposition 2), every coexponential basis $\mathcal{B}$ for the subalgebra $\mathfrak{p}$ of $\mathfrak{g}$ defines a diffeomorphism $E_\mathfrak{p}$ from $\mathbb{R}^d$ via $\mathfrak{g}/\mathfrak{p}$ onto G/P, $P = E(\mathfrak{p})$, $d = \dim G/P$. Given two such bases with diffeomorphisms E_1 and E_2, the mapping $F = E_2^{-1} \circ E_1$ is a diffeomorphism of $\mathbb{R}^d$ onto itself. If G is nilpotent, then F is bipolynomial, however, in general F is much more complicated, in particular it may behave exponentially in some and polynomially in other directions. To keep this behaviour under control we must restrict the choices of the coexponential bases carefully. We need the following well known proposition:

(63) $\mathfrak{g}$ *contains* nilpotent *subalgebras* $\mathfrak{f}$ *such that*

$$\mathfrak{g} = \mathfrak{f} + \mathfrak{g}'$$

Proof. This is trivial if $\mathfrak{g}$ is nilpotent or $\dim \mathfrak{g} \le 2$. So let us assume that $\mathfrak{g}$ is not nilpotent and the theorem holds for algebras of smaller dimension. Since $\mathfrak{g}$ is not nilpotent, there exists $q \in \mathfrak{g}$ such that $Q = \operatorname{ad} q$ is not nilpotent. Then $\mathfrak{g} = \mathfrak{g}_0 \oplus \mathfrak{g}_1$, where $\mathfrak{g}_0$ and $\mathfrak{g}_1$ are Q invariant, $Q|_{\mathfrak{g}_0}$ is nilpotent, $Q|_{\mathfrak{g}_1}$ is regular, hence $Q(\mathfrak{g}_1) = [q, \mathfrak{g}_1] = \mathfrak{g}_1 \subset \mathfrak{g}'$ and $\mathfrak{g} = \mathfrak{g}_0 + \mathfrak{g}'$. Since Q is not nilpotent, we see that $\mathfrak{g}_0$ is a proper subalgebra, hence it contains a nilpotent subalgebra $\mathfrak{f}$ with $\mathfrak{g}_0 = \mathfrak{f} + \mathfrak{g}_0'$. Clearly $\mathfrak{g} = \mathfrak{f} + \mathfrak{g}'$.

Now let $\mathfrak{p}$ be a subalgebra of $\mathfrak{g}$, let $\mathfrak{i}$ be an ideal in $\mathfrak{g}$, contained in the maximal nilpotent ideal $\mathfrak{n}$ of $\mathfrak{g}$ and containing $\mathfrak{g}'$, hence

$$\mathfrak{g} \supset \mathfrak{n} \supset \mathfrak{i} \supset \mathfrak{g}' .$$

Let $\mathfrak{f}$ be a nilpotent subalgebra of $\mathfrak{g}$ with $\mathfrak{g} = \mathfrak{f} + \mathfrak{g}'$, hence also $\mathfrak{g} = \mathfrak{f} + \mathfrak{i} = \mathfrak{f} + \mathfrak{d}$, where $\mathfrak{d} = \mathfrak{i} + \mathfrak{p}$.

Definition. A coexponential basis for $\mathfrak{p}$ in $\mathfrak{g}$ is called $\mathfrak{i}$ *special* or, if $\mathfrak{i}$ is fixed, *special*, if it is of the form $\mathcal{B}_\mathfrak{p} = \{a_1, a_2, \dots, a_r, b_1, b_2, \dots, b_s\}$ where
$\mathcal{B}_\mathfrak{d} = \{a_1, \dots, a_r\}$ is coexponential for $\mathfrak{d}$ with $a_1, a_2, \dots, a_r \in \mathfrak{f}$,
$\mathcal{B}_{\mathfrak{d},\mathfrak{p}} = \{b_1, \dots, b_s\}$ is coexponential in $\mathfrak{d}$ for $\mathfrak{p}$ with $b_1, b_2, \dots, b_s \in \mathfrak{i}$.

Let $\kappa_{\mathfrak{B}}$ denote the diffeomorphism

$$\kappa_{\mathfrak{B}} : (\xi, \vartheta) \to \kappa_{\mathfrak{B}}(\xi, \vartheta) = E(\xi_1 a_1) \ldots E(\xi_r a_r) E(\vartheta_1 b_1) \ldots E(\vartheta_s b_s) P$$

from $\mathbb{R}^r \times \mathbb{R}^s$ onto G/P.

If $\mathfrak{B}_1$ and $\mathfrak{B}_2$ are two special coexponential bases for $\mathfrak{p}$, then $T = \kappa_{\mathfrak{B}_2}^{-1} \circ \kappa_{\mathfrak{B}_1}$ is a diffeomorphism of $\mathbb{R}^r \times \mathbb{R}^s$. We will show that T belongs to the following class:

Definition: Let $\mathfrak{T}_{r,s}$ be the group of all diffeomorphisms

$$T : \mathbb{R}^r \times \mathbb{R}^s \to \mathbb{R}^r \times \mathbb{R}^s ,$$

with $T(x, y) = (T_1(x, y), T_2(x, y))$, $x \in \mathbb{R}^r$, $y \in \mathbb{R}^s$, where
(i) $T_1(x, y) = T_1(x) \in \mathbb{R}^r$ is independent of y and linear in x,
(ii) $T_2(x, y) = \sum F_\alpha(x) y^\alpha \in \mathbb{R}^s$ is polynomial in y with $F_\alpha \in \mathfrak{C}_{r,s}(c)$ for some $c = c(T) > 0$,
and where the same holds for the inverse T^{-1}.

Proposition 8. *If $\mathfrak{B}_1$ and $\mathfrak{B}_2$ are* $\mathfrak{i}$ *special coexponential bases of* $\mathfrak{g}$ *for* $\mathfrak{p}$, *then the transformation $T = \kappa_{\mathfrak{B}_2}^{-1} \circ \kappa_{\mathfrak{B}_1}$ is in $\mathfrak{T}_{r,s}$.*

Proof. Let $\mathfrak{B}_1 = \{a_1, a_2, \ldots, a_r, b_1, \ldots, b_s\}$ with $a_j \in \mathfrak{f}_1$, the nilpotent subalgebra with $\mathfrak{g} = \mathfrak{f}_1 + \mathfrak{g}'$. Since also $\mathfrak{g} = \mathfrak{f}_2 + \mathfrak{g}'$ with the $\mathfrak{f}_2$ belonging to $\mathfrak{B}_2$, we can find $c_i \in \mathfrak{g}'$ with $a_i' = a_i + c_i \in \mathfrak{f}_2$. Since $\mathfrak{g}' \subset \mathfrak{i} \subset \mathfrak{d}$ it follows that also $\mathfrak{B}_1' = \{a_1', \ldots, a_r', b_1, \ldots, b_s\}$ is special coexponential. With $\kappa_i = \kappa_{\mathfrak{B}_i}$, $i = 1, 2$, $\kappa_0 = \kappa_{\mathfrak{B}_1'}$ we have $T = (\kappa_2^{-1} \circ \kappa_0) \circ (\kappa_0^{-1} \circ \kappa_1)$, thus it suffices to prove that both factors of T are in $\mathfrak{T}_{r,s}$.

We start with $T_0 := \kappa_0^{-1} \circ \kappa_1$. Let $(x, y) \in \mathbb{R}^r \times \mathbb{R}^s$ and $(x', y') = T_0(x, y)$, i.e. $\kappa_1(x, y) = \kappa_0(x', y')$. Then, with $D = E(\mathfrak{d})$ a fortiori $E(x_1 a_1) E(x_2 a_2) \ldots E(x_r a_r) D = E(x_1' a_1') \ldots E(x_r' a_1') D$. But G/D is commutative and $E(x a_j) \equiv E(x a_j') \pmod D$, all j, because $c_j \subset \mathfrak{g}' \subset \mathfrak{d}$. Thus $x_i = x_i'$ for $1 = 1, 2, \ldots, r$.

Because $\mathfrak{f}_2$ is nilpotent, we can write

$$A' := E(x_1 a_1') E(x_2 a_2') \ldots E(x_r a_r') = E(\sum x_i a_i') E(P(x))$$

where $P(x)$ is an $\mathfrak{f}_2'$ valued polynomial in x. From theorem 11 we conclude that $E(\sum x_i a_i') = E(\sum x_i a_i) E(F(x))$ with $F \in \mathfrak{C}_{r,d}(c)$, $d = \dim \mathfrak{g}'$ (i.e. $F(x) \in \mathfrak{g}'$, all $x \in \mathbb{R}^r$). Furthermore, $E(\sum x_i a_i) = E(x_1 a_1) \ldots E(x_r a_r) E(Q(x))$ with a $\mathfrak{f}_1'$ valued, hence $\mathfrak{g}'$ valued polynomial Q. Finally, with $A = E(x_1 a_1) E(x_2 a_2) \ldots E(x_r a_r)$, we get

$$A' = A E(Q(x)) E(F(x)) E(P(x)) .$$

Since $Q(x)$, $F(x)$ and $P(x)$ are in $\mathfrak{g}'$ there exists a function $G(x)$, polynomially depending on Q, F and P, such that $E(Q)E(F)E(P) = E(-G)$ and because $F \in \mathfrak{E}_{r,d}(c)$ also $G \in \mathfrak{E}_{r,d}(c')$ for some $c' \geq c$. Thus $A' = AE(-G(x))$ and modulo P we obtain

$$AE(-G(x))E(y'_1b_1)E(y'_2b_2)\dots E(y'_sb_s) \equiv AE(y_1b_1)E(y_2b_2)\dots E(y_sb_s)$$

or $E(y'_1b_1)\dots E(y'_sb_s) = E(G(x))E(y_1b_1)\dots E(y_sb_s)w$ for some $w \in P$. Now the elements y_jb_j and $G(x)$ are all in $\mathfrak{i}$; hence, if we complete $\{b_j\}_1^s$ to a basis $\{b_j\}_1^m$ of $\mathfrak{i}$ by taking elements $b_l \in \mathfrak{p}\cap\mathfrak{i}$, $l = s+1, \dots, m$, we see that there exist polynomials $p_j(G(x), y)$ such that

$$E(G(x))E(y_1b_1)\dots E(y_sb_s) = E(p_1(G(x), y)b_1)\dots E(p_s(G(x), y)b_s)v$$

with $v \in P$. This yields $y'_i = p_i(G(x), y)$, $i = 1, \dots, s$ and proves $\kappa_0^{-1} \circ \kappa_1 \in \mathfrak{T}_{r,s}$.

Now we suppose that $\mathfrak{f}_1 = \mathfrak{f}_2 = \mathfrak{f}$ and the a_i of $\mathfrak{B}_1$, resp. a'_i of $\mathfrak{B}_2$ are chosen in $\mathfrak{f}$. Let (α_{ij}) be the real regular $r \times r$ matrix for which

$$a_i^\vee := \sum_{j=1}^{r} \alpha_{ij}a'_j \equiv a_i \pmod{\mathfrak{d} \cap \mathfrak{f}}$$

in $\mathfrak{f}/\mathfrak{f}\cap\mathfrak{d} \stackrel{\sim}{=} \mathfrak{g}/\mathfrak{d}$. Then clearly also $\mathfrak{B}_3 = \{a_1^\vee, \dots a_r^\vee, b'_1, \dots, b'_s\}$ with $b'_j \in \mathfrak{B}_2$ is special exponential. Let $S = \kappa_2^{-1} \circ \kappa_3$ be the corresponding transformation of $\mathbb{R}^r \times \mathbb{R}^s$, thus $\kappa_3(x^\vee, y^\vee) = \kappa_2(x, y)$ if $S(x^\vee, y^\vee) = (x, y)$. Using this notation, the same arguments as above show that $x_j = \sum_k x_k^\vee \alpha_{kj}$, i.e. $x^\vee = x(\alpha_{ij})^{-1}$. Since $\mathfrak{f}$ is nilpotent and $x^\vee$ and x depend linearly on each other there are polynomials Q and P such that

$$\begin{aligned} E(x_1^\vee a_1^\vee)\dots E(x_r^\vee a_r^\vee) &= E(\sum x_i^\vee a_i^\vee)E(Q(x)) = E(\sum x_i a'_i)E(Q(x)) \\ &= E(x_1a'_1)\dots E(x_ra'_r)E(P(x)) \ , \end{aligned}$$

hence $\kappa_3(x^\vee, y^\vee) = \kappa_2(x, y)$ implies

$$E(y_1^\vee b'_1)\dots E(y_s^\vee b'_s) = E(P(x))E(y_1b'_1)\dots E(y_sb'_s)w$$

with $w \in P$, and consequently $y_j^\vee = F_j(x, y)$, where the F_j now even are polynomials in x and y, because the $Q(x)$, $P(x)$ are in $\mathfrak{f}' \subset \mathfrak{i}$. It follows that S is in $\mathfrak{T}_{r,s}$.

We have reduced the proof to the case $a_i, a'_i \in \mathfrak{f}$ and $a'_i = a_i + c_i$ with $c_i \in \mathfrak{f}\cap\mathfrak{d}$. Then again $\kappa_1(x, y) = \kappa_2(x', y')$ implies $x' = x$, moreover we have in $\mathfrak{f} : E(\sum x_ia'_i) = E(\sum x_ia + \sum x_ic_i) = E(\sum x_ia_i)E(M(x))$ with a polynomial $M(x) \in \mathfrak{f}\cap\mathfrak{d}$. We proceed as before and see that also in this case $\kappa_2^{-1} \circ \kappa_1 \in \mathfrak{T}_{r,s}$.

Finally we end up in the case where $\mathfrak{f}_1 = \mathfrak{f}_2$, $a_i = a_i'$ for all i. But in this case $\kappa_1(x, y) = \kappa_2(x', y')$ implies $x = x'$ and

$$E(y_1b_1)E(y_2b_2)\dots E(y_sb_s)P = E(y_1'b_1')\dots E(y_s'b_s')P ,$$

hence, because all $E(y_ib_i)$ and $E(y_i'b_i')$ are in $I = E(\mathfrak{i})$:

$$E(y_1b_1)\dots E(y_sb_s)(P\cap I) = E(y_1'b_1')\dots E(y_s'b_s')(P\cap I)$$

Because the $\{b_i\}$ and $\{b_i'\}$ are coexponential in $\mathfrak{i}$ for $\mathfrak{i}\cap\mathfrak{p}$ and $\mathfrak{i}$, resp. I is nilpotent we conclude that $y \to y'$ is bipolynomial. This ends the proof of proposition 8.

If the group G is nilpotent, then the Schwartz space $\mathcal{S}(G)$ is an important subalgebra of $L^1(G)$, see [14] and others. $\mathcal{S}(G)$ can be defined by identifying G with $\mathbb{R}^n$, with the aid of a Malcev basis. Since different bases give raise to bipolynomial diffeomorphisms of $\mathbb{R}^n$, it is clear that $\mathcal{S}(G)$ is independent of the basis. It is well known that these classical Schwartz spaces are inadequate objects in the case of general solvable Lie groups, in fact already for the "$ax + b$" group. Thus one has to modify the definitions, see [21]. This is easily achieved by using proposition 8.

Definition. Let $\mathfrak{F}_{r,s}$ be the algebra of all functions $f : \mathbb{R}^r \times \mathbb{R}^s \to \mathbb{C}$ of the form

$$f(x, y) = \sum_{|\mu|\le m} F_\mu(x)y^\mu$$

with $F_\mu \in \mathfrak{E}_r(c)$ for some $c > 0$, $m \in \mathbb{N}$.

Thus these f are entire functions in (x, y), polynomial in y and exponentially bounded in x. Moreover, $\mathfrak{F}_{r,s}$ is stable under the partial differential operators $\frac{\partial}{\partial x_i}$, $\frac{\partial}{\partial y_j}$, $1 \le i \le r$, $1 \le j \le s$. If we set

$$D^{\alpha,\beta} = \prod_1^r \left(\frac{\partial}{\partial x_i}\right)^{\alpha_i} \prod_1^s \left(\frac{\partial}{\partial y_j}\right)^{\beta_j}$$

for $(\alpha, \beta) \in \mathbb{N}^r \times \mathbb{N}^s$, then it follows that

$$\mathfrak{D}_{r,s} = \text{ linear span } \{F(x)y^\mu D^{\alpha,\beta} ;\ F \in \mathfrak{E}_r(c),\ \alpha \in \mathbb{N}^r,\ \mu, \beta \in \mathbb{N}^s\}$$

is the subalgebra of linear endomorphisms of $\mathcal{C}^\infty(\mathbb{R}^r \times \mathbb{R}^s)$, generated by $\mathfrak{F}_{r,s}$ and the $D^{\alpha,\beta}$, more precisely by the multiplication operators, defined by functions in $\mathfrak{F}_{r,s}$, and the $D^{\alpha,\beta}$.

Now we take a transformation T in $\mathfrak{T}_{r,s}$. It follows immediately from the definitions that the elements $T_{j,k}$ of the Jacobian $\frac{\partial T}{\partial(x,y)} = (T_{j,k})$ of T are functions in

$\mathfrak{F}_{r,s}$. Let T^* be the endomorphism $f \to T^*f = f \circ T$ of $\mathscr{C}^\infty(\mathbb{R}^r \times \mathbb{R}^s)$. Clearly $T^*\mathfrak{F}_{r,s} = \mathfrak{F}_{r,s}$, moreover, if $\widetilde{F}$ denotes the multiplication operator $(\widetilde{F}g)(x, y) = F(x, y)g(x, y)$ for $F \in \mathfrak{F}_{r,s}$, then $(T^*)^{-1}\widetilde{F}T^* = (F \circ T^{-1})^\sim$. For the derivations $\frac{\partial}{\partial z_j}$ (with $z_j = x_j$ if $1 \le j \le r$, $z_j = y_{j-r}$ if $r + 1 \le j \le r + s$) we compute easily that $(T^*)^{-1} \circ \frac{\partial}{\partial z_j} \circ T^* = \sum_k (T_{jk} \circ T^{-1})\frac{\partial}{\partial z_k} \in \mathfrak{D}_{r,s}$.

From these remarks it follows that $(T^*)^{-1}\mathfrak{D}_{r,s}T^* \subset \mathfrak{D}_{r,s}$ for all $T \in \mathfrak{T}_{r,s}$, in particular for T^{-1}. This yields

(64) *For all $T \in \mathfrak{T}_{r,s}$ we have*

$$(T^*)^{-1}\mathfrak{D}_{r,s}T^* = \mathfrak{D}_{r,s} \ .$$

We define the uniform norm $|f|_\infty$ of a complex valued function f on a set X as usual as $|f|_\infty = \sup_{x \in X} |f(x)|$.

Definition. For $A \in \mathfrak{D}_{r,s}$ and $f \in \mathscr{C}^\infty(\mathbb{R}^r \times \mathbb{R}^s)$ we set

$$|f|_A = |Af|_\infty$$

and

$$\mathscr{S}_{r,s} = \left\{ f \in \mathscr{C}^\infty(\mathbb{R}^r \times \mathbb{R}^s); |f|_A < \infty \text{ for all } A \in \mathfrak{D}_{r,s} \right\}$$

It is clear that $\mathscr{S}_{r,s}$ is a complete topological vector space, in fact even a Frechet space, which contains all compactly supported smooth functions. The functions in $\mathscr{S}_{r,s}$ decay uniformly faster than any inverse polynomial in the $\mathbb{R}^s$ directions and faster then any $e^{-c|x|}$ in the $\mathbb{R}^r$ directions, in particular $\mathscr{S}_{r,s} \subset L^1(\mathbb{R}^r \times \mathbb{R}^s)$. For $r = 0$, $\mathscr{S}_{0,s}$ is the ordinary Schwartz space on $\mathbb{R}^s$. From (64) we derive

(65) *$\mathscr{S}_{r,s}$ is invariant under $\mathfrak{T}_{r,s}$, i.e. if $f \in \mathscr{S}_{r,s}$, $T \in \mathfrak{T}_{r,s}$, then $f \circ T \in \mathscr{S}_{r,s}$.*

Now we can define the relevant function spaces on the homogeneous space G/P. Let $\mathfrak{g}$ be an exponential Lie algebra with fixed nilpotent ideal $\mathfrak{i}$, containing $\mathfrak{g}'$, and let $\mathfrak{p}$ be a subalgebra. As before we set $\mathfrak{d} = \mathfrak{i} + \mathfrak{p}$ and denote the corresponding groups by G, I, P and D. We fix an $\mathfrak{i}$ special coexponential basis $\mathscr{B} = \{a_1, \dots, a_r, b_1, \dots, b_s\}$ and define the section $\sigma : \mathbb{R}^r \times \mathbb{R}^s \to G$ by

$$\sigma(x, y) = E(x_1a_1)E(x_2a_2) \dots E(x_ra_r)E(y_1b_1) \dots E(y_sb_s) \ . \tag{66}$$

Then $\kappa(x, y) = \sigma(x, y)P \in G/P$ defines the diffeomorphism $\kappa : \mathbb{R}^r \times \mathbb{R}^s \to G/P$ studied before. We take a unitary character χ of P and the induced representation $\pi = \mathrm{ind}_P^G\chi$. In our applications $\mathfrak{p}$ will be a Pukanszky polarization for some $l \in \mathfrak{g}^*$ and $\chi = \chi(l, \mathfrak{p})$ $\pi = \pi(l, \mathfrak{p})$. The Hilbert space $\mathfrak{H}(\pi) = \mathrm{ind}\, \mathfrak{H}(\chi)$ consists of measurable functions $\xi : G \to \mathbb{C} = \mathfrak{H}(\chi)$, satisfying $\xi(xq) = \overline{\chi(q)}\xi(x)$ for all $q \in P$, $x \in G$, and $\int_{G/P} |f(x)|^2 d\overline{x} < \infty$.

The diffeomorphism κ allows us to identify $\mathfrak{H}(\pi)$ with $L^2(\mathbb{R}^{r+s})$. Let $c(x)$ be the extension of $\Delta_G|_P\Delta_P^{-1}$ to G, trivial on N and I, with $d(a\overline{x}) = c(a)^{-1}d\overline{x}$ for $a \in G$. For a function $f : G \to \mathbb{C}$ with $f(xq) = \overline{\chi(q)}f(x)$ and $z = (x, y) \in \mathbb{R}^{r+s}$ we set $f^{\natural}(z) = f(\sigma(z))c(\sigma(z))^{-\frac{1}{2}}$. Then theorem 3 yields that $\xi \to \xi^{\natural}$ is unitary from $\mathfrak{H}(\pi)$ onto $L^2(\mathbb{R}^{r+s})$. Observe here, that $c(\sigma(x, y))$ is independent of $y \in \mathbb{R}^s$ and an exponential function of $x \in \mathbb{R}^r$, thus $f^{\natural} \in \mathcal{S}_{r,s}$ if and only if $f \circ \sigma \in \mathcal{S}_{r,s}$. Moreover, we conclude from proposition 8 and (64) that the fact that $f^{\natural}$ is a "Schwartz function", i.e. $f^{\natural} \in \mathcal{S}_{r,s}$, is independent of the choice of the special coexponential basis $\mathcal{B}$, which means that we can define Schwartz functions unambiguously on G/P. Similar remarks hold for definitions of Schwartz kernel operators on $L^2(\mathbb{R}^{r+s})$, resp. of $\mathfrak{H}(\pi)$. We will go into the details by computing first the kernel of the operator $\pi(f)$ for $f \in L^1(G)$.

Let Y be a connected closed subgroup of G. As before we denote the Haar moduli of G and Y by Δ_G and Δ_Y, however, different from our notation χ_Y^G in § 2, we write c_Y^G for the modulus of the relatively invariant measure $d\overline{x}$ on G/Y, thus the following general formulae hold:

$$(67)\qquad \begin{aligned} da\overline{x} &= c_Y^G(a)^{-1}d\overline{x} \qquad \text{for } a \in G \\ \int_G f(x)dx &= \int_{G/Y}\int_Y f(xy)c_Y^G(xy)dyd\overline{x} . \end{aligned}$$

Observe that for normal Y always $c_Y^G = 1$, moreover c_Y^G is always trivial on the maximal nilpotent normal subgroup N of G.

Now let H be a closed connected subgroup and τ a unitary representation of H in $\mathfrak{H}(\tau)$. Let $\pi = \mathrm{ind}_H^G\tau$ and $\mathfrak{H}(\pi) = \mathrm{ind}\ \mathfrak{H}(\tau)$, hence $(\pi(a)\xi)(x) = c_H^G(a)^{\frac{1}{2}}\xi(a^{-1}x)$ for $\xi \in \mathfrak{H}(\pi)$ and $a \in G$. For $f \in L^1(G)$ we get

$$(\pi(f)\xi)(x) = \int_G f(y)c_H^G(y)^{\frac{1}{2}}\xi(y^{-1}x)dy$$

and after the substitutions $y \to xy^{-1}$, application of (67) and the transformation properties of ξ:

$$(68)\qquad \begin{aligned} &(\pi(f)\xi)(x) = \\ &= \int_{G/H} \Delta_G(y)^{-1}c_H^G(xy)^{\frac{1}{2}}\left(\int_H c_H^G(q)^{\frac{1}{2}}f(xqy^{-1})\tau(q)dq\right)\xi(y)d\overline{y}. \end{aligned}$$

If H is normal we have $c_H^G = 1$, moreover, for $f \in L^1(G)$, $x \in G$ we define the function $\check{f}(x)$ on H by

$$\check{f}(x)(y) = f(xy), \quad y \in H$$

and for any function g on H and $x \in G$ we set $g^x(y) = g(xyx^{-1})$. Then we can write

$$f(xqy^{-1}) = f(xy^{-1} \cdot yqy^{-1}) = f(xy^{-1})^y(q) ,$$

hence $\int_H f(xqy^{-1})\tau(q)dq = \tau(f(xy^{-1})^y)$ and (68) becomes

$$(69) \qquad (\pi(f)\xi)(x) = \int_{G/H} \Delta_G(y)^{-1}\tau(f(xy^{-1})^y)\xi(y)d\overline{y} .$$

We remark that the integrand in (68) depends only on the coset $\overline{y} = yH$. This follows of course from our deduction, but it can also be computed directly.

Next assume that $H = P$, $\tau = \chi$ and $\mathfrak{H}(\tau) = \mathbb{C}$. Then $\mathfrak{H}(\pi)$ can be realized as $L^2(G/P)$ and $\pi(f)$ corresponds to a kernel operator. More precisely, we define

$$(70) \qquad K_f^\pi(x, y) = \Delta_G(y)^{-1} c_P^G(xy)^{\frac{1}{2}} \int_P c_P^G(q)^{\frac{1}{2}} f(xqy^{-1})\chi(q)dq .$$

Then (68) can be written as

$$(\pi(f)\xi)(x) = \int_{G/P} K_f^\pi(x, y)\xi(y)d\overline{y} .$$

It is easy to compute, that the kernel K_f^π satisfies the identity $K_f^\pi(xu, yv) = \chi(u^{-1}v)K_f^\pi(x, y)$ for $x, y \in G$, $u, v \in P$. If σ is the section (66), defined by the coexponential basis $\mathcal{B}$ of $\mathfrak{g}$ for $\mathfrak{p}$, then clearly K_f^π is determined by the function $K_f^\pi \circ \sigma^2$ on $\mathbb{R}^{r+s} \times \mathbb{R}^{r+s}$, with $\sigma^2(z_1, z_2) = (\sigma(z_1), \sigma(z_2)) \in G \times G$ for $z_i \in \mathbb{R}^{r+s}$.

The need to require exponential decay of the kernels in certain directions has a very serious consequence: In our proofs and in particular in certain constructions we will have to apply Fourier transformations, which – most likely – cannot be avoided, i.e. which apparently are inherent in the situation. Now clearly exponential decay of a function and all its derivatives is not stable under Fourier transformation. This forces us to impose more conditions upon the kernels to make the induction work. Actually already the simplest case, i.e. the Heisenberg groups, indicates that our conditions on the kernels may be too strong, however, we could not see how to obtain a general result without them.

Let $\mathbb{R}^k = D \oplus E$ be a direct decomposition of $\mathbb{R}^k$ into subspaces D and E. The *partial Fourier transform* $\mathcal{F}_E f$ of a function $f \in \mathcal{S}_{k,m}$ is defined by

$$(71) \qquad \mathcal{F}_E f(x \oplus x', y) = \int_E f(x \oplus u, y)e^{-i(u|x')}du .$$

If $D_1 \oplus E_1$ is another decomposition of $\mathbb{R}^r$ with $\dim D_1 = \dim D$, then we can take linear isometries $A : D \to D_1$ and $B : E \to E_1$, thus the sum $T = A \oplus B$ is an invertible linear transformation of $\mathbb{R}^k$. With B also the adjoint $B^! := (B^{-1})^*$ of the inverse transformation $B^{-1} : E_1 \to E$ is an isometry from E onto E_1, hence

$T^! = A \oplus B^!$ is invertible. An easy computation shows that for any integrable function φ on $\mathbb{R}^k$ we have

$$(72) \qquad \mathcal{F}_E(\varphi \circ T) = (\mathcal{F}_{E_1}\varphi) \circ T^!$$

where $(\varphi \circ T)(x, y) = \varphi(Tx, y)$.

Now we define:

Definition. Let $\mathcal{S}^{\mathcal{F}}_{k,m}$ be the space of all functions $f \in \mathcal{S}_{k,m}$, for which also for every decomposition $\mathbb{R}^k = D \oplus E$ with operators $\mathcal{F}_E$ and for every $T \in \mathfrak{T}_{k,m}$ the function $\mathcal{F}_E(f \circ T)$, defined by (71), is in $\mathcal{S}_{k,m}$.

If we define the partial Fourier transformation $\mathcal{F}_j$ on $\mathbb{R}^k \times \mathbb{R}^m$ by

$$(\mathcal{F}_j\varphi)(x, y) = \int_{\mathbb{R}} \varphi(x_1, \ldots, x_{j-1}, t, x_{j+1}, \ldots, x_k, y)e^{-itx_j}dt$$

then one derives easily from (72) that $f \in \mathcal{S}^{\mathcal{F}}_{k,m}$ if and only if $\mathcal{F}_j(f \circ T) \in \mathcal{S}_{k,m}$ for all $T \in \mathfrak{T}_{k,m}$ and all j with $1 \leq j \leq k$. Another simple consequence is that $\mathcal{S}^{\mathcal{F}}_{k,l}$ contains all functions of the form $f(x, y) = e^{-Q(x)} \cdot g(y)$ with $g \in \mathcal{S}_{o,m}$ and $Q(x)$ a nondegenerate positive definite quadratic form on $\mathbb{R}^k$.

Now we define smooth kernels. For technical purposes it is necessary to introduce in our next definition additional parameters.

Definition. Let k and m be natural numbers. A *smooth* (k, m) *kernel* for $G, P, \mathfrak{i}$ and the character $\chi \in \widehat{P}$ is a function

$$K : \mathbb{R}^k \times \mathbb{R}^m \times G \times G \to \mathbb{C}$$

with the following properties:

1) $K(\xi, \vartheta; xu, yv) = \chi(u^{-1}v)K(\xi, \vartheta; x, y)$ for all $(\xi, \vartheta; x, y) \in \mathbb{R}^l \times \mathbb{R}^m \times G \times G$ and all $u, v \in P$.
2) Let σ be defined by an $\mathfrak{i}$–*special* basis as in (66). With our notation as before the function

$$K^{\sigma} : (\xi, x, x'; \vartheta, y, y') \to K^{\sigma}(\xi, x, x'; \vartheta, y, y') = K(\xi, \vartheta; \sigma(x, y), \sigma(x', y',))$$

on $\mathbb{R}^{k+2r} \times \mathbb{R}^{m+2s}$ is in $\mathcal{S}^{\mathcal{F}}_{k+2r,m+2s}$.

The set of all smooth (k, m) kernels for $G, P, \mathfrak{i}$ will be denoted by

$$\mathcal{K}_{k,m}(G, P, \mathfrak{i})$$

We will show that for any $\pi \in \widehat{G}$ and a suitable class of polarizations $\mathfrak{p}$ for π the family $\mathcal{H}_{k,m}(G, P, \mathfrak{i})$ is contained in the image $\pi(L^1(G))$. To prove this fact we need more information on Vergne polarizations.

Proposition 9. *Let $\mathcal{S} = \{\mathfrak{g}_j\}_j$ be a JH sequence in the exponential Lie algebra $\mathfrak{g}$, let $\mathfrak{h}$ be a subalgebra in $\mathfrak{g}$ and $l \in \mathfrak{g}^*$. Then $\mathcal{S}^0 = \{\mathfrak{g}_j \cap \mathfrak{h}\}_j$ is a JH sequence in $\mathfrak{h}$. We set $l^0 = l|_{\mathfrak{h}}$ and*

$$\mathfrak{v} = \mathfrak{v}(l, \mathcal{S}) \ , \ \ \mathfrak{v}^0 = \mathfrak{v}(l^0, \mathcal{S}^0) \ .$$

If $\mathfrak{v} \subset \mathfrak{h}$ or $\mathfrak{g}^l + \mathfrak{h} = \mathfrak{g}$, then

$$\mathfrak{v}^0 = \mathfrak{h} \cap \mathfrak{v} \ .$$

Proof. We set $\mathfrak{h}_j = \mathfrak{g}_j \cap \mathfrak{h}$ and $l_j^0 = l|_{\mathfrak{h}_j}$. The first assertion is just (37). From the obvious inclusion $\mathfrak{g}_j^{l_j} \cap \mathfrak{h} \subset \mathfrak{h}_j^{l_j^0}$ we get $\mathfrak{g}_j^{l_j} \subset \mathfrak{h}_j^{l_j^0}$ if $\mathfrak{v} \subset \mathfrak{h}$ and consequently $\mathfrak{v} \subset \mathfrak{v}^0$. But $\mathfrak{v}$ is maximal isotropic, hence $\mathfrak{v} = \mathfrak{v}^0$.

To prove the rest we use induction on $d = \dim \mathfrak{g}/\mathfrak{h}$. For $d = 1$ there exists an element $a \in \mathfrak{g}^l$ so that $\mathfrak{g} = \mathbb{R}a \oplus \mathfrak{h}$. As $a \in \mathfrak{g}^l + \mathfrak{g}_j^{l_j}$ for all j we get $\mathfrak{g}^l + \mathfrak{g}_j^{l_j} = \mathbb{R}a \oplus \mathfrak{h} \cap (\mathfrak{g}^l + \mathfrak{g}_j^{l_j})$. From

$$l([\mathfrak{h} \cap (\mathfrak{g}^l + \mathfrak{g}_j^{l_j}), \mathfrak{h}_j]) \subset l([\mathfrak{g}^l + \mathfrak{g}_j^{l_j}, \mathfrak{g}_j]) = 0$$

we see that $\mathfrak{h} \cap (\mathfrak{g}^l + \mathfrak{g}_j^{l_j}) \subset \mathfrak{h}_j^{l_j^0}$, hence

$$\mathfrak{v} = \sum (\mathfrak{g}^l + \mathfrak{g}_j^{l_j}) = \mathbb{R}a \oplus \sum (\mathfrak{h} \cap (\mathfrak{g}^l + \mathfrak{g}_j^{l_j})) \subset \mathbb{R}a \oplus \sum \mathfrak{h}_j^{l_j^0} = \mathbb{R}a \oplus \mathfrak{v}^0 \ .$$

Since the subspace $\mathbb{R}a \oplus \mathfrak{v}^0$ is isotropic it follows that $\mathfrak{v} = \mathbb{R}a \oplus \mathfrak{v}^0 = \mathbb{R}a \oplus \mathfrak{v} \cap \mathfrak{h}$.

If $d > 1$ we take a subalgebra $\mathfrak{m}$ strictly between $\mathfrak{g}$ and $\mathfrak{h}$. Since also $\mathfrak{g} = \mathfrak{g}^l + \mathfrak{m}$ we get $\mathfrak{v}(l', \mathcal{S}') = \mathfrak{m} \cap \mathfrak{v}$ for $l' = l|_{\mathfrak{m}}$, $\mathcal{S}' = \{\mathfrak{m} \cap \mathfrak{g}_j\}$. Now

$$\mathfrak{m} = (\mathfrak{g}^l + \mathfrak{h}) \cap \mathfrak{m} = \mathfrak{g}^l \cap \mathfrak{m} + \mathfrak{h} \subset \mathfrak{m}^{l'} + \mathfrak{h} \subset \mathfrak{m} \ ,$$

i.e. $\mathfrak{m} = \mathfrak{m}^{l'} + \mathfrak{h}$. Hence we can apply our result once more to $\mathfrak{m}$, l', $\mathcal{S}'$ and $\mathfrak{h}$ and obtain $\mathfrak{v}^0 = \mathfrak{h} \cap \mathfrak{v}(l', \mathcal{S}') = \mathfrak{h} \cap \mathfrak{m} \cap \mathfrak{v} = \mathfrak{h} \cap \mathfrak{v}$.

Corollary. *Let $\mathfrak{h}$ be an ideal in $\mathfrak{g}$ and $\mathfrak{p}$ a Vergne polarization for $l \in \mathfrak{g}^*$. If $\mathfrak{p} \subset \mathfrak{h}$, then there exists a JH sequence $\mathcal{R}$ in $\mathfrak{g}$ with $\mathfrak{p} = \mathfrak{v}(l, \mathcal{R})$, $\mathfrak{h} \in \mathcal{R}$. If $\mathfrak{p}$ passes through the ideal $\mathfrak{i}$, $\mathfrak{i} \subset \mathfrak{h}$, i.e. if $\mathfrak{p} = \mathfrak{v}(l, \mathcal{S})$ and $\mathfrak{i} \in \mathcal{S}$, then one can choose $\mathcal{R}$ so that also $\mathfrak{i} \in \mathcal{R}$.*

Proof. The sequence $\mathcal{S}^0 = \{\mathfrak{g}_j \cap \mathfrak{h}\}$ is a refinement of a $\mathfrak{g}$ composition series of $\mathfrak{h}$, hence it can be completed to a JH sequence $\mathcal{R}$ of $\mathfrak{g}$. Then $\mathfrak{v}(l, \mathcal{R}) \supset \mathfrak{v}(l^0, \mathcal{S}^0) = \mathfrak{p}$, hence $\mathfrak{v}(l, \mathcal{R}) = \mathfrak{p}$. If $\mathfrak{i} \in \mathcal{S}$, $\mathfrak{i} \subset \mathfrak{h}$, then also $\mathfrak{i} \in \mathcal{S}^0 \subset \mathcal{R}$.

Now we will investigate how the structure of the polarization $\mathfrak{p} = \mathfrak{v}(l, \mathcal{S})$ depends on the nilpotent ideal $\mathfrak{i} \in \mathcal{S}$. We fix l and define

$$\mathfrak{g}_\mathfrak{i} := \{x \in \mathfrak{g}; l([x, \mathfrak{i}]) = 0\} .$$

(This is $\mathfrak{i}^l$ in our previous notation in § 4). Since $\mathfrak{g}' \subset \mathfrak{i}$ we have $[\mathfrak{g}_\mathfrak{i}, \mathfrak{g}_\mathfrak{i}] = \mathfrak{g}_\mathfrak{i}' \subset \mathfrak{g}^l \subset \mathfrak{g}_\mathfrak{i}$, in particular, $\mathfrak{g}^l$ is an ideal in $\mathfrak{g}_\mathfrak{i}$. Clearly also

$$\mathfrak{g}_0^l := \mathfrak{g}^l \cap \ker l$$

is an ideal in $\mathfrak{g}_\mathfrak{i}$. Then either $\mathfrak{g}_\mathfrak{i}' \subset \mathfrak{g}_0^l$, i.e. the quotient $\mathfrak{g}_\mathfrak{i}/\mathfrak{g}_0^l$ is abelian, or it is the direct sum of a commutative and a Heisenberg algebra. With $\mathfrak{p}$ as above we claim:

(73) *There exists a subalgebra $\mathfrak{q}$ in $\mathfrak{g}_\mathfrak{i}$ with $\mathfrak{g}_0^l \subset \mathfrak{q}$, for which $\mathfrak{q}/\mathfrak{g}_0^l$ is maximal commutative in $\mathfrak{g}_\mathfrak{i}/\mathfrak{g}_0^l$, such that*

$$\mathfrak{p} = \mathfrak{q} + \mathfrak{p} \cap \mathfrak{i}$$

Proof. For any $\mathfrak{g}_j \in \mathcal{S}$ with $\mathfrak{i} \subset \mathfrak{g}_j$ clearly $\mathfrak{g}_j^{l_j} \subset \mathfrak{g}_\mathfrak{i}$, hence

$$\mathfrak{p} = \sum_{\mathfrak{i} \subset \mathfrak{g}_j} \mathfrak{g}_j^{l_j} + \mathfrak{p} \cap \mathfrak{i} \subset \mathfrak{g}_\mathfrak{i} + \mathfrak{p} \cap \mathfrak{i}$$

which proves $\mathfrak{p} = \mathfrak{q} + \mathfrak{p} \cap \mathfrak{i}$ with $\mathfrak{q} = \mathfrak{p} \cap \mathfrak{g}_\mathfrak{i}$. Since $\mathfrak{q}' \subset \mathfrak{p}' \cap \mathfrak{g}^l \subset \mathfrak{g}_0^l$ we see that $\mathfrak{q}/\mathfrak{g}_0^l$ is commutative.

If $\mathfrak{g}^l \subset \mathfrak{r} \subset \mathfrak{g}_\mathfrak{i}$ and $\mathfrak{r}' = [\mathfrak{r}, \mathfrak{r}] \subset \mathfrak{g}_0^l$, i.e. $\mathfrak{r}/\mathfrak{g}_0^l$ is commutative, then

$$[\mathfrak{r} + \mathfrak{p} \cap \mathfrak{i}, \mathfrak{r} + \mathfrak{p} \cap \mathfrak{i}] = \mathfrak{r}' + [\mathfrak{r}, \mathfrak{p} \cap \mathfrak{i}] + (\mathfrak{p} \cap \mathfrak{i})' \subset \mathfrak{g}_0^l + [\mathfrak{g}_\mathfrak{i}, \mathfrak{i}] + \mathfrak{p}' ,$$

hence $\mathfrak{r} + \mathfrak{p} \cap \mathfrak{i}$ is isotropic. This implies $\mathfrak{r} = \mathfrak{q}$ if $\mathfrak{r} \supset \mathfrak{q}$, hence $\mathfrak{q}$ is maximal.

We remark that, using the arguments in the proof of proposition 5, one can show that $[\mathfrak{p} \cap \mathfrak{i}, \mathfrak{g}_\mathfrak{i}] \subset \mathfrak{p} \cap \mathfrak{i}$, so that $\mathfrak{g}_\mathfrak{i} + \mathfrak{p} \cap \mathfrak{i}$ is always a subalgebra of $\mathfrak{g}$.

Later the following remark will be useful:

(74) *If $\mathfrak{p}_k = \mathfrak{v}(l, \mathcal{S}_k)$, $k = 1, 2$, are Vergne polarizations through $\mathfrak{i}$, i.e. if $\mathfrak{i} \in \mathcal{S}_k$, then*

$$\dim(\mathfrak{p}_1 + \mathfrak{i}) = \dim(\mathfrak{p}_2 + \mathfrak{i})$$

in particular $\mathfrak{p}_1 + \mathfrak{i} = \mathfrak{p}_2 + \mathfrak{i}$ if $\mathfrak{p}_2 \subset \mathfrak{p}_1 + \mathfrak{i}$.

This follows from the formula $\dim(\mathfrak{p}_k + \mathfrak{i}) + \dim \mathfrak{p}_k \cap \mathfrak{i} = \dim \mathfrak{p}_k + \dim \mathfrak{i}$ and the fact that the $\mathfrak{p}_k \cap \mathfrak{i}$ are polarizations for $l|_\mathfrak{i}$ in $\mathfrak{i}$, hence $\dim \mathfrak{p}_1 \cap \mathfrak{i} = \dim \mathfrak{p}_2 \cap \mathfrak{i}$.

Finally, let $\mathcal{B}$ be an $\mathfrak{i}$ special coexponential basis for 0 in $\mathfrak{g}$ thus in particular a JH basis for $\mathfrak{g}$ and let κ be the diffeomorphism of $\mathbb{R}^{r_0} \times \mathbb{R}^{s_0}$ onto G, defined by $\mathcal{B}$, with $s_0 = \dim \mathfrak{i}$, $r_0 = \dim(\mathfrak{g}/\mathfrak{i})$. Then we write

$$\mathcal{S}_{k,m}(G, \mathfrak{i})$$

for the space of all $\mathbb{C}$ valued functions f on $\mathbb{R}^k \times \mathbb{R}^m \times G$, for which the function

$$(\xi, x; \vartheta, y) \to f(\xi, \vartheta, \sigma(x, y))$$

from $\mathbb{R}^{k+r_0} \times \mathbb{R}^{m+s_0}$ in $\mathbb{C}$ is in $\mathcal{S}^{\mathcal{F}}_{l+r_0, m+s_0}$. Again this definition is independent of $\mathcal{B}$. With the notation

$$f_{\xi,\vartheta}(x) = f(\xi, \vartheta, x)$$

for $(\xi, \vartheta) \in \mathbb{R}^k \times \mathbb{R}^m$, $x \in G$, we see that $(\xi, \vartheta) \to f_{\xi,\vartheta}$ is a mapping of $\mathbb{R}^k \times \mathbb{R}^m$ onto a dense subspace of $L^1(G)$.

The principal result of this paragraph is

Theorem 12. *Let G be a connected, simply connected exponential group with Lie algebra $\mathfrak{g}$. Let $\mathfrak{i}$ be a nilpotent ideal with $\mathfrak{g}' \subset \mathfrak{i}$, $l \in \mathfrak{g}^*$, $\mathfrak{p} = \mathfrak{v}(l, \mathcal{S})$ a Vergne polarization through $\mathfrak{i}$, $P = E(\mathfrak{p})$ and $\pi = K(\Omega_l)$ the corresponding irreducible representation of G. For every kernel $K \in \mathcal{K}_{k,m}(G, P, \mathfrak{i})$ there exists a function $f \in \mathcal{S}_{k,m}(G, \mathfrak{i})$ such that*

$$K^{\pi}_{f_{\xi,\vartheta}} = K(\xi, \vartheta)$$

for all $(\xi, \vartheta) \in \mathbb{R}^k \times \mathbb{R}^m$, with $K(\xi, \vartheta)(x, y) = K(\xi, \vartheta; x, y)$.

Proof by induction on $n = \dim \mathfrak{g}$.

There is nothing to prove if G is commutative, hence we suppose that $n > 1$ and the theorem holds for all groups of lower dimension.

Let $\mathcal{S}$ be a JH sequence with $\mathfrak{i} \in \mathcal{S}$ and $\mathfrak{p} = \mathfrak{v}(l, \mathcal{S})$ a Vergne polarization for $\pi = K(\Omega_l)$. Let $\mathfrak{k}$ be the largest ideal in $\mathfrak{g}$ with $l(\mathfrak{k}) = 0$, thus $\mathfrak{k} = \ker d\pi$ and $K = E(\mathfrak{k})$ is the e component of $\ker \pi$ in G. We set $\mathfrak{k}_0 = \mathfrak{i} \cap \mathfrak{k}$ and $\overline{\mathfrak{g}} = \mathfrak{g}/\mathfrak{k}_0$. The canonic epimorphism $\mathfrak{g} \to \overline{\mathfrak{g}}$ maps $\mathcal{S}$ onto a JH sequence $\overline{\mathcal{S}}$ of $\overline{\mathfrak{g}}$ with $\overline{\mathfrak{i}} = \mathfrak{i}/\mathfrak{k}_0 \in \overline{\mathcal{S}}$, furthermore, l defines $\overline{l} \in \overline{\mathfrak{g}}^*$ and obviously $\overline{\mathfrak{p}} := \mathfrak{v}(\overline{l}, \overline{\mathcal{S}}) = \mathfrak{p}/\mathfrak{k}_0$, $\overline{\mathfrak{p}} + \overline{\mathfrak{i}} = (\mathfrak{p} + \mathfrak{i})/\mathfrak{k}_0$. This implies that an $\mathfrak{i}$ special coexponential basis for $\mathfrak{p}$ in $\mathfrak{g}$ maps onto an $\overline{\mathfrak{i}}$ special coexponential basis for $\overline{\mathfrak{p}}$ in $\overline{\mathfrak{g}}$ and, since $K_0 = E(\mathfrak{k}_0) \lhd P$ and the character $\chi = \chi(l)$ is trivial on K_0, allows to identify $\mathcal{K}_{k,m}(G, P; \mathfrak{i})$ with $\mathcal{K}_{k,m}(\overline{G}, \overline{P}, \overline{\mathfrak{i}})$. If $\dim \overline{\mathfrak{g}} < \dim \mathfrak{g}$, i.e. of $\mathfrak{k}_0 \neq 0$, then there exists $\overline{f} \in \mathcal{S}_{k,m}(\overline{G}, \overline{P})$ with $K = K^{\overline{\pi}}_{\overline{f}}$. Since, as manifolds, $G = \overline{G} \times K_0$, we can define f on $\mathbb{R}^{k+m} \times G$

by $f(\xi, \vartheta; \overline{x}, y) = \overline{f}(\xi, \vartheta, \overline{x})h(y)$ with $h \in \mathcal{C}_0^\infty(K_0)$, such that $\int_{K_0} h dy = 1$. Then clearly $f \in \mathcal{S}_{k,m}(G, \mathfrak{i})$ and $K_f^\pi = K_{\overline{f}}^\pi = K$.

If $\mathfrak{k}_0 = 0$ and $\mathfrak{k} \neq 0$, then $\mathfrak{k} \cap \mathfrak{g}' = \mathfrak{k} \cap \mathfrak{i} = 0$ implies that $\mathfrak{g}$ is the direct sum of $\mathfrak{k}$ and an ideal $\mathfrak{g}^0$ with $\mathfrak{i} \subset \mathfrak{g}^0$, furthermore, the subalgebras $\mathfrak{g}_j^0 = \mathfrak{k} + \mathfrak{g}_j$, $\mathfrak{g}_j \in \mathcal{S}$, together with a descending chain of subalgebras in $\mathfrak{k}$, form a JH sequence $\mathcal{S}^0$ in $\mathfrak{g}$, passing through $\mathfrak{i}^0 = \mathfrak{k} + \mathfrak{i}$ and for which

$$\mathfrak{v}(l, \mathcal{S}^0) = \mathfrak{v}(l, \mathcal{S}) = \mathfrak{p}\,, \; \mathfrak{p} + \mathfrak{i}^0 = \mathfrak{p} + \mathfrak{i}\,.$$

Hence, an $\mathfrak{i}$ special basis is also an $\mathfrak{i}^0$ special basis and $\mathcal{K}_{k,l}(G, P, \mathfrak{i}) = \mathcal{K}_{k,l}(G, P, \mathfrak{i}^0)$. Since $0 \neq \mathfrak{k} \subset \mathfrak{i}^0$ we are now in the previous case with respect to $\mathfrak{i}^0$, thus the theorem is true whenever $\mathfrak{k} \neq 0$.

From now on we assume that $\mathfrak{k} = 0$, i.e. that $\pi = K(\Omega_l)$ is locally faithful, resp. that the representation $d\pi$ of $\mathfrak{g}$ is faithful. This implies that the center $\mathfrak{z}$ of $\mathfrak{g}$ has dimension ≤ 1.

Let $\mathcal{S}$ be the JH sequence with $\mathfrak{i} \in \mathcal{S}$, $\mathfrak{p} = \mathfrak{v}(l, \mathcal{S})$. We treat first the case that $\mathfrak{i}$ contains no noncentral ideal. Since $\dim \mathfrak{z} \leq 1$ this means that $\mathfrak{i} = \mathfrak{g}' = \mathfrak{z} = \mathbb{R}z$ and $\mathfrak{g}$ is a Heisenberg algebra, $\mathfrak{g}^l = \mathfrak{z}$ and $\mathfrak{p}$ is a maximal isotropic subalgebra. Thus we have the following situation: G is the $n = 2q + 1$ dimensional Heisenberg group $\mathbb{H}_q = \mathbb{R}^q \times \mathbb{R}^q \times \mathbb{R}$ with product

$$(x, y, s)(x', y', s') = (x + x', y + y', s + s' + \frac{1}{2}(yx' - xy'))$$

with $xy = \sum_1^q x_i y_i$ the usual inner product of $\mathbb{R}^q$, $P = \{0\} \times \mathbb{R}^q \times \mathbb{R} = E(\mathfrak{p})$ and the character $\chi = \chi(l, p)$ is given by $\chi(0, y, s) = e^{is}$, hence $\pi = \pi(l)$ acts on $\mathfrak{H}_\pi = L^2(\mathbb{R}^q)$ by

$$(\pi(x, y, s)\xi)(r) = e^{is} e^{i(r-x)y} \xi(r - x)\,.$$

Let $f \in L^1(\mathbb{H}_q)$. If $f_0(x, y) = \int f(x, y, s) e^{is} ds$ and $\widehat{f}_0$ is the Fourier transform for f_0 with respect to the second variable $y \in \mathbb{R}^q$, then the kernel of $\pi(f)$ is

$$K_f(x, y) = \widehat{f}_0(x - y, -y)$$

or $\widehat{f}_0(x, y) = K_f(x + y, -y)$, which shows how the function f_0 can be regained from K_f via the inverse Fourier transformation. The same argument also shows that every kernel $K(x, y)$ in the ordinary Schwartz space $\mathcal{S}(\mathbb{R}^q \times \mathbb{R}^q)$ is a K_f with $f \in L^1(G)$ and $f_0 \in \mathcal{S}(\mathbb{R}^q \times \mathbb{R}^q)$. K is in $\mathcal{K}_{k,m}(G, P, \mathfrak{i})$ if and only if $f \in \mathcal{S}_{k,m}(G, \mathfrak{i})$. So in this case theorem 12 is true. We want to point out that theorem 12 in the case of $G = \mathbb{H}_q$, $\mathfrak{i} = \mathfrak{g}' = \mathfrak{z}$, is not optimal, because we know that *all* Schwartz kernels are in the image of $L^1(G)$, a result which is true for all nilpotent groups and follows also, by the way, from our theorem, when we choose $\mathfrak{i} = \mathfrak{g}$.

Let us now assume that $\mathfrak{i}$ is not central in $\mathfrak{g}$. Then $\mathcal{S}$ contains a minimal noncentral ideal $\mathfrak{a}$ of $\mathfrak{g}$. Since $\mathfrak{a}_0 := [\mathfrak{g}, \mathfrak{a}]$ is a nonzero ideal, contained in $\mathfrak{a}$, either

$\mathfrak{a}_0 = \mathfrak{a}$ or $\mathfrak{a}_0 = \mathfrak{z} \neq 0$ or $\mathfrak{a} = \mathfrak{a}_0 \oplus \mathfrak{z}$. In this latter case $\mathfrak{a}_0 \notin \mathcal{S}$, because of the minimality of $\mathfrak{a}$. This implies that $\mathcal{S}$ refines a composition series of $\mathfrak{g}$, ending with $\mathfrak{a} \supset \mathfrak{z} \supset 0$. If we exchange $\mathfrak{z}$ in this series by $\mathfrak{a}_0$, we obtain a new JH sequence $\mathcal{S}_0$, which contains $\mathfrak{a}_0$ as minimal noncentral ideal. Since obviously also $\mathfrak{p} = \mathfrak{v}(l, \mathcal{S}_0)$ we are allowed to assume that always either $[\mathfrak{g}, \mathfrak{a}] = \mathfrak{a}$ or $[\mathfrak{g}, \mathfrak{a}] = \mathfrak{z} \neq 0$.

Let $\mathfrak{c}$ be the centralizer of $\mathfrak{a}$ in $\mathfrak{g}$ and $\mathfrak{h}$ the annihilator of $\mathfrak{a}/\mathfrak{a}\cap\mathfrak{z}$, hence $\mathfrak{g} \supset \mathfrak{h} \supset \mathfrak{c}$ and $\mathfrak{h}$ and $\mathfrak{c}$ are ideals in $\mathfrak{g}$ with $\dim(\mathfrak{g}/\mathfrak{h}) \leq 1$, $\dim(\mathfrak{g}/\mathfrak{c}) \leq 3$. Keeping in mind our last assumption we have to distinguish the following cases:

I. $\mathfrak{g} = \mathfrak{h}$, hence $[\mathfrak{g}, \mathfrak{a}] = \mathfrak{z} \neq 0$, $\mathfrak{g} = \mathbb{R}a \oplus \mathfrak{c}$.

II. $\mathfrak{g} \supset \mathfrak{h} = \mathfrak{c}$, i.e. $[\mathfrak{g}, \mathfrak{a}] \cap \mathfrak{z} = 0$, hence $\mathfrak{a} \cap \mathfrak{z} = 0$.

III. $\mathfrak{g} \underset{\neq}{\supset} \mathfrak{h} \underset{\neq}{\supset} \mathfrak{c}$, $\mathfrak{p} \subset \mathfrak{h}$.

IV. $\mathfrak{g} \underset{\neq}{\supset} \mathfrak{h} \underset{\neq}{\supset} \mathfrak{c}$, $\mathfrak{p} \not\subset \mathfrak{h}$.

In the cases I, II and III the polarization $\mathfrak{p}$ is contained in $\mathfrak{c}$: We have always $\mathfrak{a} \subset \mathfrak{p}$, hence $[\mathfrak{p}, \mathfrak{a}] \subset \mathfrak{p}'$ which in case I implies $[\mathfrak{p}, \mathfrak{a}] \subset \mathfrak{p}' \cap \mathfrak{z} = 0$.

In case II $\mathfrak{p} \not\subset \mathfrak{c} = \mathfrak{h}$ would imply $[\mathfrak{p}, \mathfrak{a}] = [\mathfrak{g}, \mathfrak{a}] = \mathfrak{a}$, because of $\mathfrak{a} \cap \mathfrak{z} = 0$. Thus $\mathfrak{a} \subset \mathfrak{p}'$, which contradicts $l(\mathfrak{a}) \neq 0$.

In case III necessarily $[\mathfrak{h}, \mathfrak{a}] = \mathfrak{z} \neq 0$, hence $[\mathfrak{p}, \mathfrak{a}] \subset \mathfrak{z} \cap \mathfrak{p}' = 0$, $\mathfrak{p} \subset \mathfrak{c}$. So we have in these cases the following chain of inclusions:

$$\mathfrak{g} \supset \mathfrak{h} \supset \mathfrak{c} \supset \mathfrak{p} \supset \mathfrak{a} \supset \mathfrak{z}$$

As usual we write G, H, C, P, A and Z for the groups, corresponding to $\mathfrak{g}, \mathfrak{h}, \mathfrak{c}$ etc. If we set

$$\mathfrak{f} = \begin{cases} \mathfrak{c} & \text{in case} \quad \text{I} \\ \mathfrak{h} & \text{in cases} \quad \text{II} \quad \text{and III} \end{cases},$$

and $F = E(\mathfrak{f})$, then

$$\tau := \operatorname{ind}_P^F \chi(l, \mathfrak{p}) \in \widehat{F}$$

and π is induced from τ, more exactly

$$\pi \tilde{=} \pi' = \operatorname{ind}_F^G \tau \ .$$

We take $a \in \mathfrak{g}$, $a \notin \mathfrak{f}$, with $l(a) = 0$. Then $\mathfrak{g} = \mathbb{R}a \oplus \mathfrak{f}$, $G = R \ltimes F$ with $R = E(\mathbb{R}a)$ and the elements in G can be written as (s, x) with $s \in \mathbb{R}$, $x \in F$, which means

$$(s, x) = E(sa)x \in G = R \ltimes F \ .$$

The usual trick, namely to write $\xi(s, x) = \xi(s)(x)$ and to consider $\xi(s)$ as an element of $\mathfrak{H}(\tau)$, hence the function $s \to \xi(s)$ as element of $\mathfrak{H}(\pi')$, permits us to identify the spaces of π and π' with $L^2(\mathbb{R}, \mathfrak{H}(\tau))$ and the quotient G/F with $\mathbb{R}$. Then formula (69) yields

(75) $$(\pi(f)\xi)(s) = \int_{\mathbb{R}} \delta(t)^{-1}\tau(f(s-t)^t)\xi(t)dt$$

with $\delta(t) = \Delta_G(t, e)$.

Let K be a kernel on $G \times G$ and $\widetilde{K}$ the corresponding operator on $\mathfrak{H}(\pi)$, i.e.

$$(\widetilde{K}\xi)(s, x) = \int_{G/P} K((s, x), (t, y))\xi(t, y)d\overline{(t, y)} .$$

If we set $K(s, t)(x, y) = K((s, x), (t, y))$, then $K(s, t)$ is a kernel on $F \times F$ and in our new notation

(76) $$(\widetilde{K}\xi)(s) = \int_{\mathbb{R}} \widetilde{K}(s, t)\xi(t)dt .$$

If $K \in \mathcal{K}_{k,m}(G, P, \mathfrak{i})$, then for every $(s, t) \in \mathbb{R}^2$ clearly $K(s, t) \in \mathcal{K}_{k,m}(F, P, \mathfrak{i})$ and K', defined by

$$\left.\begin{array}{l} K'(\xi, s, t, \vartheta; x, y) \\ K'(\xi, \vartheta, s, t; x, y) \end{array}\right\} = K(\xi, \vartheta; (s, x), (t, y)) \quad \text{if} \quad \left\{\begin{array}{l} a \notin \mathfrak{i} \\ a \in \mathfrak{i} \end{array}\right.$$

is in $\mathcal{K}_{k+2,m}(F, P, \mathfrak{i})$, resp. $\mathcal{K}_{k,m+2}(F, P, \mathfrak{i})$. Comparing (75) and (76) we see that to prove theorem 12 we have to find a function $f \in \mathcal{S}_{k,m}(G, \mathfrak{i})$ such that

(77) $$K^{\tau}_{f_{\xi,\vartheta}(s-t)^t} = \delta(t)K'(\xi', \vartheta')$$

with $\xi' = (\xi, s, t)$, $\vartheta' = \vartheta$ if $a \notin \mathfrak{i}$, $\xi' = \xi$, $\vartheta' = (\vartheta, s, t)$ if $a \in \mathfrak{i}$.

To keep our notation transparent we suppress the parameter ξ and ϑ, thus we are seeking a function f for which $K^{\tau}_{f(s-t)^t} = \delta(t)K'$, all $s, t \in \mathbb{R}$. We treat first the case II, $\mathfrak{f} = \mathfrak{h} = \mathfrak{c}$. Then $\mathfrak{a}$ is central in $\mathfrak{h}$ and $\mathfrak{a} \cap \mathfrak{z} = 0$.

The kernel $\delta K'$ is in $\mathcal{K}_{k',m'}(H, P, \mathfrak{i})$ with $(k', m') = (k+2, m)$ if $a \notin \mathfrak{i}$, $(k', m') = (k, m+2)$ if $a \in \mathfrak{i}$. Thus by induction, there exists a $g \in \mathcal{S}_{k',m'}(H, \mathfrak{i})$ with $K^{\tau}_g = \delta K'$. We have to find a function $h = h(t)$ in $\mathcal{S}_{k+1,m}(H, \mathfrak{i})$, resp. in $\mathcal{S}_{k,m+1}(H, \mathfrak{i})$, for which $\tau(h(s-t)^t) = \tau(g(s, t))$. Then the function f, defined by $f(s, x) = h(s; x)$, $s \in \mathbb{R}$, $x \in H$, is in $\mathcal{S}_{k,m}(G, \mathfrak{i})$ and satisfies (77).

Let $\mathcal{B}_{\mathfrak{a}} = \{c_1, c_2, \ldots, c_q\}$ be a coexponential basis for $\mathfrak{a}$ in $\mathfrak{h}$ and $\mathfrak{w} = \sum_1^q{}^{\oplus}\mathbb{R}c_j$, hence $\mathfrak{h} = \mathfrak{w} \oplus \mathfrak{a}$ and $H = WA$ with $W = E_{\mathfrak{a}}(\mathfrak{w}, e)$, see § 1. We identify $\mathfrak{a}$ with $\mathbb{C}$ if $\dim \mathfrak{a} = 2$ and with $\mathbb{R}$ if $\dim \mathfrak{a} = 1$, accordingly we denote the elements of A by e^z, $z \in \mathbb{C}$, resp. e^r, $r \in \mathbb{R}$. Then every element in H can uniquely be written as we^z, resp. we^r with $w \in W$.

We choose a and the basis of $\mathfrak{a}$ in such a way that $\operatorname{ad} a(z) = e^{\Lambda}z$ with $\Lambda = 1 - i\lambda$ if $\mathfrak{a} = \mathbb{C}$ and $\operatorname{ad} a(r) = r$ if $\mathfrak{a} = \mathbb{R}$. The action of $\exp \mathbb{R}a$ on A is given by $(e^z)^{-t} = e^{e^{\Lambda t}z}$, resp. $(e^r)^{-t} = e^{e^t r}$. Furthermore, for the character $\chi = \chi(l, \mathfrak{p})$ on

A we have $\chi(e^z) = e^{i\Im z}$, resp. $\chi(e^r) = e^{ir}$, where $\Im z$ denotes the imaginary part of z.

The Haar measure on H is the product measure $dw\,dz$, where dw can be understood as the Haar measure on H/A or as Lebesgue measure on $W = \exp \mathfrak{w}$ and $dz = dr$ if $A = \mathbb{R}$, $dz = dx\,dy$ if $A = \mathbb{C} \cong \mathbb{R}^2$, $z = x + iy$.

With these notations we get

$$\begin{aligned}\tau(g(s,t)) &= \int_W \int_A g(s,t;we^z)\tau(we^z)dz\,dw \\ &= \int_W \tau(w)\left(\int_A g(s,t;we^z)e^{il(z)}dz\right)dw\end{aligned}$$

with $l(z) = r$ if $z = r \in \mathbb{R}$, $l(z) = y$ if $z = x + iy \in \mathbb{C}$.

We define for $x \in H$:

$$\widehat{g}(s,t;x) = \int_A g(s,t;xe^z)e^{il(z)}dz\ , \tag{78}$$

hence $\tau(g(s,t)) = \int_W \widehat{g}(s,t;w)\tau(w)dw$ and also

$$\widehat{g}(s,t;xe^z) = \widehat{g}(s,t;x)e^{-il(z)} \tag{79}$$

for $z \in A$, $x \in H$. Observe that the function on H/A, defined by $wA \to \widehat{g}(s,t;w)$, is in $\mathcal{S}_{k+2,m}(H/A, \mathfrak{i}/\mathfrak{a})$! We set

$$\gamma(s,t;x) := \widehat{g}(s-t,-t;x^{-t})\ .$$

Then also $wA \to \gamma(s,t;w)$ is in $\mathcal{S}_{k+2,m}(H/A, \mathfrak{i}/\mathfrak{a})$. Now let

$$h_0(s;x) = \int_{\mathbb{R}} \gamma(s,t;x)dt\ .$$

We have to treat separately the cases $A = \mathbb{R}$ and $A = \mathbb{C}$. For $A = \mathbb{R}$ we get $\gamma(s,t;xe^r) = \gamma(s,t;x)e^{-ie^t r}$ and consequently

$$h_0(s;we^r) = \int_{\mathbb{R}} \gamma(s,t;w)e^{-ie^t r}dt\ .$$

The function $(s,t,w) \to e^{2t}\gamma(s,t,w)$ is also in $\mathcal{S}_{k+2,m}(H/A, \mathfrak{i}/\mathfrak{a})$, hence if we define

$$\widetilde{\gamma}(s,\xi;w) = \begin{cases} \xi^{-1}\gamma(s,\log\xi;w) & \text{for } \xi > 0 \\ 0 & \text{for } \xi \le 0 \end{cases},$$

then $\widetilde{\gamma} \in \mathcal{S}_{k+1,m+1}(H/A, \mathfrak{i}/\mathfrak{a})$ and

$$h_0(s; we^r) = \int_{\mathbb{R}} \widetilde{\gamma}(s, \xi; w)e^{-i\xi r} d\xi$$

is easily seen to belong to $\mathcal{S}_{k+1}(H, \mathfrak{i})$.

For $s, r \in \mathbb{R}$ we have with $h_0(s)(x) = h_0(s; x)$:

$$\begin{aligned}\tau(h_0(s)^r) &= \int_W \tau(w) \int_{\mathbb{R}} h_0(s; w^{-r}e^{e^r t})e^{it} dt dw \\ &= \int_W \tau(w) \int_{\mathbb{R}} \int_{\mathbb{R}} \widehat{g}(s-u, -u; w^{-(u+r)}e^{e^{u+r}t})e^{it} du dt dw \\ &= \int_W \tau(w) \int_{\mathbb{R}} \int_{\mathbb{R}} \widehat{g}(s+r-u, r-u; w^{-u})e^{-i(e^u-1)t} du dt dw\end{aligned}$$

As in the case of $\widetilde{\gamma}$ we see that the function Γ_r, defined by

$$\Gamma_r(s, \xi; w) = \\ = \begin{cases} (1+\xi)^{-1}\widehat{g}(s+r-\log(1+\xi), r-\log(1+\xi); w^{-\log(1+\xi)}) & \text{if } \xi > -1 \\ 0 & \text{if } \xi \leq -1 \end{cases}$$

is in $\mathcal{S}_{k+1,m+1}(H/A, \mathfrak{i}/\mathfrak{a})$ and

$$\tau(h_0(s)^r) = \int_W \tau(w) \int_{\mathbb{R}} \int_{\mathbb{R}} \Gamma_r(s, \xi; w)e^{-i\xi t} d\xi dt dw \ .$$

But the inner integral is the double Fourier transformation of Γ_r with respect to ξ, hence (we disregard π factors) equals $\Gamma_r(s, 0; w)$. It follows that

$$\tau(h_0(s)^r) = \int_W \tau(w)\widehat{g}(s+r, r; w)dw = \tau(g(s+r, r))$$

and consequently $\tau(f(s-r)^r) = \tau(g(s, r))$ for $f(s, x) = h_0(s)(x)$.

If $A = \mathbb{C}$ we set $h_\varphi(s; we^z) = h_0(s; we^{e^{i\varphi}z})$ and

$$\begin{aligned}h(s; we^z) &= \frac{1}{2\pi} \int_0^{2\pi} h_\varphi(s; we^z) d\varphi \\ &= \frac{1}{2\pi} \int_0^{2\pi} \int_{\mathbb{R}} \widehat{g}(s-t, -t; w^{-t}e^{e^{\Lambda t+i\varphi}z}) dt d\varphi \\ &= \frac{1}{2\pi} \int_0^{2\pi} \int_{\mathbb{R}} \gamma(s, t; w)e^{i\Im(e^{\Lambda t+i\varphi}z)} dt d\varphi \\ &= \frac{1}{2\pi} \int_0^{2\pi} \int_{\mathbb{R}} \gamma(s, t; w)e^{i\Im(e^{t+i\varphi}z)} dt d\varphi \ .\end{aligned}$$

The mapping $(t, \varphi) \to e^{t+i\varphi}$ from $\mathbb{R}\times]0, 2\pi]$ onto $\mathbb{C} \setminus \{0\}$ transforms the last integral into

$$\int_{\mathbb{R}^2} \gamma(s, \tfrac{1}{2}\log(x^2+y^2); w)(x^2+y^2)^{-1} e^{i\Im(x+iy)z} dxdy$$

which is essentially the Fourier transform with respect to (x, y) of the function

$$\widetilde{\gamma}(s, x, y; w) = \begin{cases} \gamma(s, \tfrac{1}{2}\log(x^2+y^2); w)(x^2+y^2)^{-1} & \text{if } (x, y) \neq (0,0) \\ 0 & \text{if } (x, y) = (0,0)\ . \end{cases}$$

As above one sees that $\widetilde{\gamma} \in \mathcal{S}_{k+1,m}(\mathbb{R}^2 \times W) \cong \mathcal{S}_{k+1,m}(H, \mathfrak{i})$ and consequently also $h \in \mathcal{S}_{k+1,m}(H, \mathfrak{i})$.

Now we get for $s, r \in \mathbb{R}$:

$$\tau(h(s)^r) = \int_W \tau(w) \int_{\mathbb{C}} h(s; w^{-r} e^{e^{\Lambda r}z}) e^{i\Im z} dz dw$$

with $dz = dx'dy'$ for $z = x' + iy'$. We proceed as before:

$$\int_{\mathbb{C}} h(s; w^{-r} e^{e^{\Lambda r}z}) e^{i\Im z} dz = \frac{1}{2\pi} \int_{\mathbb{C}} \int_0^{2\pi} \int_{\mathbb{R}} \widehat{g}(s-t, -t; w^{-r-t}) e^{-i\Im((e^{\Lambda(r+t)+i\varphi}-1)z)} d\varphi dt dz$$

$$= \int_{\mathbb{C}} \int_{\mathbb{R}^2} \widehat{g}(s+r-l(x,y), r-l(x,y); w^{-l(x,y)}(x^2+y^2)^{-1} e^{-i\Im((x-1+iy)z)} dxdydz$$

$$= \int_{\mathbb{R}^2} \int_{\mathbb{R}^2} \Gamma(s, r, x, y; w) e^{-i\Im(x+iy)z} dxdydz$$

with $l(x, y) = \frac{1}{2}\log(x^2 + y^2)$ and

$$\Gamma(s, r, x, y; w) = \widehat{g}(s+r+l(x+1,y), r+l(x+1,y); w^{-l(x+1,y)})((x+1)^2+y^2)^{-1}$$

for $(x, y) \neq (-1, 0)$. With now familiar arguments we see that finally

$$\tau(h(s)^r) = \int_W \tau(w)\widehat{g}(s+r, r; w) dw$$

which means $\tau(f(s-r)^r) = \tau(g(s, r))$ for $f(s, x) = h(s; x)$.

In Case I we have $\mathfrak{f} = \mathfrak{c}$, $\mathfrak{a} = \mathbb{R}y \oplus \mathbb{R}z$ and $[a, y] = z$ hence $\mathbb{R}a \oplus \mathfrak{a}$ is the Heisenberg algebra $\mathfrak{h}_1$. We identify $A = E(\mathfrak{a})$ with $\mathbb{R}^2$ via $(s_1, s_2) \to E(s_1 y + s_2 z)$. Then

$$(s_1, s_2)^t := E(-ta)(s_1, s_2)E(ta) = (s_1, s_2 - ts_1)$$

for $t \in \mathbb{R}$, $(s_1, s_2) \in A$. As before we decompose $\mathfrak{c} = \mathfrak{w} \oplus \mathfrak{a}$, $C = W \times A$, so $x \in C$ has a unique decomposition $x = w(x)(z_1(x), z_2(x))$ with $w(x) \in W$, $z_i(w) \in \mathbb{R}$.

For $w \in W$, $r \in \mathbb{R}$ we set

$$v_i(w,r) = z_i(w^{-r}), \quad w_r = w(w^{-r}) ,$$

hence

$$(w(z_1,z_2))^{-r} = w_r(z_1 + v_1(w,r), z_2 + rz_1 + v_2(w,r)) ,$$

$$z_2((w(z_1,z_2))^{-r}) = z_2 + V(w,z_1,r)$$

with $V(w,z_1,r) = rz_1 + v_2(w,r)$.

As in Case II above we take $g \in \mathcal{S}_{k',m'}(C, \mathrm{i})$ with $K_g^\tau = \delta K'$ and define $\widehat{g}$ by (78). Then we take a compactly supported smooth function Φ on $\mathbb{R}$ with $\int_{\mathbb{R}} \Phi(t)dt = 1$ and define

$$h(s;x) = \int_{\mathbb{R}} \widehat{g}(s-t,-t;x^{-t})dt\Phi(z_2(x)) .$$

If $y, z \in \mathfrak{a}$ are choosen so that $l(y) = 0$, $l(z) = 1$, then $\chi(z_1,z_2) = e^{iz_2}$ for $(z_1,z_2) \in A$ and $\tau(z_1,z_2) = e^{iz_2}\mathbf{1}$, furthermore, from (79) we obtain

$$h(s;w(z_1,z_2)) = \int_{\mathbb{R}} \widehat{g}(s-t,-t;w^{-t})e^{-itz_1}dt\, e^{-iz_2}\Phi(z_2)$$

for $w \in W$, $(z_1,z_2) \in A$. This can be written as $h(s;w(z_1,z_2)) = \widehat{\gamma}(s;w,z_1)e^{-iz_2}\Phi(z_2)$, where $\gamma(s;w,t) = \widehat{g}(s-t,-t;w^{-t})$, $\widehat{\gamma}$ the Fourier transform with respect to t. As before this implies $h \in \mathcal{S}_{k',m'}(C, \mathrm{i})$, with $k'+m' = k+m+1$.

For $r \in \mathbb{R}$ we have

$$\tau(h(s)^r) = \int_W \tau(w) \int_{\mathbb{R}^2} h(s;w^{-r}(z_1,z_2)^{-r})e^{iz_2}dz_1dz_2dw .$$

We compute the inner integrand:

$$\begin{aligned}
&\int_{\mathbb{R}^2} h(s;(w(z_1,z_2))^{-r})e^{iz_2}dz_1dz_2 = \\
&= \int_{\mathbb{R}^2}\int_{\mathbb{R}} \widehat{g}(s-t,-t;(w(z_1,z_2))^{-r-t})dt\, e^{iz_2}\Phi(z_2((w(z_1,z_2))^{-r})dz_1dz_2 \\
&= \int_{\mathbb{R}^2}\int_{\mathbb{R}} \widehat{g}(s+r-t,r-t;w^{-t})e^{-i(z_2+tz_1)}e^{iz_2}dt\Phi(z_2+V(w,z_1,r))dz_1dz_2 \\
&= \int_{\mathbb{R}}\int_{\mathbb{R}} \widehat{g}(s+r-t,r-t;w^{-t})e^{-itz_1}dt \int_{\mathbb{R}} \Phi(z_2+V(w,z_1,r))dz_2dz_1 \\
&= \int_{\mathbb{R}}\int_{\mathbb{R}} \widehat{g}(s+r-t,r-t;w^{-t})e^{-itz_1}dtdz_1 \\
&= \widehat{g}(s+r,r;w) .
\end{aligned}$$

So we have again $\tau(h(s)^r) = \int \tau(w)\widehat{g}(s+r,r;w)dw = \tau(g(s+r,r))$ and the function $f(s,x) = h(s;x)$ on $G = \mathbb{R} \ltimes C$ is what we want.

In case III again $\mathfrak{f} = \mathfrak{h}$, $F = H$, and $\mathfrak{g}$ acts non trivially on $\mathfrak{a}/\mathfrak{z}$. From the two subcases $\dim \mathfrak{a} = 2$ and $\dim \mathfrak{a} = 3$ we will discuss only the case $\dim \mathfrak{a} = 3$; it will be obvious how the proof has to be modified, which here means simplified, to cover also $\dim \mathfrak{a} = 2$.

As in § 4, case (b_2), let $\mathfrak{a} = \mathbb{R}u_1 \oplus \mathbb{R}u_2 \oplus \mathbb{R}z$ with $l(u_1) = l(u_2) = 0$, $l(z) = 1$ and ad a $(u_1 + iu_2) = \Lambda(u_1 + iu_2)$ in $\mathfrak{a}^{\mathbb{C}}$, $\Lambda = 1 - i\lambda$. We choose $b_1, b_2 \in \mathfrak{h}$ with $[b_j, u_k] = \delta_{jk}z$, hence

$$\mathfrak{g} = \mathbb{R}a \oplus \mathfrak{h}, \quad \mathfrak{h} = \mathbb{R}b_1 \oplus \mathbb{R}b_2 \oplus \mathfrak{c}$$

and $G = \mathbb{R} \ltimes H$, $H \supset C = E(\mathfrak{c}) \supset P$. As before let $\chi = \chi(l, \mathfrak{p})$ be the character of P, defined by l, and set

$$\rho = \operatorname{ind}_P^C \chi\,, \quad \check{\pi} = \operatorname{ind}_C^G \rho\ .$$

The intertwining operator U, which establishes the equivalence between $\pi = \operatorname{ind}_P^G \chi$ and $\check{\pi} = \operatorname{ind}_C^G \rho$ is given by

$$(80) \qquad (U\varphi)(x)(u) = c_P^G(x)^{-\frac{1}{2}}\varphi(xu), \quad \varphi \in \mathfrak{H}(\pi)$$

For the inverse unitary U^* from $\mathfrak{H}(\check{\pi})$ onto $\mathfrak{H}(\pi)$ we get $(U^*\psi)(x) = c_P^G(x)^{\frac{1}{2}}\psi(x)(e)$, at least for continuous $\psi \in \mathfrak{H}(\check{\pi}) = \operatorname{ind} \mathfrak{H}(\rho)$. We leave the standard computations verifying these facts to the reader.

Let K be a kernel in $\mathcal{K}_{k,m}(G, P, \mathfrak{i})$ and let $\widetilde{K}$ be the corresponding operator in $\mathfrak{H}(\pi)$. We have to find an expression for its image $U\widetilde{K}U^*$ in $\mathfrak{H}(\check{\pi})$. From (67) one derives easily the general formula

$$(81) \qquad \int_{G/P} F(\overline{y})d\overline{y} = \int_{G/C} c_P^G(y)^{-1} \int_{C/P} F(y\overline{w})d\overline{w}d\dot{y}$$

where $d\overline{y}, d\overline{w}$ and $d\dot{y}$ are the respective, suitably normalized relatively invariant measures on G/P,C/P and G/C. It follows from (80), (81) and $\psi(y)(v) = c_P^C(v)^{\frac{1}{2}}\psi(yv)(e)$ for $\psi \in \mathfrak{H}(\check{\pi})$ that

$$(82) \qquad (U\widetilde{K}U^*\psi)(x)(u) = \int_{G/C}\int_{C/P} c_P^G(xy)^{\frac{1}{2}}K(xu, yv)\psi(y)(v)d\overline{v}d\dot{y}\ .$$

For a function f on G and $x \in G$ we define the function $\check{f}(x)$ on C by

$$\check{f}(x)(u) = f(xu), \quad x \in G, \quad u \in C\ ,$$

furthermore we set $h^y(u) = h(yuy^{-1})$ for $y \in G$ and any function h on C. With this notation we derive from (82) that in order to prove theorem 12 we have to find a function $f \in \mathcal{S}_{k,m}(G, \mathfrak{i})$ for which

$$K^{\rho}_{\overset{\vee}{f(xy^{-1})^y}}(u, v) = \Delta_G(y) c_P^G(xy)^{-\frac{1}{2}} K(xu, yv)$$

or equivalently

$$K^{\rho}_{\overset{\vee}{f(x)^y}}(u, v) = \Delta_G(y) c_P^G(x)^{-\frac{1}{2}} c_P^G(y)^{-1} K(xyu, yv) \tag{83}$$

holds.

For $s = (s_1, s_2, s_0) \in \mathbb{R}^3$ we set

$$\sigma(s) = E(s_1 b_1 + s_2 b_2) E(s_0 a) ,$$

thus $(s, w) \to \sigma(s) w \in G$ is a diffeomorphism from $\mathbb{R}^3 \times C$ onto G. We define a kernel $J \in \mathcal{K}_{k',m'}(C, P, \mathfrak{i} \cap \mathfrak{c})$ by

$$J(s, t; u, v) = \Delta_G(\sigma(t)) c_P^G(\sigma(s))^{-\frac{1}{2}} c_P^G(\sigma(t))^{-1} K(\sigma(s)\sigma(t)u, \sigma(t)v) .$$

Again we remark that the Δ_G and c_P^G factor is independent of s_1, s_2, t_1 and t_2, if $b_1 \in \mathfrak{n}$ or $b_2 \in \mathfrak{n}$, moreover, from (83) we get

$$K^{\rho}_{\overset{\vee}{f(\sigma(s))^{\sigma(t)}}}(u, v) = J(s, t; u, v) . \tag{84}$$

Since $(s, t, u, v) \to (\sigma(s)\sigma(t)u, \sigma(t)v)$ maps $\mathbb{R}^3 \times \mathbb{R}^3 \times C \times C$ diffeomorphically onto $G \times G$ we see that the transition from K to J is a one–to–one correspondence, which means that for an $f \in \mathcal{S}_{k,m}(G, \mathfrak{i})$ we have $K = K_f^{\pi}$ if and only if (84) holds.

Let $\mathfrak{b} = \mathbb{R}a \oplus \mathfrak{c}$ and $B = E(\mathfrak{b}) = E(\mathbb{R}a) \ltimes C$. By induction there exists a function $g \in \mathcal{S}_{k'',m''}(B, \mathfrak{i})$ (with $k'' = k + 4$ if $b_1, b_2 \notin \mathfrak{i} + \mathfrak{p}$ etc.) with

$$K^{\rho}_{g(s_1,s_2,r_1,r_2)(\sigma_0(s_0))^{\sigma_0(r_0)}}(u, v) = J(s, r, u, v) \tag{85}$$

where $\sigma_0(s_0) = E(s_0 a)$. Basically we proceed as in the previous cases: With $\mathfrak{y} = \mathbb{R}u_1 \oplus \mathbb{R}u_2$, $Y = E(\mathfrak{y})$, $Z = E(\mathfrak{z})$ and a coexponential basis for $\mathfrak{a} = \mathfrak{y} \oplus \mathfrak{z}$ in $\mathfrak{c}$, which defines decompositions $\mathfrak{c} = \mathfrak{w} \oplus \mathfrak{y} \oplus \mathfrak{z}$, $C = W \times Y \times Z$, the elements x in C can be written as $x = we^y e^r$ with $e^y = E(y_1 u_1 + y_2 u_2) \in Y$, $e^r = E(rz) \in Z$ and $w \in W$.

In the following computations and expressions we will change our notation slightly: The r, s and t will now denote points in $\mathbb{R}^2$, hence $t = (t_1, t_2)$ etc., moreover we set $\sigma_1(t) = \sigma(t_1, t_2, 0) = E(t_1 b_1 + t_2 b_2)$, hence $\sigma(t_1, t_2, t_0) = \sigma_1(t)\sigma_0(t_0)$ etc, and for $x \in C$:

$$g(s_0, s, t)(x) = g(s_1, s_2, t_1, t_2)(\sigma_0(s_0)x) , \tag{86}$$

for the function g occurring in (85). This function, as we remember, also depends on the "silent" parameter ξ and ϑ.

With

$$\widehat{g}(s_0, s, t)(x) = \int_{\mathbb{R}} \int_{\mathbb{R}^2} g(s_0, s, t)(xe^y e^r) e^{ir} dy dr$$

we have as before

$$\rho(g(s_0, s, t)) = \int \rho(w) \widehat{g}(s_0, s, t)(w) dw$$

$$\widehat{g}(s_0, s, t)(xe^y e^r) = \widehat{g}(s_0, s, t)(x) e^{-ir} .$$

Let the function f on G be defined by

$$f(\sigma(s_1, s_2, s_0)x) = \int_{\mathbb{R}^2} \widehat{g}(s_0, s-t, -t)(\sigma_1(t) x \sigma_1(t)^{-1}) dt \Phi(z)$$

where $x = we^y e^z \in W \times Y \times Z = C$ and Φ is a compactly supported smooth function on $\mathbb{R}$ with $\widehat{\Phi}(0) = 1$. As in the other cases we see that $f \in \mathcal{S}_{k,m}(G, \mathfrak{i})$, furthermore we compute

$$\begin{aligned} & f(\sigma(s, s_0))^{\sigma(r, r_0)}(x) = \\ & = \int_{\mathbb{R}^2} \widehat{g}(s_0, s-t, -t)(\sigma_1(t)\sigma(r, r_0) we^y e^z \sigma(r, r_0)^{-1} \sigma_1(t)^{-1}) dt \Phi(z) \\ & = \int_{\mathbb{R}^2} \widehat{g}(s_0, s-t, -t)(\sigma_1(t)\sigma(r, r_0) w \sigma(r, r_0)^{-1} \sigma_1(t)^{-1}) e^{-ity} dt e^{-iry} e^{-iz} \Phi(z) \end{aligned}$$

with $ty = t_1 y_1 + t_2 y_2$. We consider the integrand as a function of t:

$$G(t) = \widehat{g}(s_0, s-t, -t)(\sigma_1(t)\sigma(r, r_0) w \sigma(r, r_0)^{-1} \sigma_1(t)^{-1}) .$$

Then $G \in \mathcal{S}(\mathbb{R})$ and

$$f(\sigma(s, s_0))^{\sigma(r, r_0)}(we^y e^z) = \widehat{G}(y) e^{-iry} e^{-iz} \Phi(z) .$$

From this expression we get (neglecting π factors)

$$\begin{aligned} \rho(f(\sigma(s, s_0))^{\sigma(r, r_0)}) &= \int_W \rho(w) \int_{\mathbb{R}^2} \widehat{G}(y) e^{-iry} dy \int_{\mathbb{R}} \Phi(z) dz = \int_W \rho(w) G(-r) dw \\ &= \int_W \rho(w) \widehat{g}(s_0, s+r, r)(\sigma_0(r_0) w \sigma_0(r_0)^{-1}) dw \\ &= \rho(g(s_0, s+r, r)^{\sigma_0(r_0)}) . \end{aligned}$$

Comparing this result with (85) and (86) we see that we have proved that the function f satisfies (84). This ends the proof of theorem 12, case III.

In case IV we can write $\mathfrak{g} = \mathbb{R}a \oplus \mathfrak{h}$, $\mathfrak{h} = \mathbb{R}b \oplus \mathfrak{c}$, resp. $\mathfrak{h} = \mathbb{R}b_1 \oplus \mathbb{R}b_2 \oplus \mathfrak{c}$ and $\mathfrak{a} = \mathbb{R}y \oplus \mathbb{R}z$, resp. $\mathfrak{a} = \mathbb{R}y_1 \oplus \mathbb{R}y_2 \oplus \mathbb{R}z$, where ad a acts irreducibly on $\mathfrak{a}/\mathfrak{z}$ and the b_i, y_j are choosen in such a way that $[b_i, y_j] = \delta_{ij}z$, $l(a) = l(b_i) = l(y_j) = 0$. It follows that $\mathfrak{i}$ is contained in $\mathfrak{h}$.

Suppose that $\mathfrak{g}^l \not\subset \mathfrak{h}$. Then we know from proposition 6 in § 4 that the restriction π_1 of π to $H = E(\mathfrak{h})$ is irreducible, with $\pi_1 = K(\Omega_{l_1})$, $l_1 = l|_{\mathfrak{h}}$; moreover, there exists $a \in \mathfrak{g}^l$ with $l(a) = 0$ and $\mathfrak{g} = \mathbb{R}a \oplus \mathfrak{h}$, $G = \mathbb{R} \ltimes H \supset P = \mathbb{R} \ltimes P_0$ with $P_0 = E(\mathfrak{p}_0)$ for $\mathfrak{p}_0 = \mathfrak{p} \cap \mathfrak{h}$. For the characters $\chi = \chi(l, \mathfrak{p})$ and $\chi_1 = \chi(l_1, \mathfrak{p}_0)$ we have $\chi(t, q) = \chi(0, q) = \chi_1(q)$ for $(t, q) \in \mathbb{R} \ltimes P_0$, so, considering P_0 as subgroup of P, we can write $\chi(t, q) = \chi(q)$. Since $(s, x) = (0, x^{-s})(s, 0)$ and $\chi(s, 0) = 1$, we see that a kernel $K \in \mathcal{K}_{0,0}(G, P)$ satisfies

$$K((s, x), (t, y)) = K((0, x^{-s}), (0, y^{-t}))$$

and similarly for (k, m) kernels. So the restriction K_1 defined by

$$K_1(x, y) = K((0, x), (0, y))$$

determines K, more precisely $K((s, x), (t, y)) = K_1(x^{-s}, y^{-t})$, and $K \to K_1$ maps $\mathcal{K}_{k,m}(G, P, \mathfrak{i})$ bijectively onto $\mathcal{K}_{k,m}(H, P_0, \mathfrak{i})$.

Now let K be a smooth (k, m) kernel on G, i.e. $K \in \mathcal{K}_{k,m}(G, P, \mathfrak{i})$. We have to find an $F \in \mathcal{S}_{k,m}(G, \mathfrak{i})$ with $K = K_F^{\pi}$. We take the Gauss function $h(x) = (2\pi)^{-\frac{1}{2}}e^{-\frac{1}{2}x^2}$ and define a kernel J on H by

$$J(\xi, r, \vartheta; x, y) = h(r)K_1(\xi, \vartheta; x^{-r}, y) \ .$$

It is easily checked that $J \in \mathcal{K}_{k+1,m}(H, P_0, \mathfrak{i})$, hence there exists by induction a function $f \in \mathcal{S}_{k+1,m}(H, \mathfrak{i})$ with $K^{\pi_1}_{f_{\xi,r,\vartheta}} = J(\xi, r, \vartheta)$.

As H is normal in G and the diffeomorphism $xP_0 \to (0, x)P$ from H/P_0 onto G/P is H equivariant, it follows that $\Delta_G|_H = \Delta_H$ and $c_P^G|_H = c_{P_0}^H$. Then the formula (70) for π_1 and f and with x^r instead of x yields

$$(87)\quad h(r)K_1(\xi, \vartheta; x, y) = \Delta_G(y)^{-1}c_P^G(xy)^{\frac{1}{2}} \int_{P_0} c_P^G(q)^{\frac{1}{2}} f(\xi, r, \vartheta; x^r qy^{-1})\chi(q)dq \ .$$

Now define F on $\mathbb{R}^k \times \mathbb{R}^m \times G$ by

$$F(\xi, \vartheta; (r, x)) = f(\xi, r, \vartheta; x) \ .$$

Then clearly $F \in \mathcal{S}_{k,m}(G, \mathfrak{i})$, hence $F_{\xi,\vartheta} \in L^1(G)$, and (70) gives

(88) $$K^{\pi}_{F_{\xi,\vartheta}}((0,x)(0,y)) = \\ = \Delta_G(y)^{-1} c_P^G(xy)^{\frac{1}{2}} \int_{\mathbb{R}} \int_{P_0} c_P^G(q)^{\frac{1}{2}} F(\xi,\vartheta;(r,x^r q y^{-1}))\chi(q)dqdr$$

because $F(\xi,\vartheta;(0,x)(r,q)(0,y^{-1})) = F(\xi,\vartheta;(r,x^r q y^{-1})) = f(\xi,r,\vartheta;x^r q y^{-1})$. Comparison of (87) and (88) shows that integrating (87) over r gives

$$K^{\pi}_{F_{\xi,\vartheta}}((0,x),(0,y)) = K_1(\xi,\vartheta;x,y) = K(\xi,\vartheta;(0,x)(0,y))$$

and proves $K = K^{\pi}_{F_{\xi,\vartheta}}$ in case $\mathfrak{g}^l \not\subset \mathfrak{h}$.

Now let $\mathfrak{g}^l \subset \mathfrak{h}$. Since $\mathfrak{a}$ and $\mathfrak{g}^l$ are in $\mathfrak{p}$ we have $[\mathfrak{g}^l,\mathfrak{a}] \subset \mathfrak{p}' \cap [\mathfrak{h},\mathfrak{a}] \subset \mathfrak{p}' \cap \mathfrak{z} = 0$, hence $\mathfrak{g}^l \subset \mathfrak{c}$. Clearly also $\mathfrak{i} \subset \mathfrak{h}$. If $\mathfrak{i}$ would be contained in $\mathfrak{c}$, then $\mathfrak{g}/\mathfrak{c}$ would be commutative, because $\mathfrak{g}' \subset \mathfrak{i}$. But the discussion of the cases (b_1) and (b_2) in § 4 shows that for $\bar{\mathfrak{g}} = \mathfrak{g}/\mathfrak{c}$ and $\bar{\mathfrak{h}} = \mathfrak{h}/\mathfrak{c}$ we have $[\bar{\mathfrak{g}},\bar{\mathfrak{h}}] = \bar{\mathfrak{h}}$ and $\bar{\mathfrak{h}}$ is the only nontrivial ideal in $\bar{\mathfrak{g}}$. So we conclude that

$$\mathfrak{h} = \mathfrak{i} + \mathfrak{c} .$$

In the following again we will discuss only the case $\dim \mathfrak{a} = 3$. Then $\mathfrak{a} = \mathfrak{a}_0 \oplus \mathfrak{z}$ with $l(\mathfrak{a}_0) = 0$. Let

$$\mathfrak{d} = \mathfrak{a}^l = \{x \in \mathfrak{g}\,;\, l([x,\mathfrak{a}]) = 0\} .$$

This $\mathfrak{d}$ is a subalgebra and contains $\mathfrak{c}$ and $\mathfrak{p}$. Let $\mathfrak{a}_0 = \mathbb{R}y_1 \oplus \mathbb{R}y_2$ and take $b_1, b_2 \in \mathfrak{i}$ with $[b_j, y_k] = \delta_{jk} z$. This is possible because $\mathfrak{h} = \mathfrak{i} + \mathfrak{c}$. If $a \in \mathfrak{p}$, $a \notin \mathfrak{c}$, then we have $\mathfrak{g} = \mathbb{R}a \oplus \mathfrak{h}$, $\mathfrak{h} = \mathbb{R}b_1 \oplus \mathbb{R}b_2 \oplus \mathfrak{c}$. Since $x = \xi_1 b_1 + \xi_2 b_2 \in \mathfrak{d}$ implies $0 = l([x,y_k]) = \xi_k l(z) = \xi_k$, we get

$$\mathfrak{d} = \mathbb{R}a \oplus \mathfrak{c} \supset \mathfrak{p} = \mathbb{R}a \oplus \mathfrak{p} \cap \mathfrak{c} .$$

Now we use (73) and take a subalgebra $\mathfrak{q} \subset \mathfrak{g}_{\mathfrak{i}}$ with $\mathfrak{p} = \mathfrak{q} + \mathfrak{p} \cap \mathfrak{i}$. Since $\mathfrak{p} \cap \mathfrak{i} \subset \mathfrak{p} \cap \mathfrak{h} \subset \mathfrak{c}$, we have $\mathfrak{p} \cap \mathfrak{c} = \mathfrak{q} \cap \mathfrak{c} + \mathfrak{p} \cap \mathfrak{i} = \mathfrak{q}_0 + \mathfrak{p} \cap \mathfrak{i}$, with $\mathfrak{q}_0 = \mathfrak{q} \cap \mathfrak{c}$.

Let $\mathcal{S} = \{\mathfrak{g}_j\}_j$ be the JH sequence of $\mathfrak{g}$ with $\mathfrak{p} = \mathfrak{v}(l,\mathcal{S})$, hence $\mathfrak{i} = \mathfrak{g}_m$ for some index m. We will define new JH sequences $\mathcal{R}$ and $\widehat{\mathcal{R}}$. For this we take elements $c_j \in \mathfrak{q}$, $0 \le j \le p$, with

$$\mathfrak{p} = \mathbb{R}c_0 \oplus \mathfrak{p} \cap \mathfrak{c}\,,\ \mathfrak{p} \cap \mathfrak{c} = \mathbb{R}c_1 \oplus \mathbb{R}c_2 \oplus \cdots \oplus \mathbb{R}c_p \oplus \mathfrak{p} \cap \mathfrak{i} .$$

The ideals

$$\mathfrak{s}_k = \mathbb{R}c_k \oplus \mathbb{R}c_{k+1} \oplus \cdots \oplus \mathbb{R}c_p \oplus \mathfrak{i}$$

form a chain between $\mathfrak{s}_0 = \mathfrak{p} + \mathfrak{i}$ and $\mathfrak{i}$, which we extend to a JH sequence $\mathcal{R}$ by taking the $\mathfrak{g}_j \in \mathcal{S}$ for $j \geq m$, i.e. $\mathfrak{g}_j \subset \mathfrak{i}$, and an arbitrary chain between $\mathfrak{g}$ and $\mathfrak{p}+\mathfrak{i}$. Then $\mathfrak{p} \cap \mathfrak{i} \subset \mathfrak{v}(l, \mathcal{R})$. But $\mathfrak{v}(l, \mathcal{R})$ contains also all c_j: Since $c_j \in \mathfrak{p} \cap \mathfrak{g}_{\mathfrak{i}}$ we have $l([c_k, c_j]) = 0$ and $l([c_k, \mathfrak{p} \cap \mathfrak{i}]) = 0$, hence $l([c_k, \mathfrak{s}_k]) = 0$, i.e. $c_k \in \mathfrak{s}_k^{l'} \subset \mathfrak{v}(l, \mathcal{R})$, $l'_k = l|_{\mathfrak{s}_k}$, and therefore $\mathfrak{p} \subset \mathfrak{v}(l, \mathcal{R})$, which means $\mathfrak{p} = \mathfrak{v}(l, \mathcal{R})$.

Next we extend the sequence $\{\mathfrak{s}_1, \mathfrak{s}_2, \ldots, \mathfrak{s}_p, \mathfrak{i}, \mathfrak{g}_{m+1}, \ldots, \mathfrak{g}_m\}$ to a JH sequence $\widehat{\mathcal{R}}$, which contains $\mathfrak{h}$, and set

$$\widehat{\mathfrak{p}} = \mathfrak{v}(l, \widehat{\mathcal{R}}) \ .$$

Then $\widehat{\mathfrak{p}} \subset \mathfrak{h}$, because $\mathfrak{g}^l \subset \mathfrak{h}$ and $\widehat{\mathfrak{p}} \cap \mathfrak{c} \supset \mathfrak{p} \cap \mathfrak{c}$. Then $\widehat{\mathfrak{p}} = \mathbb{R}b \oplus \mathfrak{p} \cap \mathfrak{c}$ for some $b \in \mathfrak{c}$ (remember that we had seen that $\widehat{\mathfrak{p}} \subset \mathfrak{h}$ implies $\widehat{\mathfrak{p}} \subset \mathfrak{c}$!). The elements a and b can both be chosen in $\mathfrak{g}_{\mathfrak{i}}$, thus $[a, b] \subset \mathfrak{g}^l \subset \mathfrak{p} \cap \widehat{\mathfrak{p}}$.

This implies that

$$\mathfrak{m} = \mathfrak{p} + \widehat{\mathfrak{p}} = \mathbb{R}a \oplus \mathbb{R}b \oplus \mathfrak{p} \cap \widehat{\mathfrak{p}}$$

is a subalgebra of $\mathfrak{g}$ and $\mathfrak{p} \cap \widehat{\mathfrak{p}}$ is an ideal in $\mathfrak{m}$. Since $\mathfrak{p}$ and $\mathfrak{p} \cap \widehat{\mathfrak{p}}$ are maximal isotropic, it follows that $l([a, b]) \neq 0$, so let us assume that

$$l([a, b]) = 1 \ .$$

Let $c = [a, b]$ and $\mathfrak{r}$ be intersection of $\mathfrak{p} \cap \widehat{\mathfrak{p}}$ with the kernel of l, hence

$$\mathfrak{m} = \mathbb{R}a \oplus \mathbb{R}b \oplus \mathbb{R}c \oplus \mathfrak{r} \ .$$

Then $\mathfrak{r}$ is an ideal in $\mathfrak{m}$ and $\mathfrak{m}/\mathfrak{r}$ is the Heisenberg algebra. As usual we denote the corresponding subgroups of G by $M, P, \widehat{P}$ and R, thus $M/R \cong \mathbb{H}_1$. With the notation

$$\langle x, y, z \rangle = E(xa)E(yb)E(zc)$$

the elements of M can uniquely be written as $\langle x, y, z \rangle r$ with $r \in R$, furthermore

$$\langle x, y, z \rangle \langle x', y', z' \rangle \equiv \langle x + x', y + y', z + z' - yx' \rangle \quad (\mathrm{mod}\ R)$$
$$\langle u, v, w \rangle^{-1} \equiv \langle -u, -v, -w - uv \rangle \quad (\mathrm{mod}\ R) \, .$$

The characters $\chi = \chi(l, \mathfrak{p})$ and $\widehat{\chi} = \chi(l, \widehat{\mathfrak{p}})$ are trivial on R and

$$\mu := \mathrm{ind}_P^M \chi \, , \ \widehat{\mu} := \mathrm{ind}_{\widehat{P}}^M \widehat{\chi}$$

are equivalent irreducible representations of M. Since μ and $\widehat{\mu}$ are trivial on R, the spaces $\mathfrak{H}(\mu)$ and $\mathfrak{H}(\widehat{\mu})$ are function spaces on M/R, i.e. the arguments of $\xi \in \mathfrak{H}(\mu)$ e.g. are the elements $\langle x, y, z \rangle$ as canonical representatives of the cosets of R in M.

The conditions for ξ, resp. ϑ to belong to $\mathfrak{H}(\mu)$, resp. $\mathfrak{H}(\widehat{\mu})$ are the identities

$$\xi\langle x, y, z\rangle = e^{-i(z+xy)}\xi\langle 0, y, 0\rangle$$
$$\vartheta\langle x, y, z\rangle = e^{-iz}\vartheta\langle x, 0, 0\rangle \ ,$$

while the intertwining operator $\Phi : \mathfrak{H}(\mu) \to \mathfrak{H}(\widehat{\mu})$ is given by

$$(\Phi\xi)\langle x, y, z\rangle = e^{-iz}\int_{\mathbb{R}} \xi\langle 0, s, 0\rangle e^{-isx} ds \ ,$$

i.e. Φ is just the Fourier transformation.

Let $\pi_1 = \operatorname{ind}_M^G \mu$, $\widehat{\pi}_1 = \operatorname{ind}_M^G \widehat{\mu}$. The intertwining operator Φ extends to an intertwining operator Θ_1, defined by

$$(\Theta_1\xi)(x) = \Phi(\xi(x))$$

for $\xi \in \mathfrak{H}(\pi_1)$. The equivalence of π and π_1, resp.$\widehat{\pi} = \operatorname{ind}_{\widehat{P}}^G \widehat{\chi}$ and $\widehat{\pi}_1$ is then defined by identifying the function $\xi \in \mathfrak{H}(\pi)$ with $\xi_1 \in \mathfrak{H}(\pi_1)$ as usually via

$$\xi_1(x)(y) = \xi(xy) \ , \ x \in G, y \in M \ .$$

No nontrivial c–factors occur, because P and $\widehat{P}$ are normal in M, hence $c_M^G|_P = c_P^G$. So we see that the intertwining operator Θ from $\mathfrak{H}(\pi)$ onto $\mathfrak{H}(\widehat{\pi})$ is given by

$$(\Theta\xi)(x\langle u, v, w\rangle y) = \int_{\mathbb{R}} \xi(x\langle 0, s, w\rangle y)e^{-isv} ds$$

for $x \in G$, $y \in R$, $\langle u, v, w\rangle = E(ua)E(vb)E(wc)$.

We claim that $b \notin \mathfrak{p}+\mathfrak{i}$. This follows from (74), because $b \in \mathfrak{p}+\mathfrak{i}$ would imply $\widehat{\mathfrak{p}} \subset \mathfrak{p}+\mathfrak{i}$, hence $\widehat{\mathfrak{p}}+\mathfrak{i} = \mathfrak{p}+\mathfrak{i}$. This contradicts the fact that $\mathfrak{h}$ contains $\widehat{\mathfrak{p}}+\mathfrak{h}$, but not $\mathfrak{p}+\mathfrak{h}$.

We will now construct $\mathfrak{i}$ special coexponential bases $\mathcal{B}_{\mathfrak{p}}$ and $\mathcal{B}_{\widehat{\mathfrak{p}}}$ for $\mathfrak{p}$ and $\widehat{\mathfrak{p}}$. Let $\{\mathfrak{i}_j\}$ be a refinement of a $\mathfrak{g}$ composition series of $\mathfrak{i}$, hence $\mathfrak{i} = \mathfrak{i}_1$, $\mathfrak{i}_j = \mathbb{R}v_j \oplus \mathfrak{i}_{j+1}$. Let $\{j_i\}$ be the jump indices of the chain $\{\mathfrak{i}_j + \mathfrak{p}\cap\widehat{\mathfrak{p}}\}$ and set $b_i = v_{j_i}$. Hence $\mathfrak{i}_{j_i} + \mathfrak{p}\cap\widehat{\mathfrak{p}} = \mathbb{R}b_i \oplus (\mathfrak{i}_{j_{i+1}} + \mathfrak{p}\cap\widehat{\mathfrak{p}})$ and $\{b_i\}$ is a coexponential basis for $\mathfrak{p}\cap\widehat{\mathfrak{p}}$ in $\mathfrak{i}+\mathfrak{p}\cap\widehat{\mathfrak{p}}$, $1 \le i \le s$ and $b_i \in \mathfrak{i}$. Next we take a coexponential basis $\{a_j\}$, $1 \le j \le r-1$, for $\mathfrak{m}+\mathfrak{i}$ in $\mathfrak{g}$. Then

$$\mathcal{B}_{\mathfrak{p}} = \{a_1, \ldots, a_{r-1}, b; b_1, \ldots, b_s\} \ , \text{ resp.}$$
$$\mathcal{B}_{\widehat{\mathfrak{p}}} = \{a_1, \ldots, a_{r-1}, a; b_1, \ldots, b_s\}$$

is an $\mathfrak{i}$ special coexponential basis for $\mathfrak{p}$, resp. $\widehat{\mathfrak{p}}$ in $\mathfrak{g}$, with $b_i \in \mathfrak{i}$. This is clear, because $\mathfrak{m}+\mathfrak{i} = \mathbb{R}b \oplus (\mathfrak{p}+\mathfrak{i}) = \mathbb{R}a \oplus (\widehat{\mathfrak{p}}+\mathfrak{i})$ and also $\mathfrak{p}+\mathfrak{i}_j = \mathbb{R}a \oplus (\mathfrak{i}_j + \mathfrak{p}\cap\widehat{\mathfrak{p}})$, $\widehat{\mathfrak{p}}+\mathfrak{i}_j = \mathbb{R}b \oplus (\mathfrak{i}_j + \mathfrak{p}\cap\widehat{\mathfrak{p}})$. We use the following notation: For $x \in \mathbb{R}^r$, $y \in \mathbb{R}^s$ we set

$$\sigma(x) = E(x_1a_1)E(x_2a_2)\dots E(x_{r-1}a_{r-1})\,,\ \sigma(y) = E(y_1b_1)E(y_2b_2)\dots E(y_sb_s)\,.$$

Hence

$$\sigma(x, y) = \sigma(x)E(x_rb)\sigma(y)\,,\ \widehat{\sigma}(x, y) = \sigma(x)E(x_ra)\sigma(y)$$

resp. $\kappa : (x, y) \to \sigma(x, y)P$, $\widehat{\kappa} : (x, y) \to \widehat{\sigma}(x, y)\widehat{P}$ are the parametrizations of G/P and $G/\widehat{P}$, defined by $\mathcal{B}_{\mathfrak{p}}$ and $\mathcal{B}_{\widehat{\mathfrak{p}}}$.

Because P and $\widehat{P}$ are normal in M, the Haar moduli Δ_P and $\Delta_{\widehat{P}}$ are just the restrictions of Δ_M. This implies that $c := c_M^G$ extends Δ_G/Δ_P and $\Delta_G/\Delta_{\widehat{P}}$. Consequently

$$U\xi(x, y) = c(\sigma(x, y))^{-\frac{1}{2}}\xi(\sigma(x, y))$$

defines a unitary operator U from $\mathfrak{H}(\pi)$ onto $L^2(\mathbb{R}^{r+s})$. Similarly one gets $\widehat{U} : \mathfrak{H}(\widehat{\pi}) \to L^2(\mathbb{R}^{r+s})$. We want an explicit expression of the operator $\widehat{U}\Theta U^{-1}$.

Let $\nu = \dim \mathfrak{i}$ and $\beta : \mathbb{R}^\nu \to I$ a canonical coordinate system for $I = E(\mathfrak{i})$. There exists a function $g : \mathbb{R} \times \mathbb{R}^\nu \to \mathbb{R}^\nu$ with coordinates in $\mathfrak{F}_{1,\nu}$ with $E(-tb)\beta(z)E(tb) = \beta(g(t, z))$ for $t \in \mathbb{R}$, $z \in \mathbb{R}^\nu$. Since the $E(tb)$ normalize P, $P \cap I$ and $P \cap R$, this implies the existence of $G_0 = (G_1, G_2) \in \mathfrak{T}_{1,s}$ and $\gamma \in \mathfrak{F}_{1,s}$ with

$$E(tb)\sigma(G_2(t, y)) \equiv \sigma(y)E(tb)E(\gamma(t, y)c) \quad (\text{mod } R \cap I)\,. \tag{89}$$

If we define the function $G : \mathbb{R}^{r+s} \to \mathbb{R}^{r+s}$ by $G(x, y) = (x, G_2(x_r, y))$, then $G \in \mathfrak{T}_{r,s}$. With $x' = (x_1, x_2, \dots, x_{r-1}, 0)$ formula (89) yields

$$\sigma(G(x, y)) = \sigma(x')\sigma(y)\langle 0, x_r, \gamma(x_r, y)\rangle \cdot w$$

with $w \in R \cap I$. It is easy to see that the determinant of the Jacobian of G equals $\delta(x) := c(E(x_rb))$, hence

$$(\Gamma\vartheta)(x, y) = \delta(x)^{\frac{1}{2}}e^{i\gamma(x_r, y)}\vartheta(G(x, y))$$

defines a unitary operator Γ of $L^2(\mathbb{R}^{r+s})$. For $\xi \in \mathfrak{H}(\pi)$, $\vartheta = U\xi$ we have with $x' = (x_1, x_2, \dots, x_{r-1}, 0)$ and $c(x) = c(\sigma(x, y))$:

$$\begin{aligned}\Gamma U\xi(x, y) &= \delta(x)^{\frac{1}{2}}c(x)^{-\frac{1}{2}}e^{i\gamma(x_r, y)}\xi(\sigma(G(x, y)) = \\ &= c(x')^{-\frac{1}{2}}e^{i\gamma(x_r, y)}\xi(\sigma(x')\sigma(y)\langle 0, x_r, \gamma(x_r, y)\rangle) \\ &= c(x')^{-\frac{1}{2}}\xi(\sigma(x')\sigma(y)\langle 0, x_r, 0\rangle)\,.\end{aligned}$$

In the very same way we obtain the formula

$$(\widehat{\Gamma}\widehat{U}\xi)(x, y) = c(x')^{-\frac{1}{2}}\xi(\sigma(x')\sigma(y)\langle x_r, 0, 0\rangle)$$

for the operators $\widehat{\Gamma}$, defined on $L^2(\mathbb{R}^{r+s})$ by a function $\widehat{G} \in \mathfrak{T}_{r,s}$, and $\widehat{U} : \mathfrak{H}(\widehat{\pi}) \to L^2(\mathbb{R}^{r+s})$.

Finally we obtain for $\Phi = \widehat{\Gamma}\widehat{U}\Theta U^{-1}\Gamma^{-1}$ and $\psi \in L^1(\mathbb{R}^{r+s}) \cap L^2(\mathbb{R}^{r+s})$:

$$\begin{aligned}(\Phi\psi)(x, y) &= c(x')^{-\frac{1}{2}}(\Theta U^{-1}\Gamma^{-1}\psi)(\sigma(x')\sigma(y)\langle x_r, 0, 0\rangle) = \\ &= c(x')^{-\frac{1}{2}} \int_{\mathbb{R}} (U^{-1}\Gamma^{-1}\psi)(\sigma(x')\sigma(y)\langle 0, s, 0\rangle e^{-isx_r} ds \\ &= \int_{\mathbb{R}} \psi(x', s, y) e^{-isx_r} ds = (\mathcal{F}_r\psi)(x, y) \ ,\end{aligned}$$

where $\mathcal{F}_r$ is the partial Fourier transformation of the r-th coordinate on $\mathbb{R}^{r+s}$. So we have proven:

$$V := \widehat{U}\Theta U^{-1} = \widehat{\Gamma}^{-1}\mathcal{F}_r\Gamma \ .$$

If $\pi_\bullet$ and $\widehat{\pi}_\bullet$ are the equivalent representations of π and $\widehat{\pi}$ in $L^2(\mathbb{R}^{r+s})$, transformed via U and $\widehat{U}$, then V is the intertwining operator: $\widehat{\pi}_\bullet = V\pi_\bullet V^{-1}$. Now the class $\mathcal{K}_{k,m}(G, P, \mathfrak{i})$ of smooth kernels had been defined by the function space $\mathcal{S}^{\mathcal{F}}_{k+2r,m+2s}$. But this space is by definition invariant under the substitutions $F \to F \circ T$ for $T \in \mathfrak{T}_{k+2r,m+2s}$, so the corresponding family of kernel operators is invariant in particular under the transformation with the Γ and $\widehat{\Gamma}$, and under partial Fourier transformations of the x coordinates. This implies that for every $K \in \mathcal{K}_{k,m}(G, P, \mathfrak{i})$ with corresponding kernel operator $\widetilde{K}$ in $L^2(\mathbb{R}^{r+s})$, there exists $K_\wedge \in \mathcal{K}_{k,m}(G, P, \mathfrak{i})$ with kernel operator $\widetilde{K}_\wedge$, so that $\widetilde{K}_\wedge = V\widetilde{K}V^{-1}$.

Now we can finish the proof of theorem 12: Since $\widehat{\mathfrak{p}} \subset \mathfrak{h}$, i.e. because with $\widehat{\mathfrak{p}}$ we are back in case III, there exists $f \in \mathcal{S}_{k,m}(G, \mathfrak{i})$ with $\widehat{\pi}_\bullet(f) = \widetilde{K}_\wedge$, hence $\pi_\bullet(f) = \widetilde{K}$, which means that $K = K^\pi_f$.

Chapter 2

Variable structures

§ 1 Variable structures

As a motivation and in order to illustrate the notion of variable structures let us consider a real vector space V with basis $\{b_1, b_2, \ldots, b_n\}$. Assume that V is an associative algebra. Then its structure is completely determined by the structural constants $c = \{c^{\gamma}_{\alpha\beta}\}$, given by

$$(1) \qquad b_\alpha \cdot b_\beta = \sum_{\gamma=1}^{n} c^{\gamma}_{\alpha\beta} b_\gamma ,$$

i.e. by the point $c \in \mathbb{R}^{n^3}$. V is associative if and only if c satisfies the polynomial identities

$$(2) \qquad \sum_{\sigma} c^{\sigma}_{\alpha\beta} c^{\delta}_{\sigma\gamma} = \sum_{\sigma} c^{\delta}_{\alpha\sigma} c^{\sigma}_{\beta\gamma}$$

for all $\alpha, \beta, \gamma, \delta$. Thus every point c of the algebraic manifold P in $\mathbb{R}^{n^3}$, which is defined by the equations (2), determines on V a structure of an associative algebra.

Let us denote the product on V, defined by $c \in P$, by $\mu_c(a, b)$, in particular (1) becomes $b_\alpha \cdot b_\beta = \mu_c(b_\alpha, b_\beta)$. It is then evident that

$$(3) \qquad \mu : (c, a, b) \longrightarrow (c, \mu_c(a, b))$$

defines a polynomial and a fortiori continuous mapping, from the real manifold $P \times V \times V$ into $P \times V$. This mapping reflects in a continuous way the dependence of multiplication of the different associative structures on V. It is obvious how one has to proceed in an analogous way in cases of other structures. We will not try here to give a general definition of a variable structure, but will essentially confine ourselves to groups, associative algebras and Lie algebras.

Definition. A *Hausdorff variable group*, abbreviated VG, is a triple

$$\mathbf{G} := (P \times G, \mu, \iota)$$

where

$P \times G$ is the topological product of the topological Hausdorff spaces P and G,
$\mu : P \times G \times G \longrightarrow P \times G$ and
$\iota : P \times G \longrightarrow P \times G$
are continuous mappings with
$\mu(p, x, y) = (p, \mu(p; x, y))$, $\iota(p, x) = (p, \iota(p; x))$
such that for every $p \in P$

$$(x, y) \longrightarrow x \underset{p}{\cdot} y := \mu(p; x, y) \ , \ x \longrightarrow {}_px^{-1} := \iota(p; x)$$

defines a topological group structure on G.

The corresponding group will always be denoted by ${}_pG$. If no confusion is possible we will usually delete the front index p of the inverse, i.e. we will write x^{-1} instead of ${}_px^{-1}$.

One may think of $\mathbf{G}$ as a family $\{{}_pG\}_{p \in P}$ of Hausdorff groups with the same underlying space G, such that the mappings

$$(p, x, y) \longrightarrow x \underset{p}{\cdot} y \quad , \quad (p, x) \longrightarrow {}_px^{-1}$$

are jointly continuous in all variables.

We call $\mathbf{G}$ locally compact, resp. compact, solvable, nilpotent etc., if all ${}_pG$, $p \in P$, are locally compact, resp. compact, solvable, nilpotent etc.

In the locally compact case some further restrictions are reasonable:

Definition. A variable group $\mathbf{G}$ is said to be *locally compact*, abbreviated LCVG, if all groups ${}_pG$ are locally compact and so that the following holds: There exists a positive Radon measure dx on G and a continuous function

$$\beta : P \times G \ni (p, x) \longrightarrow \beta_p(x) \in (0, \infty) = \mathbb{R}^+$$

such that $\beta_p(x)dx$ is a left invariant Haar measure on ${}_pG$.

A LCVG $\mathbf{G}$ defines the family $\{L^1({}_pG)\}$ of the group L^1–algebras $L^1({}_pG)$ with respect to the Haar measures $d_px = \beta_p(x)dx$. Let L^1 be the Banach space $L^1(G, dx)$ of $\mathbb{C}$–valued, dx–integrable functions on G, with norm $|f|_1 = \int |f(x)|dx$. Then for every $p \in P$

$$\Theta_p : L^1 \ni f \longrightarrow \Theta_p f = \beta_p^{-1} f \in L^1({}_pG)$$

defines a Banach space isomorphism from L^1 onto $L^1({}_pG)$. Furthermore,

$$f \underset{p}{\star} h = \Theta_p^{-1}(\Theta_p f \star \Theta_p h) \quad , \quad f^* = \Theta_p^{-1}((\Theta_p f)^*)$$

define product and involution on L^1, so that L^1 becomes an involutive Banach algebra, isomorphic under Θ_p with $L^1({}_pG)$.

Now the following definition is no surprise:

Definition. A *variable Banach algebra* (VBA) is a pair

$$\mathbf{A} := (P \times A, \mu)$$

where P is a locally compact Hausdorff space, A a Banach space with norm $|a|$ and

$$\mu : P \times A \times A \longrightarrow P \times A$$

a continuous mapping with $\mu(p, a, b) = (p, \mu(p; a, b))$ such that $(a, b) \longrightarrow \mu(p; a, b) =: a \underset{p}{\cdot} b$ defines an associative Banach algebra product on A (i.e. $|a \underset{p}{\cdot} b| \leq |a||b|$ for the norm in A). The corresponding Banach algebra is denoted by ${}_pA$.

We call $\mathbf{A}$ *involutive*, if in addition to μ an *involution*, i.e. a continuous map

$$i : P \times A \longrightarrow P \times A$$

with $i(p, a) = (p, i_p(a))$ is given, such that $a \longrightarrow i_p(a)$ defines an isometric involution on ${}_pA$ for all $p \in P$.

Now we can prove

Proposition 1. *Let* $\mathbf{G} = (P, G)$ *be a LCVG. For* $p \in P$ *and with the above notation define*

$$\mu(p; f, g) = f \underset{p}{\star} g \quad , \quad i_p(f) = f^*$$

for $f, g \in L^1 = L^1(G, dx)$ *with convolution and involution in* $L^1({}_pG)$. *Then* $\mathbf{L} = (P \times L^1, \mu, i)$ *is an involutive VBA.*

Proof. Let Δ_p be the modular function of ${}_pG$, i.e. let $d_p(x \underset{p}{\cdot} y) = \Delta_p(y) d_p x$ for $y \in {}_pG$. For every continuous compactly supported real function φ on G we have

$$\Delta_p(y) \int \varphi(x) \beta_p(x) dx = \int \varphi(x \underset{p}{\cdot} y^{-1}) \beta_p(x) dx$$

which proves continuity of $(p, y) \longrightarrow \Delta_p(y)$.

All we have to prove is the continuity of the mappings $(p, f, h) \longrightarrow f \underset{p}{\star} h$, $(p, f) \longrightarrow f^*$ and standard arguments show that it suffices to prove this for fixed, continuous, compactly supported functions f and h. Now we have for $x \in G$:

$$(f \underset{p}{\star} h)(x) = \beta_p(x) \int_G \beta_p^{-1}(x \underset{p}{\cdot} y) f(x \underset{p}{\cdot} y) \beta_p^{-1}(y^{-1}) h(y^{-1}) \beta_p(y) dy .$$

To show continuity in $p_0 \in P$ let Y be a compact neighborhood of p_0 in P and let K be the compact support of h in G. As $(x, p) \longrightarrow {}_p x^{-1}$ is continuous from $K \times Y$ into G, its image H is compact. Then $h({}_p x^{-1}) = 0$ for $p \in Y$, $x \notin H$. Similarly there is a compact set $F \subset G$ with $f(x \underset{p}{\cdot} y) = 0$ for all $p \in Y$, $y \in H$, $x \notin F$. Now the function $b(x, y, p)$, defined by

$$b(x, y, p) = \beta_p(x) \beta_p^{-1}(x \underset{p}{\cdot} y) \beta_p^{-1}({}_p y^{-1}) \beta_p(y)$$

is positive and continuous on $F \times H \times Y$ and

$$(f \underset{p}{\star} h)(x) = \int_H f(x \underset{p}{\cdot} y) h({}_p y^{-1}) b(x, y, p) dy .$$

The continuity of the integrand implies, that $f \underset{p}{\star} h$ converges for $p \longrightarrow p_0$ uniformly on F to $f \underset{p_0}{\star} h$. Thus

$$\lim_{p \to p_0} |f \underset{p}{\star} h - f \underset{p_0}{\star} h|_1 = 0 ,$$

which proves our proposition.

One has a certain amount of freedom to define the notion of *morphisms* between variable structures. For our purposes the following definition is useful:

Definition. Let $\mathbf{G}_1 = (P_1 \times A_1, \mu_1)$ and $\mathbf{G}_2 = (P_2 \times A_2, \mu_2)$ be variable structures of the same kind (i.e. both are e.g. VG's or VBA's etc.) A *morphism* from $\mathbf{G}_1$ onto $\mathbf{G}_2$ is a pair $\alpha = (\alpha_1, \alpha_2)$, where α_1 is a continuous map from P_1 into P_2 and $\alpha_2 = \{\alpha_2^p\}_{p \in P_1}$ is a family of maps from A_1 into A_2, such that α_2^p defines for every $p \in P_1$ a morphism $\alpha_2^p : {}_p A_1 \longrightarrow {}_{\alpha_1(p)} A_2$ in the usual sense and $p \longrightarrow \alpha_2^p(a)$ is continuous for every $a \in A_1$.

α, resp. $\mathbf{G}_1$, is called a *restriction* of $\mathbf{G}_2$ onto P_1, if $P_1 \subset P_2$, α_1 is the injection, and all α_2^p, are the identities.

α, resp. $\mathbf{G}_1$, is a *substructure* of $\mathbf{G}_2$, if $P_1 = P_2$, α_1 is the identity mapping, $A_1 \subset A_2$ and α_2 is the injection of A_1 into A_2.

The definition of *isomorphism* is as usual. An isomorphic α is an *equivalence*, if $P_1 = P_2$ and α_1 is the identity mapping. $\mathbf{G}_1$ and $\mathbf{G}_2$ are called *equivalent* (or *isomorphic*) if there exists an equivalence (isomorphism) α from $\mathbf{G}_1$ onto $\mathbf{G}_2$.

In a VG $\mathbf{G} = P \times G$ every group ${}_pG$ has a unit element e_p. With fixed $a \in G$ we have $e_p = a \underset{p}{\cdot} a^{-1}$. Because product and inversion are continuous in p it follows that $p \longrightarrow e_p$ is continuous. Now let $\mathcal{G}$ be an arbitrary topological group with product ab. For a fixed $z \in \mathcal{G}$ we define $a \cdot b = az^{-1}b$ and leave it to the reader to prove, that $\mathcal{G}^\bullet = (\mathcal{G}, \bullet)$, i.e. $\mathcal{G}$ with the new product $(a, b) \longrightarrow a \cdot b$ is a topological group with unit element z and that $a \longrightarrow za$ is an isomorphism from $\mathcal{G}$ onto $\mathcal{G}^\bullet$.

The above remarks imply, that we can define a new VG $\dot{\mathbf{G}} = (P \times G, \dot{\mu})$ by setting

$$\dot{\mu}_1(p; x, y) = x \underset{p}{\cdot} e^{-1} \underset{p}{\cdot} y\ , \quad \dot{\mu}_2(p; x) = e \underset{p}{\cdot} x^{-1} \underset{p}{\cdot} e$$

where e is a fixed element in G.

Proposition 2. $\mathbf{G}$ *and* $\dot{\mathbf{G}}$ *are equivalent variable groups. In* $\dot{\mathbf{G}}$ *the element* e *is the unit element in every group* ${}_pG$.

Proof. Obviously $\dot{\mathbf{G}}$ is a VG. The equivalence α from $\mathbf{G}$ onto $\dot{\mathbf{G}}$ is given by

$$\alpha(p, x) = (p, {}_pe^{-1} \underset{p}{\cdot} x)\ ,$$

i.e. $\alpha_2^p(x) = {}_pe^{-1} \underset{p}{\cdot} x$. Clearly α is a homeomorphism from $P \times G$ onto itself and, as we have seen, α_2^p is an isomorphism of $(G, \mu(p))$ onto $(G, \dot{\mu}(p))$, which proves our proposition.

Proposition 2 allows us to assume, that all groups ${}_pG$ of a VG $\mathbf{G}$ have the same unit element.

§ 2 Unitary representations of variable groups and algebras

Neither is a VG a group nor is a VBA an algebra, hence the meaning of "representations" of such objects is not immediately clear. However, there are of course obvious procedures to define groups and algebras from our variable structures:

Definition. Let $\mathbf{G}$ be a VG. The set of all continuous sections

$$f : P \ni p \longrightarrow f(p) \in {}_pG$$

is denoted by $\mathcal{C}(\mathbf{G})$. Taking pointwise operations and the topology of uniform convergence on compact subsets of P, $\mathcal{C}(\mathbf{G})$ becomes a topological group, which we will also denote by $\mathcal{C}(\mathbf{G})$.

Similarly for VBA's:

Definition. For a VBA $\mathbf{A}$ let $\mathscr{C}(\mathbf{A})$ be the algebra of all continuous sections

$$f : P \ni p \longrightarrow f(p) \in {}_pA$$

with pointwise operations. Define

$$\|f\| = \sup_P |f(p)|$$

$$\mathscr{C}_b(\mathbf{A}) = \{f \in \mathscr{C}(\mathbf{A}) ; \quad \|f\| < \infty\}$$

$$\mathscr{C}_0(\mathbf{A}) = \{f \in \mathscr{C}(\mathbf{A}) ; \quad \text{the support of } f \text{ is compact } \} .$$

It is obvious that $\mathscr{C}_b(\mathbf{A})$ with norm $\|f\|$ is a Banach algebra and that $f^*(p) = f(p)^*$ defines an isometric involution on $\mathscr{C}_b(\mathbf{A})$, if $\mathbf{A}$ is involutive. Furthermore, $\mathscr{C}_0(\mathbf{A})$ is an $(*-)$ ideal in $\mathscr{C}_b(\mathbf{A})$ and so is its closure

$$\mathscr{C}_\infty(\mathbf{A}) = \overline{\mathscr{C}_0(\mathbf{A})} .$$

In [18] we have defined the *adjoint algebra* $\mathscr{A}^\flat$ of an involutive Banach algebra $\mathscr{A}$. In all interesting cases $\mathscr{A}^\flat$ is the largest involutive Banach algebra, which contains $\mathscr{A}$ as closed ideal such that $T\mathscr{A} = 0$, $T \in \mathscr{A}^\flat$, implies $T = 0$. Every unitary (=bounded $*$–) representation of $\mathscr{A}$ extends uniquely to a unitary representation of $\mathscr{A}^\flat$. If $\mathscr{A}$ has a unit–element, then $\mathscr{A}^\flat = \mathscr{A}$. For a group L^1–algebra we have $L^1(G)^\flat = \mathscr{M}(G) =$ the algebra of all bounded Radon measures on G.

More or less obvious is

(4) *The Banach algebra $\mathscr{C}_b(P)$ of all complex valued bounded functions φ on P (with supremum norm) is contained in the center of $\mathscr{C}_\infty(\mathbf{A})^\flat$ and of $\mathscr{C}_b(\mathbf{A})^\flat$.*

Here for $\varphi \in \mathscr{C}_b(P)$, $f \in \mathscr{C}_\infty(\mathbf{A})$ the product φf is of course defined by $(\varphi f)(p) = \varphi(p) f(p)$.

Next consider a LCVG $\mathbf{G}$ and the associated variable L^1 algebra $\mathbf{L} = (P \times L^1(G), \mu, i)$, as defined in proposition 1. Analogously to the embedding of a group G in $L^1(G)^\flat$ we will show, that there exists a natural injection of $\mathscr{C}(\mathbf{G})$ into $\mathscr{C}_\infty(\mathbf{L})^\flat$: For $u \in \mathscr{C}(\mathbf{G})$ and $f \in \mathscr{C}_\infty(\mathbf{L})$ we define

(5) $$(D_u f)(p)(x) = f(p)(u(p)^{-1} \underset{p}{\cdot} x)$$

and claim

(6) *The operator D_u, defined by (5), belongs to $\mathscr{C}_\infty(\mathbf{L})^\flat$ and $u \longrightarrow D_u$ is a strongly continuous injection of $\mathscr{C}(\mathbf{G})$ into $\mathscr{C}_\infty(\mathbf{L})^\flat$.*

Proof. Let Z be a compact neighborhood of $p \in P$ and $a \in \mathscr{C}_0(G)$, i.e. let a be a continuous function on G with compact support S. Then

$$(p, x) \longrightarrow a(u(p)^{-1} \underset{p}{\cdot} x) =: a_{(p)}(x)$$

is continuous on $P \times G$. For $p \in Z$, all $a_{(p)}$ are supported on the compact image of $Z \times S$ under $(p, x) \longrightarrow u(p)^{-1} \underset{p}{\cdot} x$, thus $p \longrightarrow a_{(p)}$ maps P continuously into $L^1(G)$. Then also for every $\varphi \in \mathscr{C}_0(P)$ the function

$$p \longrightarrow \varphi(p)a_{(p)}$$

is continuous from P into $L^1(G)$, moreover $|\varphi(p)a_{(p)}|_1 = |\varphi(p)||a|_1$. But clearly

$$\varphi(p)a_{(p)} = D_u(\varphi \otimes a)(p)$$

with $(\varphi \otimes a)(p) = \varphi(p)a$, i.e. $\varphi \otimes a \in \mathscr{C}_0(P) \otimes \mathscr{C}_0(G)$. It follows that D_u maps $\mathscr{C}_0(P) \otimes \mathscr{C}_0(G)$ into $\mathscr{C}_0(\mathbf{L})$. As $\mathscr{C}_0(P) \otimes \mathscr{C}_0(G)$ is dense in $\mathscr{C}_\infty(\mathbf{L})$ and D_u is isometric, we see that D_u extends uniquely to an isometric linear automorphism of $\mathscr{C}_\infty(\mathbf{L})$, evidently contained in $\mathscr{C}_\infty(\mathbf{L})^\flat$.

The continuity of $u \longrightarrow D_u f$ follows similarly: Again it suffices to take $f = \varphi \otimes a \in \mathscr{C}_0(P) \otimes \mathscr{C}_0(G)$. For a fixed u and $\varepsilon > 0$ there exists a compact neighborhood Q of e in G with

$$|a(u(p)^{-1} \underset{p}{\cdot} x) - a(y \underset{p}{\cdot} u(p)^{-1} \underset{p}{\cdot} x)| < \varepsilon$$

for all $x \in Z =$ support φ, $y \in Q$, $x \in G$. Now

$$\Omega = \left\{ v \in \mathscr{C}(\mathbf{G});\ \ v(p) \in u(p) \underset{p}{\cdot} Q^{-1},\ \ p \in Z \right\}$$

is a neighborhood of u in $\mathscr{C}(\mathbf{G})$ and for $v \in \Omega$ we have

$$\|D_u(\varphi \otimes a) - D_v(\varphi \otimes a)\| \leq \|\varphi\| C\varepsilon$$

with a constant C depending on u and Z only.

The definition of the D_u and their properties, stated in (5) and (6), suggests the conjecture that there exist relations between the unitary representations of $\mathscr{C}(\mathbf{G})$ and of $\mathscr{C}_\infty(\mathbf{L})$ analogous to the known relations in the classical case of a locally compact group G and its algebra $L^1(G)$. First let us define:

Definition. A *general unitary representation* of the LCVG $\mathbf{G}$ is a strongly continuous homomorphism π of $\mathscr{C}(\mathbf{G})$ into the group of unitary operators of a Hilbert space $\mathfrak{H}(\pi)$. A *unitary representation* of an involutive Banach algebra $\mathscr{A}$ is a non-degenerate bounded representation π of $\mathscr{A}$ in a Hilbert space $\mathfrak{H}(\pi)$, satisfying $\pi(a^*) = \pi(a)^*$ for all $a \in \mathscr{A}$.

Now let us assume that we have a unitary representation π of $\mathscr{C}_\infty(\mathbf{L})$ for the VBA $\mathbf{L}$ of $\mathbf{G}$. It follows from the results of [18] that π extends uniquely to a representation $\pi^\flat$ of $\mathscr{C}_\infty(\mathbf{L})^\flat$, hence

$$\pi'(u) := \pi^\flat(D_u), \quad u \in \mathscr{C}(\mathbf{G})$$

defines a mapping π' of $\mathscr{C}(\mathbf{G})$ into the bounded operators of $\mathfrak{H}(\pi)$. It is easy to see that $D_u^* = D_{u^{-1}}$, hence $\pi'(u^{-1}) = \pi^\flat(D_u^*) = \pi'(u)^*$, i.e. $\pi'(u)$ is unitary. We leave it as an exercise to the reader to show that π' is also continuous. Thus (7) defines a unitary representation of the LCVG $\mathbf{G}$. In contrast to the classical case, i.e. where $\mathbf{G} = G$ is a group, $\mathscr{C}(\mathbf{G}) = G$, $\mathscr{C}_\infty(\mathbf{L}) = L^1(G)$, the mapping $\pi \longrightarrow \pi'$ defines no longer a bijection between the sets of equivalence classes of unitary representations of $\mathscr{C}_\infty(\mathbf{L})$ and of $\mathscr{C}(\mathbf{G})$: If $P = \{1, 2\}$, then $\mathscr{C}(\mathbf{G}) = {}_1G \times {}_2G$, the ordinary direct product of ${}_1G$ and ${}_2G$, while $\mathscr{C}_\infty(\mathbf{L}) = L^1({}_1G) \oplus L^1({}_2G)$. If moreover ${}_1G = {}_2G$ is commutative, with dual group $\widehat{G}$ and we consider only the irreducible representations, i.e. the homomorphisms of $\mathscr{C}_\infty(\mathbf{L})$ onto $\mathbb{C}$, resp. of $\mathscr{C}(\mathbf{G})$ into $\mathbb{T}$, then $\mathscr{C}_\infty(\mathbf{L})^\wedge$ is the disjoint union of two copies of $\widehat{G} : \mathscr{C}_\infty(\mathbf{L})^\wedge = \widehat{G} \cup \widehat{G}$, while $\mathscr{C}(\mathbf{G})^\wedge = \widehat{G} \times \widehat{G}$. The mapping $\pi \longrightarrow \pi'$ is one to one from $\mathscr{C}_\infty(\mathbf{L})^\wedge$ onto $\widehat{G} \times \{\varepsilon\} \cup \{\varepsilon\} \times \widehat{G}$, ε the unit element in $\widehat{G}$, except in $(\varepsilon, \varepsilon)$, which is the image of "the two ε's" in $\mathscr{C}_\infty(\mathbf{L})^\wedge$. Of course one could try to define a suitable notion of representation directly for the LCVG $\mathbf{G}$, something like "variable representations", which then correspond in a one to one manner to the representation of $\mathscr{C}_\infty(\mathbf{G})$, however, for our purposes we would gain nothing, because we will have to work anyway with $\mathscr{C}_\infty(\mathbf{G})$ and its C^* hull. So we continue with the following

Definition. The *C^* hull* $C^*(\mathbf{G})$ of the LCVG $\mathbf{G}$ is the C^* hull of its variable algebra $\mathscr{C}_\infty(\mathbf{L})$:

$$C^*(\mathbf{G}) := C^*(\mathscr{C}_\infty(\mathbf{L})) .$$

The *unitary dual* or shorter the *dual* of $\mathbf{G}$ is the ordinary dual of the C^* algebra $C^*(\mathbf{G})$ with its usual topology:

$$\widehat{\mathbf{G}} := C^*(\mathbf{G})^\wedge .$$

So in the future when we talk about (unitary) representations of $\mathbf{G}$ we always mean representations of $C^*(\mathbf{G})$, resp. $\mathscr{C}_\infty(\mathbf{L})$. In terms of the general representations of $\mathscr{C}(\mathbf{G})$ these are exactly those of the form π' for some representation π of $\mathscr{C}_\infty(\mathbf{L})$.

Let

$$\mathbf{A} = (P \times A, \mu, \iota)$$

be an involutive variable Banach algebra. In analogy with the definition of $C^*(\mathbf{G})$ we define the C^* hull of $\mathbf{A}$ as the C^* hull of $\mathscr{C}_\infty(\mathbf{A})$:

$$C^*(\mathbf{A}) := C^*(\mathscr{C}_\infty(\mathbf{A})) \ ,$$

and set

$$\widehat{\mathbf{A}} := C^*(\mathbf{A})^\wedge \ .$$

By a unitary representation of $\mathbf{A}$ we mean a unitary representation of $\mathscr{C}_\infty(\mathbf{A})$, resp. of $C^*(\mathbf{A})$.

We have a canonical embedding of the C^* algebra $\mathscr{C}_b(P)$ of all continuous bounded complex valued functions u on P into the center of $\mathscr{C}_\infty(\mathbf{A})^\flat$, hence of $C^*(\mathbf{A})^\flat$: For $f \in \mathscr{C}_\infty(\mathbf{A})$ and $u \in \mathscr{C}_b(P)$ the image $u^\natural \in$ center $(\mathscr{C}_\infty(\mathbf{A})^\flat)$ is given by

$$(u^\natural f)(p) = u(p)f(p) \ .$$

It follows that a representation π of $\mathbf{A}$ in a Hilbert space $\mathfrak{H}(\pi)$ defines a representation $\pi^\natural$ of $\mathscr{C}_b(P)$ such that $\pi^\natural(\mathscr{C}_b(P))$ is central in the Neumann algebra $\pi'' \subset \mathscr{L}(\mathfrak{H}(\pi))$ generated by $\pi(C^*(\mathbf{A}))$ in the bounded operators of $\mathfrak{H}(\pi)$.

Now consider a closed subset $Q \subset P$. This defines an injection

$$\rho : \mathbf{A}_Q = (Q \times A, \mu|_Q \, , \, i|_Q) \longrightarrow \mathbf{A}$$

and by means of the restriction $f \longrightarrow f|_Q$ an epimorphism

$$\rho^* : \mathscr{C}_\infty(\mathbf{A}) \longrightarrow \mathscr{C}_\infty(\mathbf{A}_Q)$$

Let $k(Q)$ be the kernel of Q in $\mathscr{C}_b(P)$, i.e. the closed ideal of all $u \in \mathscr{C}_b(P)$ vanishing on Q. We claim

$$\ker \rho^* = k(Q)\mathscr{C}_\infty(\mathbf{A}) \tag{8}$$

where the right hand side clearly means the closed ideal in $\mathscr{C}_\infty(\mathbf{A})$, generated by all $u^\natural f$ with $u \in k(Q)$, $f \in \mathscr{C}_\infty(\mathbf{A})$. To prove (8) we remark that $\ker \rho^*$ can be naturally identified with $\mathscr{C}_\infty(R, A)$, the space of continuous mappings from the complement $R = P \setminus Q$ of Q in P into A, vanishing at infinity, because every such f can be extended uniquely onto P by setting $f(p) = 0$ for $p \in Q$. Now $\mathscr{C}_\infty(R, A)$ contains the tensor product $\mathscr{C}_\infty(R) \otimes A$ as a dense subspace. But $\mathscr{C}_\infty(R) = k(Q)$, by the same argument as above, hence $\mathscr{C}_\infty(R, A) \subset k(Q)\mathscr{C}_\infty(\mathbf{A})$. As $k(Q)\mathscr{C}_\infty(\mathbf{A}) \subset \ker \rho^*$ is trivial we have proved (8).

The epimorphism ρ^* defines an injection

$$\rho_* : \widehat{\mathbf{A}}_Q \longrightarrow \widehat{\mathbf{A}} ,$$

in particular for $Q = \{p\}$, $p \in P$, we obtain an injection r_p of $\widehat{A}_{\{p\}}$ into $\widehat{\mathbf{A}}$. But clearly $\widehat{\mathbf{A}}_{\{p\}} = {}_p\widehat{A}$, because ${}_pA \cong \mathscr{C}_\infty(\mathbf{A})/k(\{p\})\mathscr{C}_\infty(\mathbf{A})$, hence we can use the r_p to identify the ${}_p\widehat{A}$ with the corresponding closed subsets of $\widehat{\mathbf{A}}$. In this sense we have

Proposition 3. $\widehat{\mathbf{A}}$ *is the disjoint union of all* ${}_p\widehat{A}$*:*

$$\widehat{\mathbf{A}} = \bigcup_{p \in P} {}_p\widehat{A} .$$

Proof. It is obvious that the union is disjoint. So let $\pi \in \widehat{\mathbf{A}}$. Because π is irreducible, it is scalar on the center of $\mathscr{C}_\infty(\mathbf{A})^\flat$, hence on the image of $\mathscr{C}_\infty(P)$ in $\mathscr{C}_\infty(\mathbf{A})^\flat$, which means that $\pi^\natural$ is scalar: $\pi^\natural(u) = u(p)\mathbf{1}$ for some $p \in P$, $\mathbf{1}$ the identity operator of $\mathfrak{H}(\pi)$. Thus $k(\{p\})\mathscr{C}_\infty(\mathbf{A}) \subset \ker \pi$, which implies $\pi \in {}_p\widehat{A}$.

Next we recall some basic facts about positive functionals on an involutive Banach algebra A. Let A' be the linear dual of A, i.e. the space of all bounded linear functionals $\varphi : A \longrightarrow \mathbb{C}$ with norm $|\varphi| = \sup_{|a| \le 1} |\varphi(a)|$. Let A'_+ be the w^* closed cone of positive functionals, i.e. $\varphi \in A'$ with $\varphi(a^*a) \ge 0$ for all $a \in A$. For $\varphi \in A'_+$ we define

$$|\varphi|_\sim = \sup\left\{|\varphi(a)|^2 ; \varphi(a^*a) = 1 , \ a \in A\right\} .$$

In [19], § 2, it is shown that $|\varphi + \psi|_\sim \le |\varphi|_\sim + |\psi|_\sim$ and $|\varphi| \le |\varphi|_\sim$ for all $\varphi, \psi \in A'_+$, furthermore that $|\varphi| = |\varphi|_\sim$ if A has an approximate identity.

The sets

$$\mathscr{P}(A) = \left\{\varphi \in A'_+ ; |\varphi|_\sim \le 1\right\} \quad , \quad \mathscr{P}_1(A) = \left\{\varphi \in A'_+ ; \ |\varphi|_\sim = 1\right\}$$

are convex and $\mathscr{P}(A)$ is w^* compact. $\mathscr{P}_1(A)$ is exactly the set of $\varphi \in \mathscr{P}(A)$ for which there exists a unitary representation π of A in a Hilbert space $\mathfrak{H}$ and a cyclic vector ε of norm 1, such that $\varphi(a) = (\pi(a)\varepsilon|\varepsilon)$.

If A has a unit element 1, then $|\varphi|_\sim = |\varphi| = \varphi(1)$ for all $\varphi \in A'_+$, in particular $\mathscr{P}(A) = \left\{\varphi \in A'_+ ; \varphi(1) = 1\right\}$ is compact und convex in A'_+. In general let

$$\tilde{A} = A \oplus \mathbb{C} \cdot 1$$

the usual extended Banach algebra with unit element $0 \oplus 1 = 1$. Then we have

$$\mathcal{P}(A) = \left\{ \tilde{\varphi}|_A ;\ \tilde{\varphi} \in \mathcal{P}_1(\tilde{A}) \right\} .$$

The elements of $\mathcal{P}_1(A)$ are usually called the *states* of A and the extremal points are the *pure states*. We denote the set of pure states of A by $\mathcal{E}(A)$, thus $\mathcal{E}(A) \cup \{0\}$ is the set of the extremal points of $\mathcal{P}(A)$.

(9) *The restriction* $r : \tilde{\varphi} \longrightarrow \tilde{\varphi}|_A$ *maps* $\mathcal{E}(\tilde{A})$ *bijectively onto* $\mathcal{E}(A) \cup \{0\}$.

Proof. Let $\omega \in \mathcal{P}_1(\tilde{A})$ be the functional $\omega(a \oplus \xi a) = \xi$. Then $r(\omega) = 0$ and ω is the only element in $\mathcal{E}(\tilde{A})$ mapped onto 0. More generally r maps $\mathcal{P}(\tilde{A}) \setminus \mathbb{R}^+\omega$ into $\mathcal{P}(A) \setminus \{0\}$. In the other direction

$$s : \varphi \longrightarrow s(\varphi) = \varphi \oplus |\varphi|_\sim \omega ,$$

i.e. $s(\varphi)(a + \xi 1) = \varphi(a) + |\varphi|_\sim \xi$, maps $\mathcal{P}(A) \setminus \{0\}$ into $\mathcal{P}(\tilde{A}) \setminus \mathbb{R}^+\omega$ and both r and s are affine, i.e. $r(u\varphi + v\psi) = u \cdot r(\varphi) + v \cdot r(\psi)$ if $\varphi, \psi \in \mathcal{P}(\tilde{A})$, $u, v \geq 0$, $u + v = 1$, similarly for s. Clearly also $rs(\varphi) = \varphi$ for all φ and $sr(\tilde{\varphi}) = \tilde{\varphi}$ for all $\tilde{\varphi} \in \mathcal{P}_1(\tilde{A})$. Now assume that $\tilde{\varphi} \in \mathcal{E}(\tilde{A})$, $\tilde{\varphi} \neq \omega$. If $r(\tilde{\varphi}) = u\psi + v\chi$ with $0 < u, v, u+v = 1$, $\psi, \chi \in \mathcal{P}(A)$, then $\tilde{\varphi} = us(\psi) + vs(\chi)$, hence $s(\psi) = s(\chi) = \tilde{\varphi}$, $\psi = \chi = r(\tilde{\varphi})$ and consequently $r(\tilde{\varphi})$ is extremal. If $\varphi \in \mathcal{E}(A)$, then necessarily $|\varphi|_\sim = 1$, hence $s(\varphi) = \tilde{\varphi} \in \mathcal{P}_1(\tilde{A})$. If $\tilde{\varphi} = u\tilde{\psi} + v\tilde{\chi}$ in $\mathcal{P}_1(\tilde{A})$, $u+v = 1$, $u, v > 0$, then $\varphi = ur(\tilde{\psi}) + vr(\tilde{\chi})$ hence $r(\tilde{\psi}) = r(\tilde{\chi}) = \varphi$, because $0 < u, v < 1$. It follows that $\tilde{\psi} = \tilde{\chi} = \tilde{\varphi}$, i.e. that $\tilde{\varphi} \in \mathcal{E}(\tilde{A}) \setminus \{\omega\}$, $\varphi = r(\tilde{\varphi})$. This proves (9). Another way to prove (9) is to use the unitary representations corresponding to the φ's. See also [33], in particular ch. IV, §6.

Now let us return to the algebras $\mathcal{C}_\infty(\mathbf{A})$ of an involutive VBA $\mathbf{A} = P \times A$. We set

$$\mathcal{E}(\mathbf{A}) := \mathcal{E}(\mathcal{C}_\infty(\mathbf{A})) .$$

As a state $\varphi \in \mathcal{P}_1(\mathcal{C}_\infty(\mathbf{A}))$ of the form $\varphi(f) = (\pi(f)\varepsilon|\varepsilon)$ is pure if and only if the representation π of $\mathcal{C}_\infty(\mathbf{A})$ is irreducible, proposition 3 implies

$$\mathcal{E}(\mathbf{A}) = \bigcup_{p \in P} \mathcal{E}({}_pA)$$

where the union is disjoint. So for every $\varphi \in \mathcal{E}(\mathbf{A})$ there is a unique

$$p = p(\varphi) \in P$$

and a $\varphi_p \in \mathcal{E}({}_pA)$ with $\varphi(f) = \varphi_p(f(p))$ for all $f \in C_\infty(\mathbf{A})$. Since we can consider $\mathcal{E}({}_pA)$ as a part of the linear dual A' of ${}_pA = A$, (10) allows us to imbed also $\mathcal{E}(\mathbf{A})$ naturally into A'.

Proposition 4. *A sequence* $\{\varphi_n\}$ *in* $\mathscr{E}(\mathbf{A})$ *converges strongly to an element* φ *in* $\mathscr{E}(\mathbf{A})$ *if and only if*
(i) $\lim_{n\to\infty} p(\varphi_n) = p(\varphi)$ *in* P *and*
(ii) $\{\varphi_n\}$ *as a sequence in the dual* A' *of the Banach space* A *converges strongly to* $\varphi \in A'$.

Proof. Let $\tilde{A} = A \oplus \mathbb{C} \cdot 1$ with norm $|a \oplus \xi \cdot 1| = |a| + |\xi|$ and for $p \in P$ define the product $\tilde{\mu}(p)$ by

$$\tilde{\mu}(p; a \oplus \xi \cdot 1\,,\ b \oplus \vartheta \cdot 1) = (\mu(p; a, b) + \vartheta a + \xi b) \oplus \xi \cdot 1$$

and $\tilde{\iota}(p; a \oplus \xi \cdot 1) = \iota(p; a) \oplus \overline{\xi} \cdot 1$ for the involution. Then $\tilde{\mathbf{A}} = (p \times \tilde{A}, \tilde{\mu}, \tilde{\iota})$ is a VBA and we have canonical injections

$$\mathbf{A} \longrightarrow \tilde{\mathbf{A}}\ ,\ \mathscr{C}_\infty(\mathbf{A}) \longrightarrow \mathscr{C}_\infty(\tilde{\mathbf{A}})\ ,$$

moreover, the φ_n and φ can be uniquely extended to functionals $\tilde{\varphi}_n$ and $\tilde{\varphi}$ in $\mathscr{E}(\tilde{\mathbf{A}})$ so that the convergence properties in proposition 4 are preserved. So we can assume that already in $\mathbf{A}$ all the algebras ${}_pA$ have a fixed element $1 \in A$ as unit element.

Now suppose that $\{\varphi_n\}$ converges strongly in $\mathscr{C}_\infty(\mathbf{A})'$ to φ. For $u \in \mathscr{C}_\infty(P)$ let $u^1 \in \mathscr{C}_\infty(\mathbf{A})$ be the function $p \longrightarrow u(p) \cdot 1 \in {}_pA = A$. Then $\varphi_n(u^1) = u(p(\varphi_n))$, $\varphi(u^1) = u(p(\varphi))$, hence $\lim_{n\to\infty} u(p(\varphi_n)) = u(p(\varphi))$ for all $u \in \mathscr{C}_\infty(P)$, which implies $\lim_{n\to\infty} p(\varphi_n) = p(\varphi)$. Then furthermore $Y = \{p(\varphi_n)\,;\ n \leq \infty\}$ (with $\varphi_\infty = \varphi$) is compact in P, thus we can find an element $v \in \mathscr{C}_0(P)$ with $v|_Y \equiv 1$. For this v and any $a \in A$ we have

$$\varphi_n(a) = v(p(\varphi_n))\varphi_n(a) = \varphi_n(v \otimes a)$$

with $(v \otimes a)(p) = v(p)a$, similarly $\varphi(a) = \varphi(v \otimes a)$. It follows that $\lim_{n\to\infty} \varphi_n(a) = \varphi(a)$ for all $a \in A$, i.e. φ_n converges strongly to φ in A'.

If, on the other hand, $\lim_{n\to\infty} p(\varphi_n) = p(\varphi)$ and $\lim \varphi_n = \varphi$ strongly in A', then $\lim_{n\to\infty} \varphi_n(u \otimes a) = \varphi(u \otimes a)$ for all $u \otimes a \in \mathscr{C}_\infty(P) \otimes A$. Because of $|\varphi_n| = |\varphi| = 1$ this implies as usual $\lim_{n\to\infty} \varphi_n(f) = \varphi(f)$ for all $f \in \mathscr{C}_\infty(\mathbf{A})$.

Before we apply proposition 4 to LCVG's and their algebras we will prove that $\mathscr{C}_\infty(\mathbf{L})$ for $\mathbf{L} = P \times L^1(G)$ has bounded approximate identities. The proof is more or less a matter of routine.

Proposition 5. *For every variable group algebra* $\mathbf{L}$ *the algebra* $\mathscr{C}_\infty(\mathbf{L})$ *has bounded approximate identities.*

Proof. Let Φ be a familiy of functions $\varphi \in \mathscr{C}_0(P)$ with $0 \leq \varphi \leq 1$ and so that for every compact $Z \subset P$ there exists $\varphi \in \Phi$ with $\varphi|_Z \equiv 1$. Let Θ be a family of functions $q \in L^1(G)$ with the following properties: $0 \leq q$, $|q|_1 = 1$, the supports of the $q \in \Theta$ are compact and form a neighborhood basis of the unit element $e \in G$. Then $\Lambda = \{\varphi \otimes q;\ \varphi \in \Phi,\ q \in \Theta\}$ is contained in $\mathscr{C}_\infty(\mathbf{L})$ and can be ordered by

$$\varphi \otimes q < \psi \otimes u\,, \quad \text{if } \varphi\psi = \varphi\,, \quad \operatorname{supp} u \subset \operatorname{supp} q\ .$$

We claim: Λ is an approximate unit, bounded by 1. The last statement is obvious. As $\mathscr{C}_\infty(P) \otimes \mathscr{C}_0(G)$ is dense in $\mathscr{C}_\infty(\mathbf{L})$ it suffices to show that for any $\vartheta \in \mathscr{C}_0(P)$, $a \in \mathscr{C}_0(G)$ and $\varepsilon > 0$ there exists $\varphi \otimes q \in \Lambda$ with $\sup_p |(\varphi \otimes q)(p) \underset{p}{\star} (\vartheta \otimes a)(p) - (\vartheta \otimes a)(p)|_1 \leq \varepsilon$. Clearly we can also assume that $|\vartheta|_\infty \leq 1$ and $|a|_1 \leq 1$. Let Z be the support of ϑ and choose $\varphi \in \Phi$ with $\varphi|_Z \equiv 1$. Then

$$|(\varphi \otimes q)(p) \underset{p}{\star} (\vartheta \otimes a)(p) - (\vartheta \otimes a)(p)|_1 \leq |q \underset{p}{\star} a - a|_1$$

for all $p \in Z$. With the definition

$$a_p(x) = \beta_p(x)^{-1} a(x)$$

we compute easily that

$$(q \underset{p}{\star} a - a)(x) = \beta_p(x) \int q(y) \left(a_p(y^{-1} \underset{p}{\cdot} x) - a_p(x) \right) dy.$$

Now as Z and $\operatorname{supp} a$ are compact and $\beta_p(x)$ and a, hence $(p, x) \longrightarrow a_p(x)$ are continuous, it follows that for any $\varepsilon' > 0$ there exists a compact neighborhood U of e in G with $|a_p(y^{-1} \underset{p}{\cdot} x) - a_p(x)| < \varepsilon'$ for $y \in U$, $p \in Z$ and all x, hence, if we choose q with $\operatorname{supp} q \subset U$ we get

$$|(q \underset{p}{\star} a - a)(x)| \leq \beta_p(x)\varepsilon'\ .$$

Again $W := \bigcup_{p \in Z} U^{-1} \underset{p}{\cdot} \operatorname{supp} a$ is compact and supports a and all $f \underset{p}{\star} a$ with $\operatorname{supp} f \subset U$. So with $B = \max_{(p,x) \in Z \times W} \beta_p(x)$ finally we obtain

$$|q \underset{p}{\star} a - a|_1 \leq B\varepsilon' = \varepsilon$$

which proves the proposition.

Now we are ready to apply proposition 4 to the case of a variable group algebra $\mathbf{L} = P \times L^1(G)$. In the classical case, i.e. P a single point, $\mathbf{L} = L^1(G)$ with a fixed locally compact group G, the positive functionals on $L^1(G)$ can be identified with the continuous positive definite functions on G. In particular, $\mathcal{P}_1(L^1(G))$ becomes a convex subset of functions $\varphi \in \mathcal{C}_b(G) \subset L^\infty(G) = L^1(G)'$ with $|\varphi|_\sim = |\varphi|_\infty = \varphi(e)$, e the identity in G. The Raikov–Yoshizawa theorem tells us, that on $\mathcal{P}_1$ the weak* topology and the compact open topology coincide. We will show that the analogue holds also in the general case. We remark first, that we can identify $\mathcal{E}(\mathbf{L})$ with a part of $\mathcal{C}_b(G)$: As $\varphi \in \mathcal{E}(\mathbf{L})$ lies in some $\mathcal{E}({}_pL^1(G))$ and ${}_pL^1(G) \stackrel{\sim}{=} L^1({}_pG)$, we have $\varphi \in \mathcal{C}_b({}_pG) = \mathcal{C}_b(G)$, $|\varphi|_\sim = |\varphi|_\infty = \varphi(e)$, where e is the common identity element for all ${}_pG$. Our proof of the following theorem is an adaption of the classical arguments to the variable case, see e.g. [7], § 13.5.

Theorem 1. *In $\mathcal{E}(\mathbf{L})$ the weak* topology coincides with the topology of uniform convergence on compacta in G.*

We start the *proof* with a

Lemma. *Let B be a continuous function on $P \times G \times G$ with complex values. For any $f \in \mathcal{C}_0(G)$ and $(p, x) \in P \times G$ let the function $C_f(p, x)$ on G be defined by*

$$C_f(p, x)(y) = f(x \underset{p}{\cdot} y^{-1}) B(p, x, y) \ ,$$

hence $C_f(p, x) \in \mathcal{C}_0(G) \subset L^1(G)$. Then

$$C_f : (p, x) \longrightarrow C_f(p, x) \in L^1(G)$$

is a continuous map from $P \times G$ into $L^1(G)$.

Proof. It suffices to prove that C_f is continuous on every compact part $Z \times K$ of $P \times G$. Let F be the compact support of f in G. Then

$$\operatorname{supp} C_f(p, x) \subset F^{-1} \underset{p}{\cdot} x \subset W := \bigcup_{q \in Z} F^{-1} \underset{q}{\cdot} K$$

for all $(p, x) \in Z \times K$. W is compact, because it is the image of the compact set $F \times Z \times K$ under the continuous mapping $(x, p, y) \longrightarrow x^{-1} \underset{p}{\cdot} y$. Now the continuity properties of the functions f and B and of product and inverse μ and ι in the variable group $\mathbf{G}$ show that if we have $\lim_\nu (p_\nu, x_\nu) = (p_0, x_0)$ in $Z \times K$, then also

$$\lim_\nu C_f(p_\nu, x_\nu) = C_f(p_0, x_0)$$

pointwise on G. But clearly for $(p, x) \in Z \times K$ the $C_f(p, x)$ are pointwise uniformly bounded on W, thus some suitable multiple $a\chi_W$, $a > 0$, of the characteristic function χ_W of W majorizes the convergence. The lemma follows now from Lebesgue dominated convergence theorem.

Now we take a function c in $\mathscr{C}_0(G)$. Then by definition of ${}_pL^1(G) \cong L^1({}_pG)$ for $\psi \in L^\infty(G)$ we get

$$(c \underset{p}{\star} \psi)(x) = \left(\Theta_p^{-1}\left(\Theta_p c \star \Theta_p \psi\right)\right)(x) =$$

$$= \beta_p(x) \int \beta_p\left(x \underset{p}{\cdot} y^{-1}\right)^{-1} c\left(x \underset{p}{\cdot} y^{-1}\right) \beta_p(y)^{-1}\psi(y)\Delta_p\left(y^{-1}\right)\beta_p(y)dy$$

$$= \int c\left(x \underset{p}{\cdot} y^{-1}\right) B(p, x, y)\psi(y)dy$$

where

$$B(p, x, y) = \beta_p(x)\beta_p\left(x \underset{p}{\cdot} y^{-1}\right)^{-1} \Delta_p\left(y^{-1}\right)$$

defines a positive continuous function on $P \times G \times G$. Thus with our previously defined mapping C_c we can write

$$(c \underset{p}{\star} \psi)(x) = \int C_c(p, x)(y)\psi(y)dy = \langle C_c(p, x), \psi\rangle$$

with the usual definition of $\langle\, ,\, \rangle$. Let $\varphi \in \mathscr{E}(\mathbf{L})$ and let $Z \times K$ be a compact subset of $P \times G$. Then the lemma has the

Corollary:

$$\left\{\psi \in \mathscr{E}(\mathbf{L})\,;\ |c \underset{p}{\star} \psi(x) - c \underset{p_0}{\star} (x)| < \varepsilon \ \ for \ \ (p, x) \in Z \times K\right\}$$

is a weak neighborhood of φ in $\mathscr{E}(\mathbf{L})$*

Now we can prove the theorem. So let

$$\varphi_0 = w^* \lim_{n\to\infty} \varphi_n \quad \text{in } \mathscr{E}(\mathbf{L})\ .$$

Then the points $p_n = p(\varphi_n)$ converge to $p_0 = p(\varphi_0)$ in P, in particular, $Z = \{p_n\,;\, n = 0, 1, 2, \dots\}$ is a compact subset of P. For simplicity in the rest of the proof we write $\underset{n}{\star}$, $a \underset{n}{\cdot} b$, etc. instead of $\underset{p_n}{\star}$, $a \underset{p_n}{\cdot} b$ etc. Furthermore it is evidently no loss of generality to assume that $\beta_0(x) \left(= \beta_{p_0}(x)\right) = 1$ for all $x \in G$ and hence that $\lim\limits_{n\to\infty} \beta_n(x) = 1$ uniformly on compacta on G.

We choose a fixed compact set K in G, ε with $0 < \varepsilon < 1$ and a compact neighborhood L of e in G with $|\varphi_0(x) - 1| < \varepsilon$ for all $x \in L$. Let $c \in \mathscr{C}_0(G)$ with

$c \geq 0$, supp $c \subset L$ and $\int c(x)dx = 1$. Then

$$|c \underset{n}{\star} \varphi_n(x) - \varphi_n(x)| =$$
$$= \left| \beta_n(x) \int_L \beta_n(y)^{-1} c(y) \beta_n \left(y^{-1} \underset{n}{\cdot} x\right)^{-1} \varphi_n \left(y^{-1} \underset{n}{\cdot} x\right) \beta_n(y) dy - \varphi_n(x) \right| =$$
$$= \left| \int_L c(y) \left(\tilde{\beta}_n(x, y) \varphi_n \left(y^{-1} \underset{n}{\cdot} x\right) - \varphi_n(x) \right) dy \right|$$

with $\tilde{\beta}_n(x, y) = \beta_n(x)\beta_n \left(y^{-1} \underset{n}{\cdot} x\right)^{-1}$. As before we see that $\bigcup_n L^{-1} \underset{n}{\cdot} K = V$ is compact, hence β converges on V uniformly to 1 and consequently $\tilde{\beta}_n$ converges on $K \times L$ uniformly to 1. If $|\tilde{\beta}_n(x, y) - 1| < \varepsilon$ for $(x, y) \in K \times L$ and $n \geq n_0$, our last inequality yields

$$|c \underset{n}{\star} \varphi_n(x) - \varphi_n(x)| \leq \varepsilon + \int_L c(y) |\varphi_n \left(y^{-1} \underset{n}{\cdot} x\right) - \varphi_n(x)| dy$$

for $n \geq n_0$ and $n = 0$. As $|\psi\left(y^{-1}z\right) - \psi(z)| \leq \sqrt{2}(1 - \operatorname{Re} \psi(y))$ for every positive definite function ψ with $\psi(e) = 1$ on an arbitrary group we see that

$$|c \underset{n}{\star} \varphi_n(x) - \varphi_n(x)| \leq \varepsilon + \sqrt{2\varepsilon} \leq 3\sqrt{\varepsilon}$$

and finally

$$|\varphi_n(x) - \varphi_0(x)| < 6\sqrt{\varepsilon} + |c \underset{n}{\star} \varphi_n(x) - c \underset{0}{\star} \varphi_0(x)|$$

uniformly on K. Now the corollary shows that there exists $n_1 = n_1(\varepsilon, K) \geq n_0$ so that

$$|c \underset{n}{\star} \varphi_n(x) - c \underset{0}{\star} \varphi_0(x)| < \sqrt{\varepsilon}$$

hence $|\varphi_n(x) - \varphi_0(x)| < 7\sqrt{\varepsilon}$ for all $n \geq n_1$ and all $x \in K$. This proves the theorem.

§ 3 Variable Lie groups and Lie algebras

It is obvious how to define an n dimensional variable Lie algebra $\mathbf{l} = (P \times L, [\cdot\,, \cdot]_p)$, so we will not explicitly state its definition. With a given basis $\{b_1, b_2, \ldots, b_n\}$ of the vector space L over the real or complex field K we have the equations

$$[b_j, b_k]_x = \sum_l \gamma^j_{kl}(x) b_l \ .$$

These yield the n^3 continuous mappings

$$\gamma^j_{kl} : P \longrightarrow K$$

respectively the mapping $\gamma : x \longrightarrow \gamma(x) = \left(\gamma^j_{kl}(x)\right)$ from P into K^{n^3}. Anticommutativity and the Jacobi identity define an algebraic submanifold $\Lambda \subset K^{n^3}$, which contains $\gamma(P)$. It is clear that we can use Λ to define a "universal" n dimensional variable Lie algebra

$$\mathbf{l}_n = (\Lambda \times L, [\cdot\ ,\ \cdot]_\lambda) \ , \tag{10}$$

over K just by setting

$$[b_j, b_k]_\lambda = \sum_l \lambda^j_{kl} b_l$$

for $\lambda = \left(\lambda^j_{kl}\right) \in \Lambda$. Then γ defines a morphism $\tilde{\gamma}$ of $\mathbf{l}$ into $\mathbf{l}_n : \tilde{\gamma}(p, v) = (\gamma(p), v)$. Although in general $\tilde{\gamma}$ is not injective it does not lose much of the structure of $\mathbf{l}$, because the algebras are constant on the fibers of $\tilde{\gamma}$, i.e. ${}_pL = {}_qL$ if $\tilde{\gamma}(p) = \tilde{\gamma}(q)$.

Definition. A *variable Lie group*, VLG, is a locally compact variable group $\mathbf{G} = (P \times G, \mu, \iota)$ where in addition to the properties of the LCVG $\mathbf{G}$ we have: G is a smooth real n manifold, $\mu(p; x, y)$, resp. $\iota(p; x)$ is twice partially differentiable with respect to (x, y), resp. to x, and the derivatives are continuous in (p, x, y), resp. in (p, x).

The last condition in this definition is necessary in order to be able to define the variable Lie algebra $\mathbf{l}(\mathbf{G})$ of a VLG $\mathbf{G}$. We assume as always that there exists a common unit element e in G for all Lie groups ${}_pG$. Let W be an euclidean neighborhood of e in G, i.e. a neighborhood which we may identify say with a cube

$$\{\xi = (\xi_1, \xi_2, \ldots, \xi_n) \in \mathbb{R}^n \, ; \, |\xi_j| < 1 \ , \ j = 1, 2, \ldots, n\}$$

in such a way that $e = 0 = (0, 0, \ldots, 0)$. We fix some p_0 in P. Then there exists a compact neighborhood V of 0 with

$$V \underset{p_0}{\cdot} V = \{\mu(p_0 \, ; \, \xi, \vartheta) \ , \ \xi, \vartheta \in V\} \subset W \ .$$

Because μ is continuous in all variables we see that $V \underset{p}{\cdot} V \subset W$ for all p in a neighborhood Y of p_0 in P. In our coordinates the components of $\mu(p;\xi,\vartheta)$ can be written as

$$\mu_j(p;\xi,\vartheta) = \xi_j + \vartheta_j + \sum_{\alpha,\beta=1}^{n} q^j_{\alpha,\beta}(p)\xi_\alpha\vartheta_\beta + Q_j(p,\xi,\vartheta)$$

where $Q_j(p,\xi,\vartheta)$ is smooth in ξ,ϑ with vanishing derivatives of order ≤ 2. From our definition of a VLG it follows that the coefficients $q^j_{\alpha,\beta}(p)$ are continuous functions on Y.

Now let $\mathfrak{g}$ be the tangent space ofG in e and let $\{b_1, b_2, \ldots, b_n\}$ be the basis of $\mathfrak{g}$ such that in our coordinates $b_j = \frac{\partial}{\partial \xi_j}$. It is well known that the vector fields

$${}_pB_j(\xi) = \sum_{k=1}^{n} \frac{\partial \mu_k}{\partial \vartheta_j}(p;\xi,0)\frac{\partial}{\partial \xi_k}$$

form a basis of the Lie algebra ${}_p\mathfrak{g}$ of ${}_pG$ (restricted to V). It is straightforward to prove that $b_j \longrightarrow {}_pB_j$ defines an isomorphism of Lie algebras when the product in $\mathfrak{g}$ is given by

$$[b_j, b_l]_p = \sum_{k=1}^{n} \left(q^k_{j,l}(p) - q^k_{l,j}(p)\right) b_k\ .$$

As the coefficients $\gamma^j_{lk}(p) = q^k_{jl}(p) - q^k_{lj}(p)$ are continuous on Y it follows that

$$\mathbf{l}(\mathbf{G}) = \left(P \times \mathfrak{g}, [\ ,\ ]_p\right)$$

defines a VLA such that ${}_p\mathfrak{g}$ is canonically isomorphic with the Lie algebra $\mathfrak{l}({}_pG)$ of ${}_pG$.

For every $p \in P$ we have the exponential mapping

$$E(p) : {}_p\mathfrak{g} \longrightarrow {}_pG\ .$$

Gluing together these $E(p)$ we obtain a mapping

$$E : \mathbf{l}(\mathbf{G}) \longrightarrow \mathbf{G}$$

defined by $E(p,x) = (p, E(p)(x)) \in \mathbf{G}$. We call E the *exponential mapping* of $\mathbf{l}(\mathbf{G})$ into $\mathbf{G}$ and claim

Theorem 2. *Let G be simply connected. E is continuous and there exist open sets $U \subset P \times \mathfrak{g}$, $V \subset P \times G$ with $P \times \{0\} \subset U$, $P \times \{e\} \subset V$, such that E maps U homeomorphically onto V.*

Proof. For $\lambda \in \mathbb{R}^n$ let $b_\lambda = \sum_1 \lambda_j b_j \in \mathfrak{g}$. Then b_λ as an element of ${}_p\mathfrak{g}$ "is" the vector field, which in the coordinates of V is given by

$$\sum_k F_k(p, \lambda; \xi) \frac{\partial}{\partial \xi_k} := \sum_k \sum_j \lambda_j \frac{\partial \mu_k}{\partial \vartheta_j}(p; \xi, 0) \frac{\partial}{\partial \xi_k} .$$

It follows that the coefficients $F_k(p, \lambda; \xi)$ and their first derivatives with respect to λ and ξ are continuous in $(p, \lambda; \xi)$ and smooth in (λ, ξ) for every $p \in P$. If $y(p, \lambda; t) = (y_k(p, \lambda; t))$ is the solution of the ordinary differential equation.

$$\dot{y}(p, \lambda; t) = F(p, \lambda; y(p, \lambda; t))$$

with initial condition $y(p, \lambda; 0) = 0$ (where "•" means of course "$\frac{\partial}{\partial t}$") then by definition $y(p, \lambda; 1) = E(p)(b_\lambda)$. With the notation

$$Y(p, \lambda) = E(p)(b_\lambda) = y(p, \lambda; 1)$$

we have to show that Y maps $P \times \mathbb{R}^n$ continuously into $P \times G$ and that there exists an open subset $U \subset P \times \mathbb{R}^n$ with $P \times \{0\} \subset U$, which Y maps homeomorphically onto an open subset of $P \times G$. First we remark that the second property implies the first one: Let $\{(p_\nu, \lambda_\nu)\}$ be a net in $P \times \mathbb{R}^n$ which converges to (p, λ). Then $(p, \frac{1}{m}\lambda) \in U$ for some $m \in \mathbb{N}$, $m > 0$, thus finally $(p_\nu, \frac{1}{m}\lambda_\nu) \in U$ and $\lim(p_\nu, \frac{1}{m}\lambda_\nu) = (p, \frac{1}{m}\lambda)$. As Y is continuous on U, it follows that $Y(p, \frac{1}{m}\lambda) = \lim Y(p_\nu, \frac{1}{m}\lambda_\nu)$ in $\mathbf{G}$. But the product is continuous on $\mathbf{G}$, hence taking the mth powers we get $Y(p, \lambda) = \lim Y(p_\nu, \lambda_\nu)$. In order to prove our claim on Y we can assume that P is compact. For p fixed the local mapping properties of $E(p)$ or in our case of $Y(p, \cdot) =: Y(p)$ are a consequence of the smoothness of $Y(p)$ and of the fact that the Jacobian $\frac{\partial Y(p)}{\partial \lambda}(\lambda)$ is regular in $\lambda = 0$. Actually in our coordinates we have $\frac{\partial Y(p)}{\partial \lambda}(0) = \mathbf{1}$, the unit matrix. In this case one usually constructs from $\frac{\partial Y(p)}{\partial \lambda}$ a family of linear transformations

$$Y'(p)(\lambda, \mu) : \mathbb{R}^n \to \mathbb{R}^n$$

continuous in (λ, μ), such that

$$Y(p, \lambda) - Y(p, \mu) = Y'(p)(\lambda, \mu)(\lambda - \mu) . \tag{11}$$

As $Y'(p)(\lambda, \lambda) = \frac{\partial Y(p)}{\partial \lambda}(\lambda) =$ Jacobian of $Y(p)$, this gives the invertibility of $Y(p)$ in some ball $B_r = \{\lambda \in \mathbb{R}^n ; |\lambda| \leq r\}$, where r depends on bounds for $\mathbf{1} - Y'(p)(\lambda, \mu)$ for small λ and μ. Thus if we can show that there are bounds independent of p and that $Y(p, \lambda)$ and $Y'(p, \lambda, \mu) := Y(p)(\lambda, \mu)$ are simultaneously continuous on $P \times B_r$, resp. $P \times B_r \times B_r$ for some $r > 0$, our theorem will follow.

Let us first consider a function F on the product $A \times Q$ of a compact topological space A and a cube $Q = \{\xi \in \mathbb{R}^n ; |\xi_j| \leq q\}$ and values in $\mathbb{R}^n$. We assume that F

has first and second derivatives with respect to $\xi \in Q$ and that these derivatives are continuous. Then F satisfies a Lipschitz condition $|F(\alpha, \xi) - F(\alpha, \vartheta)| \leq L\,|\xi - \vartheta|$ for all $\xi, \vartheta \in Q$ with a constant L *independent of* α, with norm $|\xi| = \max_j |\xi_j|$ in $\mathbb{R}^n$. Let $M = \max_{A\times Q} |F(\alpha, \xi)|$. The usual application of the Banach fixed point theorem and the fact that an estimate $0 \leq f(t) \leq ct + L\int_0^t f(\tau)d\tau$ with constant $c, L > 0$ for continuous functions f implies $f(t) \leq c\frac{e^{Lt}-1}{L}$ for $t > 0$ shows that for $0 < \gamma < \min\left\{\frac{1}{L}, \frac{q}{M}\right\}$ there exists a continuous function $y : A \times [-\gamma, \gamma] \longrightarrow Q$ such that

$$\overset{\bullet}{y}(\alpha, t) := \frac{dy(\alpha, \cdot)}{dt}(t) = F(\alpha, y(\alpha, t)), \quad y(\alpha, 0) = 0$$

in other words, the solutions of the differential equations $\overset{\bullet}{y} = F(\alpha, y)$, $y(0) = 0$, are continuous in (α, t).

We remark that the same result holds for the more general case that F depends on t, i.e. for the equation $\overset{\bullet}{y}(\alpha, t) = F(\alpha, t, y(\alpha, t))$, possibly with modified constants. This case, as usually, is reduced to the former by adding t as a new variable ξ_0, hence $(t, \xi) = (\xi_0, \xi_1, \dots, \xi_n) \in \mathbb{R}^{n+1}$, and one more equation $y_0 = F_0(\alpha, y_0, y_1, \dots, y_n) \equiv 1$.

Now let us assume that $A = P \times B$ with compact P and a compact convex neighborhood B of 0 in $\mathbb{R}^m$. Moreover, for all $p \in P$ let $F(p) : (\lambda, \xi) \longrightarrow F(p, \lambda; \xi)$ be twice differentiable in $(\lambda, \xi) \in B \times Q \subset \mathbb{R}^{m+n}$ and so that the derivatives are continuous in (p, λ, ξ). Then we claim that the solution of

$$\text{(12)} \qquad \overset{\bullet}{y}(p, \lambda; t) = F(p, \lambda; y(p, \lambda; t)), \quad y(p, \lambda; 0) = 0$$

have first order derivatives with respect to λ and that these derivatives are continuous in (p, λ, t).

Let $\lambda_0 \in B$, $\omega \in \mathbb{R}^n$ with $|\omega| = 1$ and so that $\lambda_0 + h\omega \in B$ for $0 \leq h \leq \eta$. (If $\lambda_0 \in B^0 =$ interior of B we may assume that $\lambda_0 + h\omega \in B$ for $|h| \leq \eta$.) Now we take solutions $y(p, \lambda_0, t)$ and $y(p, \lambda_0 + h\omega, t)$ with $h \neq 0$ of the equations (12). Then our assumptions on F imply the existence of a continuous map F_1 of $P \times [0, 1] \times Q$ into the linear operators of $\mathbb{R}^n$ such that

$$F(p, \lambda_0 + h\omega; \xi) - F(p, \lambda_0; \xi) = F_1(p, h, \xi)(h\omega) = hF_1(p, h, \xi)(\omega)\ .$$

Furthermore there is a continuous map F_2 of $P \times Q \times Q$ into the operators with

$$F(p, \lambda_0; \xi) - F(p, \lambda_0; \vartheta) = F_2(p; \xi, \vartheta)(\xi - \vartheta)\ .$$

For $h \neq 0$ we set

$$z(p, h; t) = \frac{1}{h}(y(p, \lambda_0 + h\omega; t) - y(p, \lambda_0; t))\ .$$

Then we derive from (12), with our functions F_1, F_2 defined above: $\dot{z}(p,h;t) = F_1(p,h,y(p,\lambda_0+h\omega;t))(\omega) + F_2(p,y(p,\lambda_0+h\omega,t),y(p,\lambda_0,t))z(p,h;t) =: G(p,h;t,z(p,h;t))$.

This function $G(p,h;t,y)$ clearly is continuous on $P \times [0,\eta] \times [-\gamma,\gamma] \times \mathbb{R}^n$ and satisfies a Lipschitz condition in (t,y) with a constant independent of (p,h). Thus $\dot{u}(p,h;t) = G(p,h;t,u(p,h;t))$ with boundary conditions $u(p,h;0) = 0$ has solutions $u(p,h;t)$ for all $(p,h) \in P \times [0,1]$ and $|t| \le \delta$ for some δ independent of (p,h) and ω. However, for $h \ne 0$ also z is a solution, hence $z(p,h;t) = u(p,h,t)$ for $h \ne 0$. As y is continuous also in $h = 0$ it follows that

$$u(p,0;t) = \lim_{h \to 0} z(p,h;t) = y_\omega(p,\lambda_0;t)$$

where y_ω is the directional derivative of $y(p,\lambda,t)$ along ω. Now letting λ_0 vary we see that y_ω solves the system

$$\dot{y}_\omega(p,\lambda;t) = F_\omega(p,\lambda,y(p,\lambda,t))(\omega) + F'(p,y(p,\lambda;t))(y_\omega)$$

where F_ω is the directional derivative of F along ω with respect to the variable λ and F' is the Jacobian of F with respect to the variable ξ. From our first general theorem now it follows, that there exists $\delta > 0$ such that the derivatives y_ω are continuous on $P \times B \times [-\delta,\delta]$.

With our previous notation for the variable exponential mapping we see that

$$Y(p,\lambda) = y(p,\lambda;1)\,.$$

Here we have to justify that we can put $t = 1$ in the right hand side, i.e. that the solutions y exists uniformly in (p,λ) for all $t \in (-\gamma,\gamma)$ for some $\gamma > 1$. As we have mentioned above we can take any $\gamma < \min\left\{\frac{1}{L}, \frac{q}{M}\right\}$ where L is the Lipschitz constant and M the upper bound of F. But here

$$F_k(p,\lambda;\xi) = \sum_{j=1}^{n} \lambda_j \beta_{jk}(p,\xi)$$

with $\beta_{jk}(p,\xi) = \frac{\partial\mu_k}{\partial\vartheta_j}(p,\xi,0)$. Hence taking $B_r = \{\lambda; |\lambda_j| \le r\}$ with $r > 0$, small enough, we can obviously choose $\gamma > 1$.

Furthermore, because also the Jacobian $Y'(p,\lambda) = \left(\frac{\partial Y_j(p,\gamma)}{\partial\lambda_k}\right)$ is continuous on $P \times B$, it follows that we can assume that also the transformations $Y'(p)(\lambda,\mu)$ in (11) are continuous on $P \times B \times B$. Consequently there exists s, $0 < s \le r$, with

$$|\mathbf{1} - Y'(p)(\lambda,\mu)| = |Y'(p)(p,p) - Y'(p)(\lambda,\mu)| \le \frac{1}{2}$$

for all $(p, \lambda, \mu) \in P \times B_s \times B_s$, in particular $T(p, \lambda, \mu) := (Y'(p)(\lambda, \mu))^{-1}$ exists, is bounded by 2 and defines a continuous mapping on $P \times B_s \times B_s$. This implies immediately that $Y : (p, \lambda) \longrightarrow Y(p, \lambda)$ defines a homeomorphism of the interior of $P \times B_s$ onto an open superset of $P \times \{e\}$ in $\mathbf{G}$.

If the variable group $\mathbf{G}$ is exponential, i.e. if all ${}_pG$, $p \in P$, are exponential Lie groups, then by definition every $E(p)$ is a diffeomorphism from ${}_p\mathfrak{g}$ onto ${}_pG$, hence E is a continuous and bijective mapping from $P \times \mathfrak{g}$ onto $P \times G$. We claim

Theorem 3. *If $\mathbf{G}$ is exponential, then E is a homeomorphism from $P \times \mathfrak{g}$ onto $P \times G$.*

Proof. For the matter of simplicity we assume that P is first countable, so that we can work with ordinary sequences. Because we know already that E is bijective and continuous we only have to show that a sequence $\{(p_j, a_j)\}$ in $P \times \mathfrak{g}$ converges to (p_∞, a_∞), if the $E(p_j, a_j) = (p_j, g_j)$ converge to some (p_∞, g_∞). Our theorem will be a consequence of the following

Lemma. *Let*

$$T = \left\{(x, y) \in \mathbb{R} \times \mathbb{R}^n \, ; \, y \in \mathbb{R}^n \, , \, x = 0 \quad or \quad x = 1 \, , \, \frac{1}{2}, \frac{1}{3}, \dots \right\}$$

and let f be a bijective mapping from T onto T with $f(\frac{1}{\nu}, y) = (\frac{1}{\nu}, f_\nu(y))$. If f is continuous, then f is also open, i.e. f is a homeomorphism of the closed subset T of $\mathbb{R}^{n+1}$.

Proof. We have to show: If U is a neighborhood of $p = (\frac{1}{\nu}, y) \in T$ (with $\frac{1}{\infty} = 0$) then $f(U)$ is a neighborhood of $f(\frac{1}{\nu}, y) = (\frac{1}{\nu}, f_\nu(y))$. This is clear for finite ν, because the continuity of f implies the continuity of all f_ν. Hence all f_ν are homeomorphisms of $\mathbb{R}^n$, by the invariance of domain theorem. So it suffices to consider the case $p = (0, y_0)$ and $U = \{\frac{1}{\nu}\}_{\nu < \infty} \times B$ with $B = \{x \in \mathbb{R}^n \, , \, |x - y_0| \leq \rho\}$. There is also no loss of generality to assume that $y_0 = 0$ and $\rho = 1$, hence $B = \{x; |x| \leq 1\}$.

Let f_∞^{-1} be the homeomorphism of $\mathbb{R}^n$, inverse to f_∞. Then

$$h(\frac{1}{\nu}, y) = (\frac{1}{\nu}, f_\infty^{-1}(y))$$

defines a homeomorphism h of T. Another homeomorphism t is defined by

$$t(\frac{1}{\nu}, y) = (\frac{1}{\nu}, y - a_\nu)$$

with $a_\nu = f_\nu(f_\infty^{-1}(0))$, because $\lim a_\nu = a_\infty = 0$. If we put $g = t \circ f \circ h$, then f is a homeomorphism if and only g is, and consequently we only have to show that $g(U)$ is a neighborhood of $g(p) = (0, 0)$, i.e. that there exists some $\delta > 0$ and $\nu_0 \in \mathbb{N}$ with

$$(\frac{1}{\nu}, y) \in g(U)$$

for all $\nu \geq \nu_0$ and $y \in \mathbb{R}^n$ with $|y| \leq \delta$.

If as usually $g(\frac{1}{\nu}, y) = (\frac{1}{\nu}, g_\nu(y))$, then the g_ν are homeomorphisms with $g_\nu(0) = 0$, thus by the general Jordan curve theorem the image $g_\nu(S)$ of the $(n-1)$ sphere $S = \{x \in \mathbb{R}^n; |x| = 1\}$ is the boundary of the open image $g_\nu(B)$ of the open unit ball B which implies that

$$\rho_\nu = \min\{|x|;\ x \in g_\nu(S)\}$$

is positive for all $\nu < \infty$. We claim that $r := \inf_{\nu<\infty} \rho_\nu > 0$. Otherwise we could find a subsequence $\{\nu_j\}$, and $y_j \in S$ with $\lim g_{\nu_j}(y_j) = 0$. As S is compact we may also assume that $\{y_j\}$ converges: $y_\infty := \lim y_j \in S$. Then $\lim_{j\to\infty} (\frac{1}{\nu_j}, y_j) = (0, y_\infty)$ and the continuity of g implies

$$g(0, y_\infty) = (0, g_\infty(y_\infty)) = \lim g(\frac{1}{\nu_j}, y_j) = (0, 0) ,$$

hence $g_\infty(y_\infty) = 0$. This is a contradiction because g_∞ is the identity, hence $g_\infty(y_\infty) = y_\infty \neq 0$. So we see that $r > 0$. Then for any δ with $0 < \delta < r$ there exists $\nu(\delta)$ such that

$$\delta B = \{x; |x| < \delta\} \subset g_\nu(B) \text{ for all } \nu \geq \nu(\delta) ,$$

hence $g(U)$ contains $\bigcup_{\nu(\delta)\leq\nu\leq\infty} \{\frac{1}{\nu}\} \times \delta B$, which is a neighborhood of $g(p) = (0, 0)$. This proves the lemma and the theorem.

The obvious question, whether the functor $\mathbf{G} \longrightarrow \mathbf{l}(\mathbf{G})$ is onto the category of variable Lie algebras has an obvious answer: The universal n dimensional VLA's $\mathbf{l}_n$ cannot be the algebras of a VLG, because this would imply that there exists a n manifold M such that every n dimensional Lie group is locally isomorphic to a group, which is diffeomorphic to M. So the problem arises to which VLA's $\mathbf{l}$ one can find VLG's $\mathbf{G}$ such that $\mathbf{l} = \mathbf{l}(\mathbf{G})$. The next proposition gives a partial answer to this question.

We start with a VLG $\mathbf{H} = (P \times H, \nu)$ with VLA $\mathbf{l}(\mathbf{H}) = (P \times \mathfrak{h}, [\cdot, \cdot]_p)$. Now let D be a continuous mapping from P into the linear operators $L(\mathfrak{h})$ such that for every $p \in P$ the image D_p is a derivation of the Lie algebra ${}_p\mathfrak{h}$. Hence for every $p \in P$ we can define a Lie product on $\mathfrak{g} := \mathbb{R} \oplus \mathfrak{h}$ by

$$[(\xi, b), (\vartheta, c)]_p = -\vartheta D_p(b) + \xi D_p(c) ,$$

In this way we obtain a VLA

$$\mathbf{g} := (P \times \mathfrak{g}, [\cdot, \cdot]_p) .$$

Now assume that H is simply connected and connected. Then the same holds for $G := \mathbb{R} \times H$, consequently for every $p \in P$ we can define a product on G such that the Lie group ${}_pG$ obtained in this way is the semidirect product

$${}_pG = \mathbb{R} \ltimes {}_pH$$

of $\mathbb{R}$ and ${}_pH$ and such that $\mathfrak{l}({}_pG) = {}_p\mathfrak{g}$. For every $p \in G$, $\mathbb{R}$ acts as an automorphism group on ${}_pH$. If we write x^{ϑ_p} for the image of $x \in H$ under the automorphism corresponding to $\vartheta \in \mathbb{R}$ we have with the notation of theorem 2

$$E(p, A)^{\vartheta_p} = E(p, e^{\vartheta D_p}(A))$$

for every $A \in \mathfrak{h}$. From this formula and theorem 2 it follows immediately that $(p, \vartheta, h) \longrightarrow h^{\vartheta_p}$ is a continuous mapping of $P \times \mathbb{R} \times H$ into H. But then in turn also

$$(p, (\xi, x), (\vartheta, y)) \longrightarrow (p, (\xi + \vartheta, x^{\vartheta_p} \underset{p}{\cdot} y))$$

maps $P \times G \times G$ continuously into $P \times G$. Similarly $(p, (\xi, x)) \longrightarrow (p, (-\xi, (x^{-1})^{-\xi_p})$ is continuous. If the $\beta_p(x)dx$ are the Haar measures on the ${}_pH$ in the sense of the second definition in § 1, then we take the product measure $d\xi dx$ as Radon measure on $G = \mathbb{R} \times H$ and the $\beta_p(x)d\xi dx$ as Haar measures on the ${}_pG$. We have proved

Proposition 6. *With definitions and notations as above there exists a simply connected variable Lie group* $\mathbf{G} = (P \times G, \nu)$ *such that*

$$\mathbf{g} = \mathbf{l}(\mathbf{G}) .$$

The injection $H \longrightarrow (0, H) \subset \mathbb{R} \times H$ *defines a natural morphism* $\iota : \mathbf{H} \longrightarrow \mathbf{G}$ *which identifies* $\mathbf{H}$ *with a variable subgroup of* $\mathbf{G}$ *and gives rise to a natural commutative diagram*

$$\begin{array}{ccc} \mathbf{l}(\mathbf{H}) & \xrightarrow{E} & \mathbf{H} \\ \downarrow & & \downarrow \iota \\ \mathbf{g} = \mathbf{l}(\mathbf{G}) & \xrightarrow{E} & \mathbf{G} \end{array} .$$

We apply this result to a certain subalgebra of the "universal" VLA $\mathbf{l}_n$. Let

$$\Delta_n = \left\{\lambda = \left(\lambda^j_{kl}\right) \in \Lambda;\ \lambda^j_{kl} = 0 \text{ if } k < j \le l\right\} .$$

Thus $\lambda \in \Delta_n$ just means that $\{b_j\}_{j=1,\dots,n}$ is a *Malcev basis* of the Lie algebra ${}_\lambda\mathfrak{l} = (L, [\ ,\ ]_\lambda)$, i.e. that the ${}_\lambda\mathfrak{l}_k = \sum_{j=k}^{n} \mathbb{R}b_j$ form a descending chain of subalgebras of ${}_\lambda\mathfrak{l}$. Let $\mathbf{s}_n$ be the restriction of $\mathbf{l}_n$ onto Δ_n, i.e.

$$\mathbf{s}_n = (\Delta_n \times L, [\ ,\ ]_\lambda) .$$

Our last theorem yields the following

Corollary. *For every n there exists a solvable simply connected variable Lie group*

$$\mathbf{S}_n = (\Delta_n \times \mathbb{R}^n, \mu)$$

with

$$\mathbf{l}(\mathbf{S}_n) = \mathbf{s}_n .$$

Proof. There is almost nothing to show for $n = 1$: Here $\Delta = \Delta_1 = \{0\}$, $\mathbf{S}_1 = \mathbb{R}$, $\mathbf{s}_1 = \mathbb{R}$. Assume that $\mathbf{S}_n$ for $n \ge 1$ exists. Then let

$$\mathbf{S}_n^{\sim} = (\Delta_{n+1} \times \mathbb{R}^n, \nu)$$

with $\nu_1(\lambda, a, b) = \mu_{n,1}(\lambda', a, b)$, where λ' is the image of $\lambda \in \Delta_{n+1}$ under the restriction $\Delta_{n+1} \longrightarrow \Delta_n$, defined by $\lambda' = (\lambda^j_{k,l}; j, k, l \ge 2)$. Similarly we define $\mathbf{s}_n^{\sim} = \mathbf{l}(\mathbf{S}_n^{\sim})$. Now we apply the theorem to the case $P = \Delta_{n+1}$, $H = \mathbb{R}^n$, $\mathfrak{h} = \sum_{j=2}^{n+1} \mathbb{R}b_j$ and D_λ the derivation

$$D_\lambda(b_j) = \sum_{k=2}^{n+1} \lambda^1_{jk} b_k\ ,\quad j = 2, 3, \dots, n+1$$

of ${}_\lambda\mathfrak{h} = {}_{\lambda'}\mathfrak{h}$. Then $\mathbf{G} = (\Delta_{n+1} \times \mathbb{R}^{n+1}, \mu) = \mathbf{S}_{n+1}$ is the desired VLG in dimension $n+1$.

Our construction of the solvable variable Lie group $\mathbf{S}_n$ shows, that we can use $\mathbb{R}^n$ in a natural way as a global coordinate system simultaneously for all groups ${}_\lambda\mathbb{R}^n$, $\lambda \in \Delta_n$. Of course these are just the canonical coordinates of the second kind, defined by the Malcev basis $\{b_j\}$ of ${}_\lambda\mathfrak{s}_n$. We will study how these coordinates are related to the canonical coordinates of the first kind, which locally are given by the exponential mapping E_n from $\mathbf{s}_n$ into $\mathbf{S}_n$.

Let us write

(13) $$E_n(\lambda, x) = (F_1(\lambda, x)\ ,\ F_2(\lambda, x), \dots\ ,\ F_n(\lambda, x)) \in \mathbb{R}^n\ ,$$

thus the F_j are continuous real functions on $\Delta_n \times \mathfrak{s}_n$, which are smooth in x. We also introduce the usual notations:

For $\alpha = (\alpha_1, \alpha_2, \dots, \alpha_n) \in \mathbb{N}^n$, $\xi = (\xi_1, \xi_2, \dots, \xi_n) \in \mathbb{R}^n$, resp. $z \in \mathbb{C}^n$ we write

$$|\alpha| = \sum_{j=1}^{n} \alpha_j\ ,\ |\xi| = \sum_{j=1}^{n} |\xi_j|\ ,\ |z| = \sum_{j=1}^{n} |z_j|\ ,\ \xi^\alpha = \prod_{j=1}^{n} \xi_j^{\alpha_j}\ .$$

Similarly $|\lambda| = \sum_{j,k,l} |\lambda_{k,l}^j|$ for $\lambda \in \Lambda \subset \mathbb{R}^{n^3}$ and $|x| = \sum_{j=1}^{n} |x_j|$ for $x = \sum_1^n x_j b_j \in \mathfrak{s}_n$. Observe that $|\xi|$, $|x|$, $|\lambda|$ are norms and that $|[x, y]_\lambda| \le |\lambda||x||y|$ for $x, y \in {}_\lambda\mathfrak{s}_n$. Our first aim is to prove by induction the following.

Theorem 4. *There exist families $\{p_{j,\alpha}(\lambda)\}_{\alpha\in\mathbb{N}^n}$, $j = 1, 2, \dots, n$, of real polynomials on $\mathbb{C}^{n^3}$ and for every $T > 0$ a constant $C_n(T) > 0$ such that*

(14) $$F_j(\lambda, x) = \sum_{\alpha} p_{j,\alpha}(\lambda) x^\alpha$$

holds for all $(\lambda, x) \in \Delta_n \times \mathbb{R}^n$ with $|\lambda| \le T$, $|x| < C_n(T)$, i.e. the power series $\sum_{\alpha} p_{j,\alpha}(\lambda) z^\alpha$ converge for $z \in \mathbb{C}^n$ with $|z| < C_n(T)$ and represents $F_j(\lambda, x)$ for real arguments in this region.

Remark: In general on Δ_n the functions $p_{j,\alpha}$ are not unique!

Proof. The theorem is true and trivial for $n = 1$. Let us assume that it holds for n. Let $\{b_0, b_1, \dots, b_n\}$ be a basis of $\mathfrak{s}_{n+1}$, so that we can identify $\mathfrak{s}_n$ with the subspace spanned by $b_1, b_2, \dots, b_n$. Similarly we write $\mathbb{R}^{n+1} = \mathbb{R} \times \mathbb{R}^n$ and $(\xi_0, \xi_1, \dots, \xi_n)$, resp. (ξ_0, ξ) for the elements of $\mathbb{R}^{n+1}$, with $\xi \in \mathbb{R}^n$.

If $\lambda = (\lambda_{k,l}^j; 0 \le j, k, l \le n) \in \Delta_{n+1}$ then $\lambda' = (\lambda_{k,l}^j; 1 \le j, k, l \le n) \in \Delta_n$ and $\lambda \longrightarrow \lambda'$ projects Δ_{n+1} onto Δ_n. With this notation we can write

$${}_\lambda\mathfrak{s}_{n+1} = \mathbb{R}b_0 \oplus {}_{\lambda'}\mathfrak{s}_n\ ,\quad {}_\lambda\mathbb{R}^{n+1} = \mathbb{R} \ltimes ({}_{\lambda'}\mathbb{R}^n)\ .$$

The derivation D, which ad b_0 induces on ${}_{\lambda'}\mathfrak{s}_n$ is given by the matrix (λ_{jk}^0), $1 \le j, k \le n$. Finally let

$$E_{n+1}(\lambda, x) = (F_0(\lambda, x), F_1(\lambda, x), \dots, F_n(\lambda, x))$$

for $x = \sum_{0}^{n} x_j b_j \in \mathfrak{s}_{n+1}$ and

$$E_n(\lambda, y) = (F'_1(\lambda, y), F'_2(\lambda, y), \dots, F'_n(\lambda, y))$$

for $y \in \mathfrak{s}_n$. Then obviously

$$\begin{aligned} F_0(\lambda, x) &= x_0 \qquad \text{for} \qquad x \in \mathfrak{s}_{n+1} \\ F_j(\lambda, y) &= F'_j(\lambda', y) \qquad \text{for} \qquad y \in \mathfrak{s}_n\,,\ \ j = 1, 2, \dots, n\ . \end{aligned}$$

We have to determine the $F_j(\lambda, x)$ for $j \geq 1$. With $F(\lambda, x) = (F_1(\lambda, x), \dots, F_n(\lambda, x)) \in {}_{\lambda'}\mathbb{R}^n$ we have

$$E_{n+1}(\lambda, x) = (x_0, F(\lambda, x)) = (x_0, 0) \underset{\lambda}{\cdot} (0, F(\lambda, x)) = E_{n+1}(\lambda, x_0 b_0) \underset{\lambda}{\cdot} (0, F(\lambda, x)),$$

thus $(0, F(\lambda, x)) = E_{n+1}(\lambda, -x_0 b_0) \underset{\lambda}{\cdot} E_{n+1}(\lambda, x)$. For further computation we use the Campbell–Baker–Hausdorff series (see [5], ch II, § 6)

$$u \underset{\lambda}{\Diamond} v = u + v + \frac{1}{2}[u, v]_\lambda + \cdots = \sum_{r,s \geq 1} {}_\lambda H_{r,s}(u, v)$$

in ${}_\lambda\mathfrak{s}_{n+1}$. Here ${}_\lambda H_{r,s}(u, v)$ is the (r, s)–homogeneous term in (u, v), see [5], p 56. As we have

$$|[u, v]|_\lambda \leq |\lambda||u||v|$$

the results of [5], p 62, imply the estimate

$$|{}_\lambda H_{r,s}(u, v)| \leqq \eta_{r,s} |\lambda|^{r+s-1} |u|^r |v|^s \tag{15}$$

where the constants $\eta_{r,s}$ are the coefficients of the power series expansion in two variables

$$-\log(2 - e^{X+Y}) = \sum_{r,s \geq 0} \eta_{r,s} X^r Y^s\ . \tag{16}$$

The series (16) converges for all $X, Y \in \mathbb{C}$ with $|X| + |Y| < \log 2$.

Now we see that for small x we have

$$(0, F(\lambda, x)) = E_{n+1}(\lambda, (-x_0 b_0) \underset{\lambda}{\Diamond} x)$$

hence, because $(-x_0 b_0) \underset{\lambda}{\Diamond} x \in {}_{\lambda'}\mathbb{R}^n$:

$$F(\lambda, x) = E_n(\lambda', (-x_0 b_0) \underset{\lambda}{\Diamond} x)\ .$$

This implies

$$F_j(\lambda, x) = F'_j(\lambda', (-x_0 b_0) \underset{\lambda}{\Diamond} x)$$

for $j = 1, 2, \ldots, n$.

We will compute the element $x' = (-x_0 b_0) \underset{\lambda}{\Diamond} x$ in ${}_{\lambda'}\mathfrak{s}_n$. First we observe that ${}_\lambda H_{r,s}(u, v)$ is a linear combination of Lie monomials of degree (r, s) in (u, v). Each such monomial is a sum of terms of the form $u^\alpha v^\beta [b_{i_1}[b_{i_2}[\ldots[b_{i_{m-1}}, b_{i_m}]_\lambda \ldots]_\lambda]_\lambda]_\lambda$ with $|\alpha| = r$, $|\beta| = s$, $m = r + s$, $0 \le i_j \le n$. As the monomial in the b_j is of the form $\sum_0^n S_j(\lambda) b_j$ with homogeneous polynomials S_j we see that we can express ${}_\lambda H_{r,s}$ in the following form:

$$(17) \qquad {}_\lambda H_{r,s}(u, v) = \sum_{j=0}^{n} \left(\sum_{\substack{|\alpha|=r \\ |\beta|=s}} P_{j,\alpha,\beta}(\lambda) u^\alpha v^\beta \right) b_j$$

with universal polynomials $P_{j,\alpha,\beta}$, homogeneous of degree $m - 1$ in λ.

For $u = -x_0 b_0$, $v = x$ we get $u^\alpha = \pm x_0^r$ if $\alpha = r$, $u^\alpha = 0$ if $\alpha < r$, hence $u^\alpha v^\beta = \pm x_0^{r+\beta_0} x_1^{\beta_1} \ldots x_n^{\beta_n} = \pm x^\gamma$ with $|\gamma| = r + |\beta| = m$, $\gamma \ge r$. From (17) we derive immediately that

$${}_\lambda h_m(x) := \sum_{r+s=m} {}_\lambda H_{r,s}(-x_0 b_0, x) = \sum_{j=1}^{n} \left(\sum_{|\gamma|=m} P_{j,\gamma}(\lambda) x^\gamma \right) b_j$$

where again the $P_{j,\gamma}$ are homogeneous polynomials of degree $|\gamma| - 1$. Observe that because of $(-x_0 b_0) \underset{\lambda}{\Diamond} x = -x_0 b_0 + x - \frac{x_0}{2}[b_0, x] + \cdots \in {}_{\lambda'}\mathfrak{s}_n$ the last summation starts with $j = 1$, i.e. we have $P_{0,\gamma} = 0$.

From the estimate (15) we get

$$|{}_\lambda h_m(x)| \le \sum_{r+s=m} \eta_{r,s} |\lambda|^{m-1} |x_0|^r |x|^s \leqq \eta_m |\lambda|^{m-1} |x|^m$$

with $\eta_m = \sum_{r+s=m} \eta_{r,s}$. As

$$\sum_{m=0}^{\infty} \eta_m X^m = -\log(2 - e^{2X})$$

we have proved: The series

$$h(\lambda, x) := \sum_{m=1}^{\infty} {}_{\lambda}h_m(x) = \sum_{j=1}^{n} h_j(\lambda, x) b_j$$

with

$$h_j(\lambda, x) = \sum_{m=1}^{\infty} \left(\sum_{|\gamma|=m} P_{j,\gamma}(\lambda) x^{\gamma} \right)$$

converges for all $x \in \mathfrak{s}_{n+1}$ with $|x| < \frac{1}{2|\lambda|} \log 2$. For every c with $0 < c < \frac{1}{2|\lambda|} \log 2$ the series converges absolutely and uniformly on $|x| \le c$.

It follows (see e.g. [12]) that the $h_j(\lambda, \cdot)$ are restrictions of holomorphic functions with power series expansion

$$h_j(\lambda, z) = \sum P_{j,\gamma}(\lambda) z^{\gamma}$$

convergent for $|z| < \frac{1}{2|\lambda|} \log 2$. Furthermore we see that for $|x| < \frac{1}{2|\lambda|} \log 2$ we get

$$|x'| = |(-x_0 b_0) \underset{\lambda}{\Diamond} x| \leqq \sum_{1}^{\infty} |{}_{\lambda}h_m(x)| \leqq -\frac{1}{|\lambda|} \log\left(2 - e^{2|\lambda||x|}\right)$$

hence, if

$$F'_j(\lambda', y) = \sum_{\alpha} p'_{j,\alpha}(\lambda') y^{\alpha}$$

converges for all $y \in \mathfrak{s}_n$, $\lambda' \in \Delta_n$ with $|\lambda'| \le T$, $|y| < C_n(T)$, it follows that

$$F_j(\lambda, x) = F'_j(\lambda', x') = \sum_{\alpha} p'_{j,\alpha}(\lambda')(x')^{\alpha}$$

converges and is holomorphic for all $\lambda \in \Delta_{n+1}$, $x \in \mathfrak{s}_{n+1}$ with $|\lambda| \le T$, $|x| \le C_{n+1}(T)$, where

$$C_{n+1}(T) = \min_{0<t\le T} \frac{1}{2t} \log\left(2 - e^{-tC_n(T)}\right) \quad .$$

Finally, for every $\alpha = (\alpha_1, \alpha_2, \ldots, \alpha_n)$ we have

$$(x')^{\alpha} = \prod_{j=1}^{n} h_j(\lambda, x)^{\alpha_j} = \sum_{\beta} \psi_{\alpha,\beta}(\lambda) x^{\beta}$$

where $\psi_{\alpha,\beta}(\lambda)$ is the sum over all products $\prod_{j=1}^{n}\prod_{i=1}^{\alpha_j} P_{j,\gamma(j,i)}$ with $\sum_{j,i}\gamma(j,i) = \beta$. Because of $|\gamma(j,i)| \geq 1$ it follows that $|\beta| \geq |\alpha|$. Thus with

$$p_{j,\beta}(\lambda) = \sum_{|\alpha|\leq|\beta|} p'_{j,\alpha}(\lambda')\psi_{\alpha,\beta}(\lambda)$$

we finally found a family of polynomials $\{p_{j,\beta}\}$ in λ with

$$F_j(\lambda, x) = \sum_{\beta} p_{j,\beta}(\lambda)x^{\beta}$$

for $j = 1, 2, \ldots, n$. With $p_{0,\beta}(\lambda) = 1$ for $\beta = (1, 0, \ldots, 0)$, $p_{0,\beta}(\lambda) = 0$ otherwise, we see that we have proved theorem 4 in dimension $n+1$, hence in general.

Theorem 4 is the basis of our next result.

Theorem 5. *There exist families $\{q_{\alpha,\beta}(\lambda)\}_{\alpha,\beta\in\mathbb{N}^n}$ and $\{r_\alpha(\lambda)\}_{\alpha\in\mathbb{N}^n}$ of polynomials and for every $T > 0$ constants $C'_n(T)$ and $C''_n(T)$ such that the multiplication $\mu(\lambda;\xi,\vartheta)$ and the inversion $\iota(\lambda;\xi)$ in $\mathbf{S}_n$ have power series expansions*

$$\mu(\lambda;\xi,\vartheta) = \sum_{\alpha,\beta} q_{\alpha,\beta}(\lambda)\xi^{\alpha}\vartheta^{\beta}$$

for all λ, ξ, ϑ with $|\lambda| \leq T$, $|\xi|, |\vartheta| \leq C'_n(T)$ and

$$\iota(\lambda,\xi) = \sum_{\alpha} r_{\alpha}(\lambda)\xi^{\alpha}$$

for all λ, ξ with $|\lambda| \leq T$, $|\xi| < C''_n(T)$.

We will only sketch the proof. Again we assume that the theorem holds for $\mathbf{S}_n$, so we may write

$$\xi \underset{\lambda'}{\cdot} \vartheta = \sum_{\alpha,\beta} q'_{\alpha,\beta}(\lambda')\xi^{\alpha}\vartheta^{\beta} \tag{18}$$

for $\lambda' \in \Delta_n$, $|\lambda'| \leq T$, $\xi, \vartheta \in \mathbb{R}^n$, $|\xi|, |\vartheta| < C'_n(t)$.

Now (18) implies that we have a similar convergent expansion for a product of $k \geq 2$ elements $\xi, \vartheta, \ldots, \omega \in {}_{\lambda'}\mathbb{R}^n$ with polynomial coefficients:

$$\xi \underset{\lambda'}{\cdot} \vartheta \underset{\lambda'}{\cdot} \cdots \underset{\lambda'}{\cdot} \omega = \sum_{\alpha,\beta,\ldots,\delta\in\mathbb{N}^n} Q_{\alpha,\beta,\ldots,\delta}(\lambda')\xi^{\alpha}\vartheta^{\beta}\ldots\omega^{\delta} \tag{19}$$

say for $|\xi|, |\vartheta|, \ldots, |\omega| < \gamma_n(T)$, $|\lambda| \leq T$.

Now take $\lambda \in \Delta_{n+1}$ and let D_λ be the corresponding derivation of ${}_{\lambda'}\mathfrak{s}_n$. Let $\xi \longrightarrow \xi^t$ be the automorphism of ${}_{\lambda'}\mathbb{R}^n$ defined by conjugation with

$$E_{n+1}(\lambda, tb) = (t, 0) \in {}_\lambda\mathbb{R}^{n+1} = \mathbb{R} \ltimes {}_{\lambda'}\mathbb{R}^n \quad ,$$

i.e. $(0, \xi^t) = (-t, 0) \underset{\lambda}{\cdot} (0, \xi) \underset{\lambda}{\cdot} (t, 0)$. Then

$$E_n(\lambda', x)^t = E_n\left(\lambda', e^{tD_\lambda}(x)\right)$$

for $x \in \mathfrak{s}_n$, where e^X for $X \in L(\mathfrak{s}_n)$ denotes the usual exponential series of the linear operator X. For $x = \xi_k b_k$ we get

$$e^{tD_\lambda}(\xi_k b_k) = \xi_k \sum_{\nu=0}^{\infty} \varphi_{k,\nu}(\lambda) t^\nu$$

with $\mathfrak{s}_n$ valued homogeneous polynomials $\varphi_{k,\nu}$ of degree ν. Now we use theorem 4 and see that

$$(20) \qquad E_n(\lambda', \xi_k b_k)^t = \sum_\alpha p_\alpha(\lambda') \xi_k^{|\alpha|} \left(\sum_\nu \varphi_{k,\nu}(\lambda) t^\nu \right)^\alpha = \sum_{l+1, m=1}^{\infty} p_{k,l,m}(\lambda) t^l \xi_k^m$$

with polynomials $p_{k,l,m}(\lambda)$ and convergence for $|t|, |\xi_k| < \gamma_n'(T)$, $|\lambda| \le T$.

Finally let $(\xi_0, \xi), (\vartheta_0, \vartheta) \in {}_\lambda\mathbb{R}^{n+1}$, i.e. $\xi, \vartheta \in {}_\lambda\mathbb{R}^n$. Then

$$(\xi_0, \xi) \underset{\lambda}{\cdot} (\vartheta_0, \vartheta) = (\xi_0 + \vartheta_0, \xi^{\vartheta_0} \underset{\lambda}{\cdot} \vartheta)$$

so in order to prove our theorem for multiplication in ${}_\lambda\mathbb{R}^{n+1}$ it suffices to show that ξ^{ϑ_0} has a suitable power series expansion in (ϑ_0, ξ). But

$$\xi = E_n(\lambda', \xi_1 b_1) \underset{\lambda'}{\cdot} \dots \underset{\lambda'}{\cdot} E_n(\lambda', \xi_n b_n)$$

hence

$$\xi^{\vartheta_0} = f(\lambda, \vartheta_0, \xi_1) \underset{\lambda'}{\cdot} f(\lambda, \vartheta_0, \xi_2) \underset{\lambda'}{\cdot} \dots \underset{\lambda'}{\cdot} f(\lambda, \vartheta_0, \xi_n)$$

where $f(\lambda, \vartheta_0, \xi_k)$ is the series right hand side of (20). Using (19), this implies that ξ^{ϑ_0} has itself a power series expansion in (ϑ_0, ξ) with polynomial coefficients, converging for $|\vartheta_0|, |\xi| \leqq C'''(T)$, $|\lambda| \le T$. This proves the multiplication part of Theorem 5. As $(\xi_0, \xi)^{-1} = (-\xi_0, (\xi^{-\xi_0})^{-1})$, the inversion part can be shown in the very same way.

One may ask whether every solvable connected and simply connected variable Lie group of dimension n is — in a suitable sense — equivalent to some variable subgroup of $\mathbf{S}_n$. This leads essentially to the following question: Let $\mathbf{G} = P \times G$ be such a solvable VLG and $\mathbf{g} = \mathbf{l}(\mathbf{G}) = P \times \mathfrak{g}$ its Lie algebra. Does there exist a linear basis $\mathfrak{b} = \{b_1, \dots, b_n\}$ in $\mathfrak{g}$ and a continuous mapping Φ from P into the linear group $GL(\mathfrak{g})$ of $\mathfrak{g}$ such that ${}_p\mathfrak{b} := \Phi(p)(\mathfrak{b})$ is a Malcev basis of ${}_p\mathfrak{g}$ or, in other words, can one select a Malcev basis ${}_p\mathfrak{b} = \{b_1(p), \dots, b_n(p)\}$ in every ${}_p\mathfrak{g}$ in such a way, that the $b_j(p)$ depend continuously on p? If this would be the case one could define with the help of $\{\Phi(p)\}$ an equivalent VLG $\mathbf{G}' = (P \times G, \mu', \iota')$ with VLA $\mathbf{g}'$ such that $\mathbf{g}'$ contains a basis $\mathfrak{b}$, which is a Malcev basis for all ${}_p\mathfrak{g}$. Clearly for such a VLG $\mathbf{G}'$ there exist canonical morphisms $\mathbf{g}' \longrightarrow \mathbf{s}_n$, $\mathbf{G}' \longrightarrow \mathbf{S}_n$.

Unfortunately such a "continuous selection" of a Malcev basis is not always possible. Already in dimension 2 we have the following counterexample: Let $D = \{z \in \mathbb{C}; |z| \leq 1\}$, $\mathfrak{g} = \mathbb{R}a \oplus \mathbb{R}b$ and set for $z = x + iy \in D$

$$[a, b]_z = xa + yb \quad ,$$

thus ${}_z\mathfrak{g} \cong \mathfrak{a}_2$, the "$ax + b$" algebra, if $z \neq 0$, and ${}_0\mathfrak{g}$ the abelian 2 dimensional algebra. Evidently $\mathbf{d} = (D \times \mathfrak{g}, [\ ,\ ]_z)$ is a solvable VLA. If $c(z)$ would be the second element of a continuous Malcev basis of $\mathbf{d}$, normalized so that $|c(z)| = 1$, for all $z \in D$ (this time with 2 norm $|\alpha a + \beta b| = (\alpha^2 + \beta^2)^{1/2}$ in $\mathfrak{g}$,) then necessarily

$$c(z) = \cos\varphi \cdot a + \sin\varphi \cdot b \quad \text{if} \quad \varphi = \arg z \quad ,$$

for all $z \neq 0$, if $c(i) = b$, which, of course, we can always assume. But this implies that every element $u \in \mathfrak{g}$ of length 1 is a limit of a sequence $c(z_j)$ for $z_j \longrightarrow 0$, hence for no $c(0)$ the family $\{c(z); z \in D\}$ can be continuous in 0.

We will say that a basis $\mathfrak{b} = \{b_j\}$ of $\mathfrak{g}$ is a *Malcev basis of the VLA $\mathbf{g}$* if $\mathfrak{b}$ is a Malcev basis in every Lie algebra ${}_p\mathfrak{g}$, $p \in P$. If $\mathfrak{b}$ is such a Malcev basis, let $\mathfrak{g}_h = \sum_{j=h}^{n} \mathbb{R}b_j$, hence $\mathfrak{g} = \mathfrak{g}_1 \supset \mathfrak{g}_2 \supset \cdots \supset \mathfrak{g}_n$ and the $\mathfrak{g}_h$ are subalgebras of ${}_p\mathfrak{g}$ for all $p \in P$; furthermore, the $\mathbf{g}_h$ are variable subalgebras of $\mathbf{g}$. Another application of proposition 6 is

(21) *Let $\mathbf{G} = (P \times G, \mu, \iota)$ be a connected, simply connected solvable VLG and assume that $\mathbf{g} = \mathbf{l}(\mathbf{G}) = P \times \mathfrak{g}$ has a Malcev basis $\mathfrak{b} = \{b_j\}$. Then we can chose $\mathbb{R}^n$ as a global coordinate system for G so that the following holds: For $h = 1, 2, \dots, n$ let*

$$G_h = \left\{\xi = \{\xi_j\} \in \mathbb{R}^n;\ \xi_1 = \xi_2 = \cdots = \xi_{h-1} = 0\right\} \subset G \quad .$$

Then the $\mathbf{G}_h = (P \times G_h, \mu_h, \iota_h)$ with the restriction μ_h and ι_h of μ and ι to $G_h \times G_h$, resp. G_h, are VLG with $\mathbf{l}(\mathbf{G}_h) = \mathbf{g}_h = P \times \mathfrak{g}_h$. The $\mathbf{G}_h$ are variable subgroups of $\mathbf{G}$ and the $\mathbf{g}_h$ are variable subalgebras of $\mathbf{g}$. If $\mathbf{G}$ is exponential, then the restrictions of the exponential map E to the $\mathbf{g}_h$ are homeomorphisms from $\mathbf{g}_h$ onto $\mathbf{G}_h$ for $h = 1, 2, \dots, n$. Finally, the Lebesgue measures on the $\mathbf{g}_h$ define the Haar measures on the $\mathbf{G}_h$.

All statements of (21) follow by induction from proposition 6, theorem 3 and the fact that the left invariant Haar measure of a semidirect product $R \ltimes H$ is the product measure of the left invariant Haar measures of R and H.

Chapter 3

The duals of exponential variable Lie groups

§ 1 The continuity of the Kirillov map

For a connected, simply connected exponential solvable group G with Lie algebra $\mathfrak{g}$ we know (ch. I, theorem 8) that the Kirillov map K defines a bijection between the orbit space $\Omega(\mathfrak{g}^*)$ of the linear dual $\mathfrak{g}^*$ of $\mathfrak{g}$ under the coadjoint action of G and the unitary dual $\widehat{G}$ of G. With the usual topologies of $\Omega(\mathfrak{g}^*)$ and $\widehat{G}$ the map K is even continuous. The *Kirillov conjecture* says that K is not only continuous but actually a homeomorphism between $\Omega(\mathfrak{g}^*)$ and $\widehat{G}$.

It is easy to see how K and the conjecture generalize to the variable case. To formulate the Kirillov conjecture adequately let us define:

Definition. Let $\mathbf{G} = P \times G$ be a locally compact variable group. A *variable* **G** *module* **M** is a pair $\mathbf{M} = (P \times M, T)$, where

i) $P \times M$ is the cartesian product of P and a normed linear space M and
ii) T is a map from $P \times G$ into the bounded operators $\mathcal{L}(M)$ such that $(p, x, m) \to T(p, x)m$ is continuous from $P \times G \times M$ into M and $T(p) : x \to T(p, x)$ is a representation of ${}_pG$ in M for every $p \in P$.

A variable **G** module **M** is in a natural way always a $\mathcal{C}(\mathbf{G})$–space: For $f \in \mathcal{C}(\mathbf{G})$ and $(p, m) \in P \times M$ we define

$$f(p, m) = (p, T(p, f(p))m) .$$

Then $\mathcal{C}(\mathbf{G})$ acts continuously on $\mathbf{M} = P \times M$.

Let M^* be the dual Banach space of M and for $A \in \mathcal{L}(M)$ let $A^* \in \mathcal{L}(M^*)$ be the adjoint operator. If we set

$$T^*(p, x) = T(p, {}_px^{-1})^*$$

with the inverse ${}_px^{-1}$ of x in ${}_pG$, then

$$\mathbf{M}^* = (P \times M^*, T^*)$$

is also a variable **G**–module.

Our main interest lies in the case in which $\mathbf{G}$ is a variable Lie group and $\mathfrak{g}$ the underlying space of the Lie algebras ${}_p\mathfrak{g}$ of the groups ${}_pG$, $p \in P$. Of course the $T(p, x)$ in this case are the adjoint operators Ad ${}_px \in \mathcal{L}({}_p\mathfrak{g}) = \mathcal{L}(\mathfrak{g})$. We will denote this module also by $\mathbf{l}(\mathbf{G})$, although in a strict sense one has to distinguish it from the variable Lie algebra of $\mathbf{G}$. Then $\mathbf{l}(\mathbf{G})^*$ is the *coadjoint module* of $\mathbf{G}$.

Now again we consider an exponential VLG $\mathbf{G}$. We know already (ch. II, proposition 3) that $\widehat{\mathbf{G}} = \bigcup_{p\in P} {}_p\widehat{G}$. Furthermore, for every $p \in P$ the Kirillov map K_p is a bijection from $\Omega({}_p\mathfrak{g}^*)$ onto ${}_p\widehat{G}$, hence an injection into $\widehat{\mathbf{G}}$. Now let $(p, l) \in \mathbf{l}(\mathbf{G})^*$, i.e. $l \in \mathfrak{g}^*$. Then the $\mathcal{C}(\mathbf{G})$ orbit $\Omega_{(p,l)}$ of (p, l) in $\mathbf{l}(\mathbf{G})^*$ obviously is given by

$$\Omega_{(p,l)} = (p, \Omega_l)$$

where Ω_l is the ${}_pG$–orbit of l in ${}_p\mathfrak{g}^*$. It follows that

$$(1) \qquad \Omega(\mathbf{G}) := \bigcup_{p\in P} \Omega_p(\mathbf{G}), \quad \text{where } \Omega_p(\mathbf{G}) := \{(p, \Omega_l)\,;\, \Omega_l \in \Omega({}_p\mathfrak{g}^*)\}$$

is the orbit space of the variable coadjoint module $\mathbf{l}(\mathbf{G})^*$. (Here (p, A) for a subset A of $\mathfrak{g}^*$ of course means the subset $\{(p, a); a \in A\}$ of $\mathbf{l}(\mathbf{G})^*$). Now we define

$$(2) \qquad K(p, \Omega_l) = K_p(\Omega_l) \in {}_p\widehat{G} \subset \widehat{\mathbf{G}}\,.$$

Definition. The map K from $\Omega(\mathbf{G})$ into $\widehat{\mathbf{G}}$ defined by (2), is called the *Kirillov map* from $\Omega(\mathbf{G})$ into $\widehat{\mathbf{G}}$.

As K is defined "fiberwise" it follows from the classical case that K maps $\Omega(\mathbf{G})$ bijectively onto $\widehat{\mathbf{G}}$. Also as in the classical case both spaces $\widehat{\mathbf{G}}$ and $\Omega(\mathbf{G})$ are topological T_0 spaces: For $\widehat{\mathbf{G}}$ this is clear because $\widehat{\mathbf{G}}$ is the dual of the type I C^* algebra $C^*(\mathbf{G})$, for $\Omega(\mathbf{G})$ this follows easily from the fact that all $\Omega({}_p\mathfrak{g}^*)$ are T_0 spaces. Now we can state our main result:

Theorem 1. *K is a homeomorphism.*

As an immediate consequence we see:

Corollary. *The Kirillov conjecture is true for all connected (ordinary) exponential solvable Lie groups.*

The corollary is just the case where P reduces to a single point.

In the following we assume that the space P satisfies the first axiom of countability, thus to prove continuity of functions etc. we can work with sequences. Our assumption is not really necessary but it simplifies substantially the notation and arguments, moreover it is hard to imagine interesting natural examples where it is not satisfied.

We start our proof with a simple remark: If Γ is a Hausdorff topological group, acting continuously on a Hausdorff space Σ, then a sequence $\{\delta_\nu\}_{\nu\in\mathbb{N}}$ of orbits $\delta_\nu \subset \Sigma$ converges in the quotient space Σ/Γ to an orbit δ_∞ if and only if Σ contains a sequence $\{s_\nu\}$, converging to some $s_\infty \in \Sigma$, such that $\delta_\nu = \Gamma s_\nu$ for $\nu \le \infty$. Hence using the following notation

$$\overline{\mathbb{N}} = \mathbb{N} \cup \{\infty\} = \{\infty, 0, 1, 2, \dots\}$$

we have to show:

Let $\{(x_\nu, l_\nu); \nu \in \overline{\mathbb{N}}\}$, resp. $\{\pi_\nu; \nu \in \overline{\mathbb{N}}\}$ be families in $P \times \mathfrak{g}^*$, resp. in $\widehat{\mathbf{G}}$ such that

$$\pi_\nu = K\left(\Omega_{(p_\nu, l_\nu)}\right), \quad \nu \in \overline{\mathbb{N}}\ .$$

(A) If $\lim\limits_{\nu\to\infty} (p_\nu, l_\nu) = (p_\infty, l_\infty)$, then a subsequence of $\{\pi_\nu\}_{\nu<\infty}$ converges to π_∞.

(B) If $\lim\limits_{\nu\to\infty} \pi_\nu = \pi_\infty$, then one can choose the $l_\nu \in \mathfrak{g}^*$, $\nu \le \infty$, so that a subsequence of $\{l_\nu\}_{\nu<\infty}$ converges to l_∞.

Before we start the proof of (A) and (B) we will rearrange the situation in such a way that it becomes more manageable. First we prove a simple

Proposition 1. *Let $\{(p_j, l_j)\}_{j\in\mathbb{N}}$ be an arbitrary sequence in $P \times \mathfrak{g}^*$ and for every $j \in \mathbb{N}$ let a subalgebra $\mathfrak{p}_j$ of $\mathfrak{g}_j := {}_{p_j}\mathfrak{g}$ be given. Then we can find a subsequence $\{j_k\}_{k\in\mathbb{N}}$ and a fixed $h \in \mathbb{N}$ with $1 \le h \le n = \dim \mathfrak{g}$ and the following properties:*

(i) *For every $k \in \mathbb{N}$ there exists a basis $\{b_{k,i}\}_i$ in $\mathfrak{g}_{j_k}$ such that $b_{\infty,i} = \lim\limits_{k\to\infty} b_{k,i}$ exists and $\{b_{\infty,i}\}$ is a basis of $\mathfrak{g}$.*

(ii)
$$\mathfrak{p}_{j_k} = \sum_{i=h}^{n} \mathbb{R} b_{k,i}\ , \quad k \in \mathbb{N}j$$

Now assume that $\lim(p_j, l_j) = (p_\infty, l_\infty)$ exists in $P \times \mathfrak{g}^*$. Then $\{j_k\}_k$ can be choosen so that

(iii) *$\{b_{\infty,i}\}_i$ is a Malcev basis of $\mathfrak{g}_\infty := {}_{p_\infty}\mathfrak{g}$, if the $\{b_{k,i}\}_i$ are Malcev bases in $\mathfrak{g}_{j_k}$.*

(iv)
$$\mathfrak{p}_\infty := \sum_{i=h}^{n} \mathbb{R} b_{\infty,i}$$

is a subalgebra of $\mathfrak{g}_\infty$ and it is subordinate to l_∞, if the $\mathfrak{p}_{j_k}$ are subordinate to l_{j_k}.

Proof. We choose a fixed euclidean inner product on $\mathfrak{g}$ and for every $j \in \mathbb{N}$ an orthonormal basis $\{b'_{j,i}\}$ with $\mathfrak{p}_j = \sum_{i=h}^{n} \mathbb{R} b'_{j,i}$. Then there is obviously a subsequence

$\{j_k\}_k$ for which all $\mathfrak{p}_{j_k}$ have the same dimension $n-h+1$ and all limits $b_{\infty,i} = \lim b_{k,i}$, $b_k := b'_{j_k}$, exist and the family $\{b_{\infty,i}\}_i$ is orthonormal, hence a basis of $\mathfrak{g}$. This proves (i) and (ii).

Now assume that $p_\infty = \lim p_j$ in P. Let

$$(3) \qquad [b_{k,l}, b_{k,m}]_{p_{j_k}} = \sum_{i=1}^{n} \beta^l_{k;m,i} b_{k,i}$$

then the convergence of the $\{b_{k,l}\}_k$ and the continuity properties of $[\ ,\ ]_p$ imply that all

$$\beta^l_{\infty;m,i} := \lim_{k\to\infty} \beta^l_{k;m,i}$$

exist and (3) holds also for $k=\infty$. Now $\mathfrak{p}_{j_k}$ is a subalgebra if and only if

$$(4) \qquad \beta^l_{k;m,i} = 0 \quad \text{for} \quad i < h, \quad \text{all} \quad l, m \geq h$$

so if all $\mathfrak{p}_{j_k}$ are subalgebras, then (4) holds also for $k=\infty$, which means that in this case $\mathfrak{p}_\infty$ is a subalgebra. Similarly, $\{b_{k,j}\}_j$ is Malcev if and only if $\beta^l_{k;m,i} = 0$ for all m, i, l with $i \leq \min\{m, l\}$. Hence in this case also $\{b_{\infty,j}\}_j$ is a Malcev basis in $\mathfrak{g}_\infty$. Finally, as

$$l_\infty([b_{\infty,l}, b_{\infty,m}]_{p_\infty}) = \lim_{k\to\infty} l_{j_k}([b_{k,l}, b_{k,m}]_{p_k})$$

if $\lim l_j = l_\infty$, it follows that $l_\infty([\mathfrak{p}_\infty, \mathfrak{p}_\infty]) = 0$, if $l_{j_k}([\mathfrak{p}_{j_k}, \mathfrak{p}_{i_k}]) = 0$ for all k.

We will use these data to construct a new variable Lie algebra on $\overline{\mathbb{N}} \times \mathfrak{g}$: We take the $b_{\infty,j} = b_j$ as linear basis for $\mathfrak{g}$. For every $\nu \in \mathbb{N}$ let $\tau^\nu = \left(t^\nu_{j,k}\right)$ be the matrix, defined by

$$b_{\nu,j} = \sum_k \tau^\nu_{j,k} b_k\ ,$$

thus $\lim_{\nu\to\infty} \tau^\nu_{j,k} = \delta_{j,k}$, resp. $\lim_{\nu\to\infty} \tau^\nu = 1$, the unit matrix. The matrix τ^ν defines a linear automorphism of $\mathfrak{g}$, which also will be denoted by τ^ν. Then

$$[x, y]_\nu = \tau_\nu^{-1}\left([\tau_\nu x, \tau_\nu y]_{p_\nu}\right)$$

defines a Lie product on $\mathfrak{g}$ and if ${}_\nu\mathfrak{g}$ is the corresponding Lie algebra, then τ_ν induces an isomorphism from ${}_\nu\mathfrak{g}$ onto ${}_{p_\nu}\mathfrak{g}$, in particular

$$\tau_\nu[b_j, b_k]_\nu = [\tau_\nu b_j, \tau_\nu b_k]_{p_\nu} = [b_{\nu,j}, b_{\nu,k}]_{p_\nu}$$

and

$$\lim_{\nu\to\infty} [b_j, b_k]_\nu = [b_j, b_k]_\infty = [b_{\infty,j}, b_{\infty,k}]_{p_\infty}\ .$$

This shows that

$$\overline{\mathfrak{g}} := (\overline{\mathbb{N}} \times \mathfrak{g}, [\ ,\ ]_\nu)$$

indeed is a variable Lie algebra. Starting with Malcev bases in the $\mathfrak{g}_{j_k}$, proposition 1 shows that the b_j form a Malcev basis in every ${}_\nu\mathfrak{g}$, hence ch. II, proposition 6, shows, that we can impose a variable Lie group structure on $\overline{\mathbb{N}} \times G$, so that

$$\overline{\mathbf{G}} = (\overline{\mathbb{N}} \times G, \overline{\mu}, \overline{\iota})$$

becomes a VLG with $\mathbf{l}(\overline{\mathbf{G}}) = \overline{\mathfrak{g}}$. Now let us assume that the sequence $\{p_\nu\} \subset P$ converges to p_∞. Under this condition we define a mapping $t : \overline{\mathbb{N}} \times \mathfrak{g} \to P \times \mathfrak{g}$ by

$$t(\nu, y) = (p_\nu, \tau_\nu(y))\,,\ \nu \in \overline{\mathbb{N}}\ .$$

Obviously t is a morphism of $\overline{\mathfrak{g}}$ into $\mathbf{l}(\mathbf{G})$. As the exponential maps are homeomorphisms from $P \times \mathfrak{g}$, resp. $\mathbb{N} \times \mathfrak{g}$ onto $P \times G$, resp. $\overline{\mathbb{N}} \times G$, it follows that there exists a unique continuous mapping T such that the diagram

$$\begin{array}{ccc} \overline{\mathbb{N}} \times \mathfrak{g} & \xrightarrow{\ t\ } & P \times \mathfrak{g} \\ {\scriptstyle E}\big\downarrow & & {\scriptstyle E}\big\downarrow \\ \overline{\mathbb{N}} \times G & \xrightarrow{\ T\ } & P \times G \end{array}$$

commutes, in particular $T(\nu, g) = (p_\nu, T_\nu(g))$ with

$$T_\nu(\exp_\nu x) = \exp_{p_\nu} t_\nu(x)\ .$$

As τ_ν is an isomorphism from ${}_\nu\mathfrak{g}$ onto ${}_{p_\nu}\mathfrak{g}$ and G is simply connected it follows from the general functorial properties of the exponential mapping that all T_ν are isomorphisms from ${}_\nu G$ onto ${}_{p_\nu}G$, hence also T is a morphism from $\overline{\mathbf{G}}$ into $\mathbf{G}$. The sequences $\{(p_\nu, l_\nu); \nu \le \infty\}$ in $P \times \mathfrak{g}^*$, resp. $\{\pi_\nu; \nu \le \infty\}$ in $\widehat{\mathbf{G}}$ of (A) and (B) define uniquely sequences $\{(\nu, l_\nu)\}$ in $\overline{\mathbb{N}} \times \mathfrak{g}^*$ and $\{\overline{\pi}_\nu\} \subset \widehat{\overline{\mathbf{G}}}$ and if $p_\infty = \lim p_\nu$ this correspondence preserves convergence. This is obvious for the $\{(p_\nu, l_\nu)\}$ and follows easily from theorem 1, because convergence in $\widehat{\mathbf{G}}$ and $\widehat{\overline{\mathbf{G}}}$ can be expressed in terms of positive definite functions on the ${}_\nu G$, resp. ${}_{p_\nu}G$. So we see that for proving theorem 1 it suffices to consider only the cases with $P = \overline{\mathbb{N}}$ and sequences of the form $\{(\nu, l_\nu); \nu \le \infty\}$, resp. $\{\pi_\nu \in {}_\nu\widehat{G}; \nu \le \infty\} \subset \widehat{\overline{\mathbf{G}}}$. We emphasize that this turns out to be the essential advantage of the method even in the "simplest" case where P is a point, i.e. where $\mathbf{G} = G$ is an ordinary exponential group.

It is now not difficult to prove the continuity of the Kirillov mapping K. Our discussion shows that we may assume that $\mathbf{G} = \overline{\mathbb{N}} \times G$, $\mathbf{g} = \overline{\mathbb{N}} \times \mathfrak{g}$ and that a sequence $\{l_\nu\}_{\nu \in \mathbb{N}}$ in $\mathfrak{g}^*$ is given, which converges to $l_\infty \in \mathfrak{g}^*$. Furthermore we may assume that $\mathfrak{g}$ has a basis $\{b_1, b_2, \ldots, b_n\}$ so that $\mathfrak{p} = \sum_{j=h}^{n} \mathbb{R}b_j$ is a subalgebra in every ${}_\nu\mathfrak{g}$, subordinate to l_ν and a Pukanszky polarization for all finite ν. It follows from (21) in ch. II that $E(\overline{\mathbb{N}} \times \mathfrak{p}) = \overline{\mathbb{N}} \times P$ is a variable subgroup of $\mathbf{G}$ and that P for every $\nu \in \overline{\mathbb{N}}$ is a closed subgroup of ${}_\nu G$, more precisely $P = {}_\nu P = E(\nu, \mathfrak{p})$.

Let χ_ν be the unitary character of ${}_\nu P$ defined by l_ν, i.e. $\chi_\nu(\exp_\nu x) = e^{il_\nu(x)}$ for $x \in \mathfrak{p}$ (with $\exp_\nu x = E(\nu, x)$ for $x \in \mathfrak{g}$). Then $\pi_\nu := K(\Omega_{(\nu,l_\nu)}) = \mathrm{ind}_{{}_\nu P}^{{}_\nu G} \chi_\nu \in \widehat{\mathbf{G}}$ for finite ν, while $\pi_\infty = K(\Omega_{(\infty,l_\infty)})$ is weakly contained in $\omega := \mathrm{ind}_{{}_\infty P}^{{}_\infty G} \chi_\infty$, see theorem 10, ch. I. So continuity of K will follow when we can show that for a total subset of elements ξ from the Hilbert space $\mathfrak{H}(\omega)$ of ω the positive definite functions $x \to (\omega(x)\xi|\xi)$ are uniform limits on compacta of sequences of positive definite functions Φ_ν belonging to the π_ν, $\nu \in \mathbb{N}$.

For this purpose we define for every $\varphi \in \mathscr{C}_0(G)$ and $\nu \in \overline{\mathbb{N}}$ a function $\varphi_{(\nu)}$ on G by

$$\varphi_{(\nu)}(x) = \int_P \varphi(x \underset{\nu}{\cdot} y) c_\nu(y)^{\frac{1}{2}} \chi_\nu(y) dy$$

where $c_\nu(y) = \Delta_{{}_\nu G}(y)\Delta_{{}_\nu P}(y)^{-1} = e^{-\mathrm{trace}\, \mathrm{ad}_{\mathfrak{g}/\mathfrak{p}} y}$. This c_ν extends to the real character, again denoted by c_ν, on ${}_\nu G$, occurring in the definition of $\pi_\nu = \mathrm{ind}_{{}_\nu P}^{{}_\nu G} \chi_\nu$. We claim that the function

$$\tilde{\varphi}_{(\nu)} = c_\nu^{\frac{1}{2}} \varphi_{(\nu)}$$

belongs to $\mathfrak{H}(\pi_\nu)$. Obviously $\tilde{\varphi}_{(\nu)}$ is measurable and satisfies $\tilde{\varphi}_{(\nu)}(x \underset{\nu}{\cdot} y) = \overline{\chi_\nu(y)} \tilde{\varphi}_{(\nu)}(x)$ for all $x \in {}_\nu G$, $y \in {}_\nu P$. If also $\psi \in \mathscr{C}_0(G)$ then

$$\begin{aligned}
(\tilde{\varphi}_{(\nu)}|\tilde{\psi}_{(\nu)}) &= \int_{{}_\nu G/{}_\nu P} \int_{P \times P} \varphi(xy)\overline{\psi(xz)} c_\nu(x) c_\nu(yz)^{\frac{1}{2}} \chi_\nu(yz^{-1}) dy dz d_\nu \bar{x} = \\
&= \int_{{}_\nu G/{}_\nu P} \int_{P \times P} \varphi(xzy) c_\nu(y)^{\frac{1}{2}} \chi_\nu(y) dy \overline{\psi(xz)} c_\nu(xz) dz d_\nu \bar{x} = \\
&= \int_{{}_\nu G/{}_\nu P} \int_P \varphi_{(\nu)}(xz)\overline{\psi(xz)} c_\nu(xz) dz d_\nu \bar{x}
\end{aligned}$$

In ch. I, § 2 we saw that for any function $F \in L^1({}_\nu G)$ the formula $\int_{{}_\nu G} F(x) c_\nu(x)^{-1} dx = \int_{{}_\nu G/{}_\nu P} \int_P F(xy) dy d_\nu \bar{x}$ holds. For $F(x) = \varphi_{(\nu)}(x)\overline{\psi(x)} c_\nu(x)$ we obtain

$$(\tilde{\varphi}_{(\nu)}|\tilde{\psi}_{(\nu)}) = (\varphi_{(\nu)}|\psi) \tag{5}$$

where the right hand side is the ordinary inner product in $L^2({}_\nu G) = L^2(G)$.

Since $(\pi_\nu(a)\tilde{\varphi}_{(\nu)})(x) = c_\nu(a)^{\frac{1}{2}} \tilde{\varphi}_{(\nu)}(a^{-1}x) = c_\nu(x)^{\frac{1}{2}} \varphi_{(\nu)}(a^{-1}x) = c_\nu(x)^{\frac{1}{2}} (\Lambda(a)\varphi)_{(\nu)}(x) = (\Lambda(a)\varphi)\tilde{}_{(\nu)}(x)$, with the left transformation operator $\Lambda(a)$, (5) implies

$$\Phi_\nu(x) := (\pi_\nu(x)\varphi^{\sim}_{(\nu)}|\varphi^{\sim}_{(\nu)}) = ((\Lambda(x)\varphi)_{(\nu)}|\varphi) =$$
$$= \int_G \int_P \varphi(x^{-1} \underset{\nu}{\cdot} y \underset{\nu}{\cdot} z)\overline{\varphi(y)}c_\nu(z)^{\frac{1}{2}}\chi_\nu(z)dzdy$$

Let $\gamma^j_{k,l}(\nu)$ be the structure constants of ${}_\nu\mathfrak{g}$ for the basis $\{b_j\}$. Then

$$\text{trace ad}_{{}_\nu\mathfrak{g}/\mathfrak{p}} b_j = \sum_{k=1}^{h-1} \gamma^j_{kk}(\nu) =: \gamma_j(\nu)$$

for $j \geq h$, hence for $u = \sum_{j=h}^{n} u_j b_j \in \mathfrak{p}$:

$$c_\nu(\exp_\nu u) = e^{-\sum_{j=h}^{n} u_j \gamma_j(\nu)}$$

which shows that $\lim_{m\to\infty} c_m(z) = c_\infty(z)$ uniformly on compacta in P. Similarly $\lim_{m\to\infty} \chi_m(z) = \chi_\infty(z)$ uniformly on compacta, because $\chi_\nu(\exp_\nu u) = e^{il_\nu(u)}$ and $\lim_{m\to\infty} l_m = l_\infty$. So finally we see that the sequence $\{\Phi_m\}_{m\in\mathbb{N}}$ converges to Φ_∞, which proves the continuity of K.

In our last proof we have used implicitly another realization of the induced representation $\text{ind}_P^G\chi$. We will describe this realization, which is due to Blattner [3], in the special case of a connected closed subgroup H of the solvable Lie group G and a representation σ of H in the Hilbert space $\mathfrak{H}$. As before we assume that $\Delta_G|_H\Delta_H^{-1}$ extends to a real character c of G and that $d\overline{x}$ is the relatively invariant measure on $\overline{G} = G/H$ with respect to c. Let $\pi = \text{ind}_H^G\sigma$ and $\mathfrak{H}(\pi) = \text{ind}\ \mathfrak{H}$ the space of π as defined in Ch. I, § 3.

We choose a coexponential basis of $\mathfrak{g}$ with respect to the Lie algebra $\mathfrak{h}$ of H and adapt the notation of ch. I, § 2, in particular we write (ξ, y) for the elements of G with $\xi \in \mathbb{R}^d$, $y \in H$. Instead of (ξ, e), $e = e^2 \in H$, or $(0, y)$ we will also write just ξ or y.

Now let $\mathfrak{H}'$ be the space of all measurable functions $f : G \to \mathfrak{H}$ with

$$f(xy) = c(y)^{-\frac{1}{2}}\sigma(y)^* f(x)$$

for almost all $x \in G$ and all $y \in H$ and

$$|f|' := \left(\int_{\mathbb{R}^d} |f(\xi)|^2 d\xi\right)^{\frac{1}{2}} < \infty ,$$

Clearly $\mathfrak{H}'$ is a Hilbert space and there is an obvious unitary operator from $\mathfrak{H}'$ onto $L^2(\mathbb{R}^d, \mathfrak{H})$. It is similarly obvious that

$$(Uf)(x) = c(x)^{\frac{1}{2}} f(x)\ , \quad f \in \mathfrak{H}'$$

defines a function Uf in $\mathfrak{H}(\pi)$: Certainly $|Uf(x)|^2$ depends only on $\overline{x} = xH$, hence theorem 3 of ch. I gives

$$\int_{\overline{G}} |Uf(\overline{x})|^2 d_c\overline{x} = \int_{\mathbb{R}^d} |f(\xi)|^2 c(\xi) c(\xi)^{-1} d\xi = |f|'^2\ .$$

This proves that U is unitary from $\mathfrak{H}'$ onto $\mathfrak{H}(\pi)$. A trivial computation shows that U establishes an equivalence between π and the left translation representation Λ of G on $\mathfrak{H}'$, hence in this sense $\Lambda = \operatorname{ind}_H^G \sigma$. In the sequel we will freely use both forms of $\operatorname{ind}_H^G \sigma$.

§ 2 The continuity of the inverse Kirillov map, part I

As before let $\mathfrak{g}$ be an exponential solvable Lie algebra with dual $\mathfrak{g}^*$. For $l \in \mathfrak{g}^*$ we define

$$\begin{aligned} \mathfrak{k}(l) &= \ker l = \{x \in \mathfrak{g}; l(x) = 0\} \\ \mathfrak{g}^l &= \{x \in \mathfrak{g}; l([x, y]) = 0 \text{ for all } y \in \mathfrak{g}\} \\ \mathfrak{a}(l) &= \text{ the largest ideal in } \mathfrak{g} \text{ contained in } \mathfrak{g}^l \end{aligned}$$

Observe that any ideal of $\mathfrak{g}$ contained in $\mathfrak{k}(l)$ is also containd in $\mathfrak{a}(l)$.

Before we start with the proof of the continuity of the inverse Kirillov map K^{-1}, we prove the

(6) **Lemma.** *If no proper ideal of* $\mathfrak{g}$ *contains* $\mathfrak{g}^l$ *and if no nonzero ideal is contained in* $\mathfrak{k}(l)$, *then* $\mathfrak{a}(l)$ *is the center of* $\mathfrak{g}$, $\dim \mathfrak{a}(l) = 1$ *and every nontrivial ideal of* $\mathfrak{g}$ *contains* $\mathfrak{a}(l)$.

Proof. Let $\mathfrak{d} = \mathfrak{k}(l) \cap \mathfrak{a}(l)$. From

$$[\mathfrak{g}, \mathfrak{d}] \subset [\mathfrak{g}, \mathfrak{a}(l)] \subset \mathfrak{a}(l) \cap [\mathfrak{g}, \mathfrak{g}^l] \subset \mathfrak{d}$$

it follows that $\mathfrak{d}$ is an ideal, hence $\mathfrak{d} = 0$ and consequently $\dim \mathfrak{a}(l) \leq 1$. Clearly $\mathfrak{a}(l)$ contains the center $\mathfrak{z}$ of $\mathfrak{g}$. On the other hand $[\mathfrak{g}, \mathfrak{a}(l)] = 0$ implies $\mathfrak{a}(l) \subset \mathfrak{z}$, thus $\mathfrak{a}(l) = \mathfrak{z}$.

Let $\mathfrak{m} \neq 0$ be a minimal ideal in $\mathfrak{g}$ and $\mathfrak{i} = \{x \in \mathfrak{g}; [x, \mathfrak{m}] = 0\}$ its centralizer. Then the quotient $\mathfrak{g}/\mathfrak{i}$ is commutative, hence $\mathfrak{g}' \subset \mathfrak{i}$ and $\mathfrak{g}^l + \mathfrak{i}$, as an ideal containing

$\mathfrak{g}^l$ equals $\mathfrak{g}$. It follows that

$$[\mathfrak{g}, \mathfrak{m}] = [\mathfrak{g}^l, \mathfrak{m}] + [\mathfrak{i}, \mathfrak{m}] = [\mathfrak{g}^l, \mathfrak{m}] \subset \mathfrak{k}(l) ,$$

consequently $[\mathfrak{g}, \mathfrak{m}] = 0$, i.e. $\mathfrak{m} \subset \mathfrak{z}$ and finally $\mathfrak{a} = \mathfrak{z} = \mathfrak{m} \neq 0$.

Now we start with the proof of (B) § 1. As we remarked before, we only need to consider the VLG $\mathbf{G} = \overline{\mathbb{N}} \times G$ with VLA $\mathbf{g} = \overline{\mathbb{N}} \times \mathfrak{g}$ and representations $\pi_\nu \in {}_\nu\widehat{G}$, $\nu \in \overline{\mathbb{N}}$. Moreover, we can assume that theorem 1 holds for all exponential VLG's of dimension less than n. For all $\nu \in \overline{\mathbb{N}}$ let l_ν be an element in $\mathfrak{g}^*$ with $\pi_\nu = K(\Omega_{(\nu, l_\nu)})$.

Suppose first that infinitely many $\mathfrak{k}(l_\nu)$ contain nonzero ideals for $\mathfrak{n}_\nu \subset {}_\nu\mathfrak{g}$. The same arguments used for proposition 1 show that in this case there is no loss of generality in assuming that $\mathfrak{n} = \mathbb{R}b_n$, resp. $\mathfrak{n} = \mathbb{R}b_{n-1} \oplus \mathbb{R}b_n$ is an ideal in all ${}_\nu\mathfrak{g}$, contained in $\mathfrak{k}(l_\nu)$, $\nu \in \mathbb{N}$. Then $\mathfrak{n}$ is also an ideal in ${}_\infty\mathfrak{g}$ and we leave it to the reader to construct a commutative diagram

$$\begin{array}{ccc} \overline{\mathbb{N}} \times \mathfrak{g} & \xrightarrow{\ \gamma\ } & \overline{\mathbb{N}} \times \overline{\mathfrak{g}} =: \overline{\mathbf{g}} \\ {\scriptstyle E}\downarrow & & {\scriptstyle E}\downarrow \\ \mathbb{N} \times G & \xrightarrow{\ \Gamma\ } & \overline{\mathbb{N}} \times \overline{G} =: \overline{\mathbf{G}} \end{array}$$

where $\overline{\mathfrak{g}} = \mathfrak{g}/\mathfrak{n}$ is the quotient space modulo $\mathfrak{n}$, γ the epimorphism of $\overline{\mathbb{N}} \times \mathfrak{g}$ onto the canonically defined quotient VLA $\overline{\mathbb{N}} \times \overline{\mathfrak{g}}$, given by $\gamma(\nu, x) = (\nu, \gamma_\nu(x))$ with

$$\gamma_\nu(x) = x + {}_\nu\mathfrak{n} \in {}_\nu\overline{\mathfrak{g}} := {}_\nu\mathfrak{g}/{}_\nu\mathfrak{n} ,$$

E the exponential maps and Γ the epimorphism of $\mathbf{G}$ onto $\overline{\mathbf{G}}$.

The dual $\overline{\mathfrak{g}}^*$ of $\overline{\mathfrak{g}}$, resp. ${}_\nu\overline{\mathfrak{g}}^*$, can be identified with the annihilator $\mathfrak{n}^\perp$ of $\mathfrak{n}$ in $\mathfrak{g}^*$, resp. with ${}_\nu\mathfrak{n}^\perp$. Thus for $\nu \in \mathbb{N}$ all l_ν lie in ${}_\nu\overline{\mathfrak{g}}^*$ and define representations $\overline{\pi}_\nu \in {}_\nu\overline{G}$ with $\pi_\nu = \overline{\pi}_\nu \circ \Gamma_\nu$, where Γ_ν is the epimorphism from ${}_\nu G$ onto ${}_\nu\overline{G}$, i.e. $\Gamma(\nu, x) = (\nu, \Gamma_\nu(x))$ for $\nu \in \mathbb{N}$. As $\pi_\infty = \lim \pi_\nu$ it follows that also $\pi_\infty = \overline{\pi}_\infty \circ \Gamma_\infty$, with $\overline{\pi}_\infty \in {}_\infty\overline{G}$, and $\lim_{\nu\to\infty} \overline{\pi}_\nu = \overline{\pi}_\infty$. But then by induction we conclude that we can choose the $\lambda_\nu \in {}_\nu\overline{\mathfrak{g}}^* \subset {}_\nu\mathfrak{g}^* = \mathfrak{g}^*$ so that $\overline{\pi}_\nu = K(\overline{\Omega}_{(\nu, l_\nu)})$, $l_\infty = \lim l_\nu$ in $\overline{\mathfrak{g}}^*$, $\overline{\Omega}_{(\nu, l_\nu)}$ the $\overline{G}$ orbit in $\overline{\mathfrak{g}}^*$. However, $\overline{\mathfrak{g}}^*$ is a G invariant subspace of $\mathfrak{g}^*$ on which the $\overline{G}$ action coincides with the G action. Thus $\pi_\nu = K(\Omega_{(\nu, l_\nu)})$ and $l_\infty = \lim l_\nu$ also for $\mathbf{G}$, which proves (B) in this case.

Next we assume that infinitely many $\mathfrak{g}^{l_\nu}$ are contained in proper ideals $\mathfrak{h}_\nu \subset {}_\nu\mathfrak{g}$. Then again it is no loss of generality to assume that $\mathfrak{h} := \sum_{j=2}^n \mathbb{R}b_j$ is an ideal in all ${}_\nu\mathfrak{g}$, $\nu \in \overline{\mathbb{N}}$, and that all $\mathfrak{g}^{l_\nu}$, $\nu \in \mathbb{N}$, are contained in $\mathfrak{h}$. In this case we consider the $(n-1)$ dimensional variable group $\mathbf{H} = \overline{\mathbb{N}} \times H$ with algebra $\mathbf{h} = \overline{\mathbb{N}} \times \mathfrak{h}$. We have the diagram

$$\begin{array}{ccccc}
\mathbf{h} & \longrightarrow & \mathbf{g} & & \\
{\scriptstyle E}\big\downarrow & & \big\downarrow{\scriptstyle E} & & \\
\mathbf{H} & \underset{\iota}{\longrightarrow} & \mathbf{G} & \underset{\pi_\nu}{\longrightarrow} & \mathfrak{H}(\pi_\nu)
\end{array} \; .$$

with injective ι. If $\mathbf{L}$, resp. $\mathbf{L}_H$ is the variable L^1–algebra of $\mathbf{G}$, resp. $\mathbf{H}$, then the bottom of the last diagram gives us the following (see [18], in particular Satz 5):

$$\begin{array}{ccccc}
\mathscr{C}_\infty(\mathbf{L}_H) & \longrightarrow & \mathscr{C}_\infty(\mathbf{L})^\flat & & \\
\big\downarrow & & \big\downarrow & & \\
C^*(\mathbf{H}) & \underset{\iota_*}{\longrightarrow} & C^*(\mathbf{G})^\flat & \underset{\pi_\nu^\flat}{\longrightarrow} & \mathfrak{H}(\pi_\nu)
\end{array} \; .$$

Here the representations $\pi_\nu^\flat \circ \iota_*$ of $C^*(\mathbf{H})$ are just the integrated representations of the restrictions $\pi'_\nu = \pi_\nu|_\mathbf{H}$ of the π_ν to the variable group $\mathbf{H}$ of $\mathbf{G}$. We write also π'_ν for $\pi_\nu^\flat \circ \iota_*$. Our assumption $\pi_\infty = \lim\limits_{\nu\to\infty} \pi_\nu$ means that $\ker \pi_\infty \supset \bigcap\limits_{j=1}^{\infty} \ker \pi_{\nu_j}$ for every infinite subsequence $\{\pi_{\nu_j}\}_j$. This implies the same relation for the kernels of the $\pi_\nu^\flat$ and in turn for the π'_ν:

$$\ker \pi'_\infty \supset \bigcap_j \ker \pi'_{\nu_j} \; .$$

In the topological language of the dual $\widehat{\mathbf{H}}$ this means that the hull $h(\ker \pi'_\infty)$ of $\ker \pi'_\infty$ is contained in the closure of the union of the hulls $h(\ker \pi'_{\nu_j})$, in particular for $\nu_j = j$:

$$h(\ker \pi'_\infty) \subset \left(\bigcup_{j=1}^{\infty} h(\ker \pi'_j) \right)^- \; .$$

The hulls $h(\ker \pi'_j)$ for $j \in \mathbb{N}$ can be described as follows: Let l'_ν be the restriction of l_ν onto $\mathfrak{h} = {}_\nu\mathfrak{h} \subset \mathfrak{g}$ and $\sigma_\nu = K(\Omega_{(\nu,l'_\nu)}) \subset {}_\nu\widehat{H} \subset \widehat{\mathbf{H}}$. As ${}_\nu\mathfrak{g}^{l_\nu} \subset \mathfrak{h}$ for $\nu \in \mathbb{N}$, it follows from theorem 9, ch. I, that $h(\ker \pi'_\nu) = \{\sigma^t_\nu; t \in \mathbb{R}\}^-$, where $\sigma^t_\nu = K(\Omega_{(\nu,l^t_\nu)})$, $l^t_\nu =$ restriction of $\mathrm{Ad}^*_\nu \exp t b_1(l_\nu)$ onto $\mathfrak{h}$.

Similarly $h(\ker \pi'_\infty) = \{\sigma^t_\infty; t \in \mathbb{R}\}^-$ if $\mathfrak{h} \subset {}_\infty\mathfrak{g}^{l_\infty}$ or $\pi'_\infty = \sigma_\infty = K(\Omega_{(\infty, l'_\infty)})$ if $\mathfrak{h} \not\subset {}_\infty\mathfrak{g}^{l_\infty}$. In any case we have

$$\sigma_\infty \in \{\sigma^t_\nu\,;\, k \leq \nu < \infty\,,\, t \in \mathbb{R}\}^-$$

for all $k \in \mathbb{N}$. This implies the existence of sequences $\{t_j\} \subset \mathbb{R}$ and $\{\nu_j\}_j$, $\nu_j < \nu_{j+1} \in \mathbb{N}$, with $\lim \sigma^{t_j}_{\nu_j} = \sigma_\infty$ in $\widehat{\mathbf{H}}$. Clearly we may assume $\nu_j = j$ and $t_j = 0$ for all j, thus $\lim \sigma_\nu = \sigma_\infty$. But then our induction hypothesis yields $\lim\limits_{\nu\to\infty} l'_\nu = l'_\infty$ for suitably chosen l'_ν in their orbits $\Omega_{(\nu, l'_\nu)}$. However, $\Omega_{(\nu,l_\nu)} = \Omega_{(\nu,l_\nu)} + \mathfrak{h}^\perp$, so we can also choose the l_ν in $\Omega_{(\nu,l_\nu)}$, $\nu \in \mathbb{N}$, thus finally $\lim\limits_{\nu\to\infty} l_\nu = l_\infty$, which proves our assertion in this case.

Left over is the case in which no l_ν vanishes on nonzero ideals in ${}_\nu\mathfrak{g}$ and $\mathfrak{g} = \mathfrak{g}^{l_\nu} + {}_\nu\mathfrak{g}'$ for all ν. It follows from lemma (6) that then we can assume, that for all $\nu \in \mathbb{N}$

$$\mathfrak{z} = \mathbb{R}b_n \text{ is the center of } {}_\nu\mathfrak{g}$$
$$\lambda_\nu := l_\nu(b_n) \neq 0\,.$$

Now consider the sequence of the quotients ${}_\nu\mathfrak{g}/\mathfrak{z}$, $\nu \in \mathbb{N}$. If $n \geq 2$ then we can choose a minimal nonzero ideal in ${}_\nu\mathfrak{g}/\mathfrak{z}$, i.e. we can find an ideal $\mathfrak{y}_\nu \subset {}_\nu\mathfrak{g}$ such that the ideal $\mathfrak{y}_\nu/\mathfrak{z}$ is nonzero and minimal in ${}_\nu\mathfrak{g}/\mathfrak{z}$. We have either $\dim \mathfrak{y}_\nu = 2$ for infinitely many $\nu \in \mathbb{N}$ or $\dim \mathfrak{y}_\nu = 3$ for infinitely many $\nu \in \mathbb{N}$.

The subsequence argument allows us to assume, that all $\mathfrak{y}_\nu$ have the same dimension, moreover, proposition 1 shows, that we can also assume that all $\mathfrak{y}_\nu$ are equal. So the problem is reduced to the following two cases:

Case 1. $\mathfrak{y} = \mathbb{R}b_{n-1} \oplus \mathbb{R}b_n$ is a minimal noncentral ideal in all ${}_\nu\mathfrak{g}$.

Case 2. $\mathfrak{y} = \mathbb{R}b_{n-2} \oplus \mathbb{R}b_{n-1} \oplus \mathbb{R}b_n$ is a minimal noncentral ideal in all ${}_\nu\mathfrak{g}$.

Here "minimal noncentral" means that the quotients $\mathfrak{y}/\mathfrak{z}$, $\mathfrak{z} = \mathbb{R}b_n =$ the center, are minimal in ${}_\nu\mathfrak{g}/\mathfrak{z}$.

First we will treat Case 1. For $x \in \mathfrak{g}$ we can write

$$[x, b_{n-1}]_\nu = \alpha_\nu(x)b_{n-1} + \beta_\nu(x)b_n$$

where α_ν is a homomorphism, i.e. $\alpha_\nu({}_\nu\mathfrak{g}') = 0$ and β_ν a linear form on $\mathfrak{g}$. The centralizer $\mathfrak{k}_\nu$ of $\mathfrak{y}$, i.e. $\mathfrak{k}_\nu = \ker \alpha_\nu \cap \ker \beta_\nu$, is a proper ideal in ${}_\nu\mathfrak{g}$, hence ${}_\nu\mathfrak{g}^{l_\nu} \not\subset \mathfrak{k}_\nu$ and $[{}_\nu\mathfrak{g}^{l_\nu}, b_{n-1}]$ is a nonzero subspace of $\mathfrak{y}$, not containing $\mathfrak{z}$ (because of $l_\nu(b_n) \neq 0$). This implies $\alpha_\nu \neq 0$. If $\beta_\nu = a\alpha_\nu$ for some real a, we would have$[x, (b_{n-1} + ab_n)] = \alpha_\nu(x)(b_{n-1} + ab_n)$, thus $\mathbb{R}(b_{n-1} + ab_n)$ would be a minimal ideal different from $\mathfrak{z}$, which was excluded. It follows that α_ν and β_ν are independent and $\ker \alpha_\nu$ is an $(n-1)$ dimensional ideal.

Proposition 1 implies now that we can choose a basis $\{b_1, b_2, \dots, b_n\}$ of $\mathfrak{g}$ with the following properties: For $\nu \in \mathbb{N}$

(7)
(i) $\mathbb{R}b_n = \mathfrak{z}$ is the center of ${}_\nu\mathfrak{g}$

(ii) $\mathbb{R}b_{n-1} \oplus \mathbb{R}b_n = \mathfrak{y}$ is a minimal noncentral ideal in ${}_\nu\mathfrak{g}$,

(iii) $\sum_{i=3}^{n} \mathbb{R}b_i = \mathfrak{k}$ is the centralizer of $\mathfrak{y}$ in ${}_\nu\mathfrak{g}$,

(iv) $[b_1, b_{n-1}]_\nu = \alpha_\nu b_{n-1} + \beta_\nu b_n$,

(v) $[b_2, b_{n-1}]_\nu = \gamma_\nu b_n$,

(vi) $[b_1, b_2]_\nu + \alpha_\nu b_2 \in \mathfrak{k}$.

Here (vi) follows from (iv) and (v) via the Jacobi identity. The α_ν, β_ν and γ_ν in (iv), (v) and (vi) are now real constants with

(8) $$\alpha_\nu\gamma_\nu \neq 0 \text{ for } \nu \in \mathbb{N}\,,\ \alpha_\infty = \lim_{\iota\to\infty} \alpha_\iota\,,\ \beta_\infty = \lim_{\iota\to\infty} \beta_\iota\,,\ \gamma_\infty = \lim_{\iota\to\infty} \gamma_\iota$$

Let

$$u_\nu = \left(\alpha_\nu^2 + \beta_\nu^2\right)^{-\frac{1}{2}} (\alpha_\nu b_{n-1} + \beta_\nu b_n) \in \mathfrak{y}\,.$$

It is easy to see that for $\nu \in \mathbb{N}$ and $t = -(\alpha_\nu\gamma_\nu\lambda_\nu)^{-1}(\alpha_\nu^2 + \beta_\nu^2)^{\frac{1}{2}} l_\nu(u_\nu)$ we have $l_\nu(\mathrm{Ad}_\nu \exp_\nu tb_2(u_\nu)) = 0$, hence taking $\mathrm{Ad}_\nu^* \exp_\nu(-tb_2)(l_\nu) \in \Omega_{(\nu,l_\nu)}$ instead of l_ν we can assume that $[b_1, u_\nu] = \alpha_\nu u_\nu$ and

(9) $$l_\nu(u_\nu) = 0 \quad \text{for all} \quad \nu \in \mathbb{N}\,.$$

As $\mathfrak{k}$ is an ideal in all ${}_\nu\mathfrak{g}$ it follows that

$$\mathfrak{h} = \mathbb{R}b_1 \oplus \mathfrak{k} = \sum_{j\neq 2} \mathbb{R}b_j$$

is a subalgebra ${}_\nu\mathfrak{h}$ in every ${}_\nu\mathfrak{g}$, $\nu \in \overline{\mathbb{N}}$. Thus $\mathbf{h} = \overline{\mathbb{N}} \times \mathfrak{h}$ is a variable subalgebra of $\mathbf{g}$ and $E(\mathbf{h}) = \overline{\mathbb{N}} \times H$ with $H = \{x \in \mathbb{R}^n\,;\, x_2 = 0\}$ is the corresponding variable subgroup of $\mathbf{G}$. Let

$$l'_\nu = l_\nu|_{\mathfrak{h}}\,,\ \tau_\nu = K\left(\Omega_{(\nu,l'_\nu)}\right) \in \widehat{\mathbf{H}}\,.$$

We claim that

(10) $$\pi_\nu = \mathrm{ind}_{{}_\nu H}^{{}_\nu G} \tau_\nu \quad \text{for all} \quad \nu \in \mathbb{N}\,.$$

To prove (10) it suffices to show that ${}_\nu\mathfrak{h}$ contains a polarization with Pukanszky condition for all $\nu \in \mathbb{N}$. For this we choose a "good sequence" $\{\mathfrak{a}_i\}_{i=1}^n$ of subalgebras

$\mathfrak{a}_i \subset {}_\nu\mathfrak{g}$ containing the ideals $\mathbb{R}b_2 \oplus \mathfrak{k}$, $\mathfrak{k}$, $\mathfrak{y}$ and form the Vergne polarization

$$\mathfrak{p}_\nu = \sum_{i=1}^{n} \mathfrak{a}_i^{l_{\nu,i}} \ , \ l_{\nu,i} = l_\nu|_{\mathfrak{a}_i} .$$

Then $u_\nu \in \mathfrak{p}_\nu$ for all $\nu \in \mathbb{N}$. Furthermore, as ${}_\nu\mathfrak{g}^{l_\nu}$ and a fortiori $\mathfrak{p}_\nu$ lies in no proper ideal it follows that $\mathfrak{p}_\nu$ contains an element $w = b_1 + \xi b_2 + c$ with $c \in \mathfrak{k}$. Then

$$[w, u_\nu]_\nu = \alpha_\nu u_\nu + \xi\gamma_\nu\alpha_\nu(\alpha_\nu^2 + \beta_\nu^2)^{-\frac{1}{2}} b_n \in \mathfrak{p}'_\nu ,$$

hence $l_\nu([w, u_\nu]_\nu) = \xi\alpha_\nu\gamma_\nu\lambda_\nu(\alpha_\nu^2 + \beta_\nu^2)^{-\frac{1}{2}} = 0$, i.e. $\xi = 0$.

The same argument shows that $\vartheta_1 b_1 + \vartheta_2 b_2 + c \in \mathfrak{p}^\nu$, $c \in \mathfrak{k}$, implies $\vartheta_2 = 0$. Hence

$$\mathfrak{p}_\nu \subset \mathfrak{p}_\nu + \mathfrak{k} = \mathbb{R}b_1 \oplus \mathfrak{k} = \mathfrak{h}$$

which proves (10). Finally we have

$$(11) \qquad l_\nu + \mathfrak{h}^\perp \subset \Omega_{(\nu, l_\nu)} \text{ for all } \nu \in \mathbb{N} .$$

This is clear because $\mathfrak{p}_\nu$ satisfies Pukanszky's condition, hence $l_\nu + \mathfrak{h}^\perp \subset l_\nu + \mathfrak{p}_\nu^\perp \subset \Omega_{(\nu, l_\nu)}$.

Summing up we see that we have to study the following situation: $\mathbf{g} = \overline{\mathbb{N}} \times \mathfrak{g}$ contains the variable subalgebra $\mathbf{h} = \overline{\mathbb{N}} \times \mathfrak{h}$, where $\mathfrak{h} = \mathbb{R}b_1 \oplus \mathfrak{k}$, $\mathfrak{k} = \sum_3^n \mathbb{R}b_j$, and $\mathfrak{k}$ is an ideal in all ${}_\nu\mathfrak{g}$, $\nu \in \overline{\mathbb{N}}$. Correspondingly we have the variable group $\mathbf{G} = \overline{\mathbb{N}} \times G$ with subgroup $\mathbf{H} = \overline{\mathbb{N}} \times H$. Here we can identify G with $\mathbb{R}^n$ (via canonical coordinates of the second kind with respect to the Malcev basis $\{b_j\}$ of $\mathfrak{g}$) and H with the subspace of all $x \in \mathbb{R}^n$ with $x_2 = 0$.

Then a sequence $\{l_\nu\}_{\nu \in \overline{\mathbb{N}}} \subset \mathfrak{g}^*$ is given with $l'_\nu = l_\nu|_\mathfrak{h} \in \mathfrak{h}^*$ and the following properties: $l_\nu(b_n) = \lambda_\nu \neq 0$ for $\nu < \infty$, $\lambda_\infty = \lim\limits_{\nu \to \infty} \lambda_\nu$, $l_\nu(u_\nu) = 0$ for the elements $u_\nu \in \mathfrak{y}$ defined above, and $l_\nu + \mathfrak{h}^\perp \subset \Omega_{(\nu, l_\nu)}$ for $\nu < \infty$.

If we define

$$\Omega_\nu := \Omega_{(\nu, l_\nu)} \in \Omega(\mathbf{G}) \ , \ \Omega'_\nu := \Omega_{(\nu, l'_\nu)} \in \Omega(\mathbf{H}) ,$$

$$\pi_\nu = K(\Omega_\nu) \in \widehat{\mathbf{G}} \ , \ \tau_\nu = K(\Omega'_\nu) \in \widehat{\mathbf{H}} ,$$

then $\lim\limits_{\nu \to \infty} \pi_\nu = \pi_\infty$ and $\pi_\nu = \operatorname{ind}_{{}_\nu H}^{{}_\nu G} \tau_\nu$ for all finite ν.

We have to show that for suitably chosen l_ν in Ω_ν and subsequence $\{\nu'\} \subset \mathbb{N}$ we have $\lim\limits_{\nu' \to \infty} l'_{\nu'} = l'_\infty$. If we take the new l_ν in the same $\mathrm{Ad}^*_\nu H$ orbit, then (9) and (11) still are true and $\lim \Omega_{\nu'} = \Omega_\infty$ follows.

Let $\mathfrak{H}_\nu$ be the representation space of π_ν. Then $\pi_\infty = \lim \pi_\nu$ means that for $\omega_\infty \in \mathfrak{H}_\infty$ with $|\omega_\infty| = 1$ there exist a sequence of unit vectors $\omega_\nu \in \mathfrak{H}_\nu$, $\nu \in \mathbb{N}$, such that

$$\lim_{\nu\to\infty}(\pi_\nu(x)\omega_\nu|\omega_\nu)_\nu = (\pi_\infty(x)\omega_\infty|\omega_\infty)_\infty$$

for all $x \in G$ uniformly on compacta.

To compute the coefficients $(\xi_\nu|\vartheta_\nu)_\nu$ for $\xi_\nu, \vartheta_\nu \in \mathfrak{H}_\nu$ we use the realization of $\mathfrak{H}_\nu$ described in § 1. We introduce the notation

$$e_j^t := E(\nu, tb_j) = (0,\dots,0,\underset{j}{t},0,\dots,0) \in G$$

for $j = 1, 2, \dots, n$, $t \in \mathbb{R}$, thus $e_j^{\mathbb{R}} = \{e_j^t; t \in \mathbb{R}\}$ is the one parameter subgroup of ${}_\nu G$ generated by $b_j \in \mathfrak{g}$. Then ${}_\nu G = e_2^{\mathbb{R}}{}_\nu H$ and $\{b_2\}$ is a coexponential basis for ${}_\nu\mathfrak{h}$ in ${}_\nu\mathfrak{g}$ for all $\nu \in \overline{\mathbb{N}}$. From § 1 we derive that $(\xi|\vartheta)_\nu = \int_{\mathbb{R}}(\xi(e_2^t)|\vartheta(e_2^t))_{\nu'}dt$ and in particular for $\xi = \pi_\nu(x)\omega_\nu$, $\vartheta = \omega_\nu$ we get

$$\varphi_\nu(x) := (\pi_\nu(x)\omega_\nu|\omega_\nu) = \int_{\mathbb{R}}(\omega_\nu(x^{-1}\underset{\nu}{\cdot} e_2^t)|\omega_\nu(e_2^t))_\nu\, dt \tag{12}$$

for the positive definite function φ_ν.

For our purposes we have to select the sequence $\{\varphi_\nu\}_{\nu\in\mathbb{N}}$ more carefully. Let $\mathbf{L}$ be the variable group algebra of $\mathbf{G}$, i.e. $\mathbf{L} = \overline{\mathbb{N}} \times L^1(G)$ as defined before and let $\mathbf{C}_0^\infty$ be the subalgebra of compactly supported smooth functions in $\mathbf{L}$, i.e. $f : \overline{\mathbb{N}} \times G \to \mathbb{C}$ with compact support and such that all $f(\nu) : x \to f(\nu, x)$ are smooth.

For $\xi, \vartheta \in \mathfrak{H}_\nu$ we define the continuous functions $\langle\xi,\vartheta\rangle_\nu$ on G by

$$\langle\xi,\vartheta\rangle_\nu(x) = (\pi_\nu(x)\xi|\vartheta)_\nu\,,$$

thus $\varphi_\nu = \langle\omega_\nu, \omega_\nu\rangle_\nu$. Furthermore, for $f \in \mathbf{C}_0^\infty$ (or more generally in $\mathbf{C}_0$) we set

$$\langle\xi,\vartheta\rangle'_\nu(f) = \int f(\nu,x)\langle\xi,\vartheta\rangle_\nu(x)dx = (\pi_\nu(f)\xi|\vartheta)_\nu\,.$$

From theorem 1 in ch. II we see that for a bounded sequence $\{\xi_\nu\}_\nu$, $\xi_\nu \in \mathfrak{H}_\nu$, the sequence $\{\langle\xi_\nu,\xi_\nu\rangle_\nu\}_{\nu\in\mathbb{N}}$ converges uniformly on compacta to $\langle\xi_\infty,\xi_\infty\rangle_\infty$ if and only if $\{\langle\xi_\nu,\xi_\nu\rangle'_\nu\}$ converges weak* to $\langle\xi_\infty,\xi_\infty\rangle'_\infty$.

Let us now start with a sequence $\{\xi_\nu\}_{\nu\in\overline{\mathbb{N}}}$ of unit vectors such that $\lim\langle\xi_\nu,\xi_\nu\rangle_\nu = \langle\xi_\infty,\xi_\infty\rangle_\infty$ uniformly on compacta. Here we may assume that for finite ν all ξ_ν have supports which are compact modulo H. This is equivalent to the fact that all the mappings $t \to \xi_\nu(e_2^t)$ from $\mathbb{R}$ into $\mathfrak{J}_\nu$, the space of τ_ν, have compact supports.

Let $h, g \in \mathbf{C}_0^\infty$. Then we have for every $f \in \mathbf{C}_0^\infty$:

$$\begin{aligned}(13)\qquad \lim_{\nu\to\infty} \langle \pi_\nu(h)\xi_\nu, \pi_\nu(h)\xi_\nu\rangle'_\nu(f) &= \lim_{\nu\to\infty} \langle \xi_\nu, \xi_\nu\rangle'_\nu(h^* \underset{\nu}{\star} f \underset{\nu}{\star} b) = \\ &= \langle \xi_\infty, \xi_\infty\rangle'_\infty(h^* \underset{\infty}{\star} f \underset{\infty}{\star} h) = \\ &= \langle \pi_\infty(h)\xi_\infty, \pi_\infty(h)\xi_\infty\rangle'(f)\ ,\end{aligned}$$

$$(14)\qquad \lim_{\nu\to\infty} |\pi_\nu(h)\xi_\nu|^2 = \lim_{\nu\to\infty} \langle \xi_\nu, \xi_\nu\rangle'_\nu(h^* \underset{\nu}{\star} h) = |\pi_\infty(h)\xi_\infty|^2\ ,$$

and consequently $w^*\ \lim\langle \pi_\nu(h)\xi_\nu, \pi_\nu(h)\xi_\nu\rangle'_\nu = \langle \pi_\infty(h)\xi_\infty, \pi_\infty(h)\xi_\infty\rangle'_\infty$, similarly with g.

The usual polarization argument together with theorem 1, ch. II, yields: For all $g, h \in \mathbf{C}_0^\infty$ we have

$$w^* \lim_{\nu\to\infty} \langle \pi_\nu(h)\xi_\nu, \pi_\nu(g)\xi_\nu\rangle'_\nu = \langle \pi_\infty(h)\xi_\infty, \pi_\infty(g)\xi_\infty\rangle'_\infty$$

$$\lim_{\nu\to\infty} \langle \pi_\nu(h)\xi_\nu, \pi_\nu(g)\rangle_\nu = \langle \pi_\infty(h)\xi_\infty, \pi_\infty(g)\xi_\infty\rangle_\infty \text{ uniformly on compacta}\quad .$$

The underlying space $\mathfrak{g}$ of the VLA $\mathbf{g}$ acts from both sides on the module $\mathbf{C}^\infty$ of all continuous functions on $\overline{\mathbb{N}} \times G$, which are smooth in $x \in G$: For $a \in \mathfrak{g}$ we define the (variable) left and right derivations L_a, R_a of $\varphi \in \mathbf{C}^\infty$ by

$$\begin{aligned}(L_a\varphi)(\nu, x) &= \frac{d}{dt}\varphi(\nu \exp_\nu(-ta) \underset{\nu}{\cdot} x)|_{t=0} \\ (R_a\varphi)(\nu, x) &= \frac{d}{dt}\varphi(\nu, x \underset{\nu}{\cdot} \exp_\nu ta)|_{t=0}\end{aligned}$$

Clearly the subspace $\mathbf{C}_0^\infty$ is stable under $\mathfrak{g}$, moreover for $g, h \in \mathbf{C}_0^\infty$ the functions $\langle \pi_\nu(g)\xi_\nu, \pi_\nu(h)\xi_\nu\rangle_\nu$ are smooth and define a function $\langle \pi(g)\xi, \pi(h)\xi\rangle$ in $\mathbf{C}^\infty$ for which we get

$$\begin{aligned}(15)\qquad & L_a\langle \pi(g)\xi, \pi(h)\xi\rangle(\nu, x) = \\ &= \frac{d}{dt}(\pi_\nu(\exp_\nu(-ta) \underset{\nu}{\cdot} x)\pi_\nu(g)\xi_\nu|\pi_\nu(b)\xi_\nu)_\nu|_{t=0} = \\ &= \frac{d}{dt}(\pi_\nu(x)\pi_\nu(g)\xi_\nu|\pi_\nu(\lambda_{\exp_\nu ta}h)\xi_\nu)_\nu|_{t=0} = \\ &= (\pi_\nu(x)\pi_\nu(g)\xi_\nu|\pi_\nu(L_a h)\xi_\nu)_\nu\end{aligned}$$

with the left translation operator $\lambda_z h(\nu, z) = h(\nu, z^{-1} \underset{\nu}{\cdot} x)$. Similarly for the right derivations

$$R_a\langle \pi(g)\xi, \pi(h)\xi\rangle(\nu, x) = (\pi_\nu(x)\pi_\nu(L_a g)\xi_\nu|\pi_\nu(h)\xi_\nu)_\nu$$

Now we fix an h in $\mathbf{C}_0^\infty$ with $|\pi_\infty(h)\xi_\infty| = 1$. Because of (14) we can also assume that all

$$\omega_\nu = \pi_\nu(h)\xi_\nu$$

have norm 1 in $\mathfrak{H}_\nu$, $\nu \in \mathbb{N}$. We claim:

(16) *The ω_ν for $\nu \in \mathbb{N}$ have compact supports modulo H and the smooth, positive definite functions $\varphi_\nu = \langle \omega_\nu, \omega_\nu \rangle_\nu$ converge with all their partial derivatives uniformly on compacta to φ_∞.*

Proof. Let $Q \subset G$ be such that $Q \underset{\nu}{\cdot} H = Q$, Q/H is compact in ${}_\nu G/H$ and $\operatorname{supp} \xi(\nu, \cdot) \subset Q$. Let $K \subset G$ be compact with $\operatorname{supp} h(\nu, \cdot) \subset K$. Then

$$\begin{aligned}\omega_\nu(x) = \pi_\nu(h)\xi_\nu(x) &= \int_G h(\nu, y)\xi_\nu(y^{-1} \underset{\nu}{\cdot} x)dy = \\ &= \int_G h(\nu, x \underset{\nu}{\cdot} y)\xi_\nu(y^{-1})dy = \int_Q h(\nu, x \underset{\nu}{\cdot} y^{-1})\Delta_\nu(y)^{-1}\xi_\nu(y)dy\end{aligned}$$

If $x \notin KQ$, then $y \in Q$ implies $x \underset{\nu}{\cdot} y^{-1} \notin K$, hence $h(\nu, x \underset{\nu}{\cdot} y^{-1}) = 0$. Thus $\operatorname{supp} \omega_\nu \in KQ$, which proves the first part of (16). Obviously all φ_ν are smooth, positive definite functions on $\mathbf{G}$ with $\varphi_\nu(e) = 1$ for the common unit element $e \in {}_\nu G$.

Let D be a polynomial in the differential operators L_a, $a \in \mathfrak{g}$. Then it follows from (15) that

$$D\varphi_\nu = \langle \omega_\nu, \pi_\nu(Dh)\xi_\nu \rangle_\nu \; .$$

As we have seen this implies $\lim\limits_{\nu \to \infty} D\varphi_\nu = D\varphi_\infty$ uniformly on compacta.

Because the partial differential operators $\frac{\partial^m}{\partial x_1^{m_1} \ldots \partial x_n^{m_n}}$, $m = \sum m_j$, are linear combinations of the D's with coefficients which are continuous on $\overline{\mathbb{N}} \times G$ and smooth on G, the second statement of (16) follows.

Another simplification concerns the $\lambda_\nu = l_\nu(b_n)$. If $\lambda_\infty \neq 0$ then it is easy to see that there is an equivalence from $\mathbf{G}$ onto some $'\mathbf{G}$, for which all $'\lambda_\nu = 1$. We will show that also for $\lambda_\infty = 0$ we can shift to another VLG $'\mathbf{G}$, a convergent sequence $\{'\pi_\nu\} \subset {}'\widehat{\mathbf{G}}$ with a corresponding sequence $\{'l_\nu\} \subset {}'\mathfrak{g}$ with $'l_\nu('b_n) = 1$ for all $\nu \in \overline{\mathbb{N}}$ such that $\lim\limits_{\nu \to \infty} {}'l_\nu = {}' l_\infty$ implies $\lim\limits_{\nu \to \infty} l_\nu = l_\infty$.

We start with the definition of the VLA $'\mathbf{g} = \overline{\mathbb{N}} \times {}'\mathfrak{g}$. Let $\{c_{k,l}^j(\nu)\}$ be the structure constants of $\mathbf{g}$, i.e. $[b_j, b_k]_\nu = \sum\limits_l c_{k,l}^j(\nu) b_l$, $\nu \in \overline{\mathbb{N}}$. We take a basis $\{'b_j\}$ of $'\mathfrak{g}$ and define for $\nu \in \overline{\mathbb{N}}$:

$$'c_{k,l}^j(\nu) = c_{k,l}^j(\nu) \;\text{ for }\; l < n \, , \; 'c_{k,n}^j(\nu) = c_{k,n}^j(\nu)\lambda_\nu$$

$$['b_j, 'b_k]_\nu = \sum_l {}'c^j_{k,l}(\nu)\, 'b_l \ .$$

Thus $'\mathbf{g} = (\overline{\mathbb{N}} \times '\mathfrak{g}, [\ ,\ ]_\nu)$ is a VLA. Next we define a homomorphism $dT : \mathbf{g} \to '\mathbf{g}$ by

$$dT(\nu, b_j) = (\nu, b_j)\,,\ j < n\,, \quad dT(\nu, b_n) = (\nu, \lambda_\nu\, 'b_n)\ .$$

If $\lambda_\infty \neq 0$, then evidently dT is an equivalence from $\mathbf{g}$ onto $'\mathbf{g}$. If $\lambda_\infty = 0$, then

$$_\infty{}'\mathfrak{g} = \mathfrak{f} \oplus \mathbb{R}\, 'b_n \quad \text{with} \quad \mathfrak{f} = \sum_{j=1}^{n-1} \mathbb{R}\, 'b_j\ ,$$

i.e. $_\infty{}'\mathfrak{g}$ is the direct sum of the ideals $\mathfrak{f}$ and $\mathbb{R}'b_n$. In this case dT defines a homomorphism from $_\infty\mathfrak{g}$ onto $\mathfrak{f}$. If $\{b_j\}_j$ is a Malcev basis for $\mathbf{g}$ (i.e. for all $_\nu\mathfrak{g}$), then $\{'b_j\}_j$ is one for $'\mathbf{g}$, hence the VLG $'\mathbf{G}$ with $l('\mathbf{G}) = {}'\mathbf{g}$ exists and we also have a homomorphism $T : \mathbf{G} \to '\mathbf{G}$ such that

$$\begin{array}{ccc} \mathbf{g} & \xrightarrow{dT} & '\mathbf{g} \\ {\scriptstyle E}\downarrow & & \downarrow{\scriptstyle E} \\ \mathbf{G} & \xrightarrow{T} & '\mathbf{G} \end{array}$$

commutes. As $T(\exp_\nu sb_j) = \exp_\nu sdT('b_j)$ we see, using as before canonical coordinates of the second kind, that we have

$$T_\nu(x_1, x_2, \ldots, x_n) = (x_1, x_2, \ldots, \lambda_\nu x_n)$$

where T_ν is the homomorphism induced on $_\nu G$. Again for all $\nu \in \mathbb{N}$ the T_ν are isomorphisms. For all $\nu \in \overline{\mathbb{N}}$ with $\lambda_\nu \neq 0$, thus in particular for all $\nu \in \mathbb{N}$, we define $'\pi_\nu \in {}_\nu{}'\widehat{G}$ by

$$'\pi_\nu(x) = \pi_\nu(T_\nu^{-1} x)$$

i.e. in coordinates $'\pi_\nu(x_1, \ldots, x_n) = \pi_\nu(x_1, \ldots, \lambda_\nu^{-1} x_n)$. If we define $'l_\nu \in {}'\mathfrak{g}^*$ for $\lambda_\nu \neq 0$ by

$$'l_\nu('b_j) = l_\nu(b_j) \quad \text{for} \quad j < n\,,\ 'l_\nu('b_n) = 1\ ,$$

then obviously $l_\nu = (dT_\nu)^*('l_\nu)$ and $'\pi_\nu = K(\Omega_{(\nu, 'l_\nu)})$. Now assume that $\lambda_\infty = 0$, hence $_\infty{}'G = F \times \mathbb{R}$ is the direct product of $F = \exp_\infty \mathfrak{f}$ and $\mathbb{R} (= \exp_\infty \mathbb{R}\, 'b_n)$ In this case we define $'\pi_\infty$ for $x = (x_1, x_2, \ldots, x_n) \in \mathbb{R}^n$ by

$$'\pi_\infty(x) = \pi_\infty(x) e^{ix_n}\ .$$

Here the x on the left hand side is understood as an element in $_\infty{}'G = \mathbb{R}^n$ in our previous canonical coordinates and the right hand x similarly as element in

${}_\infty G = \mathbb{R}^n$. Observe that now π_∞ is trivial on $\exp_\infty \mathbb{R}b_n$, hence it lives on the quotient ${}_\infty G/\exp_\infty \mathbb{R}b_n$, i.e. essentially on F. We leave it as an easy exercise to the reader that

$$'\pi_\infty = K(\Omega_{(\infty,'l_\infty)}) \text{ with } 'l_\infty('b_j) = l_\infty(b_j) \ , \ j < n \, , \ 'l_\infty('b_n) = 1 \ .$$

For $x \in \mathbb{R}^n$ let $\tilde{x} = (x_1, x_2, \dots, x_{n-1}, 0)$, thus $x = \tilde{x} \cdot e_n^{x_n}$ with $e_n^{x_n} = (0, 0, \dots, 0, x_n)$ in every ${}_\nu G$ and ${}_\nu{}'G$. From $\pi_\nu(x) = \pi_\nu(\tilde{x})e^{i\lambda_\nu x_n}$, $'\pi_\nu(x) = {}'\pi_\nu(\tilde{x})e^{ix_n}$ for all $\nu \in \overline{\mathbb{N}}$, it follows easily that for the positive definite functions φ_ν resp. $'\varphi_\nu(x) = ('\pi_\nu(x)\omega_\nu|\omega_\nu)_\nu$, we have

$$'\varphi_\nu(x) = \varphi_\nu(\tilde{x})e^{ix_n} \text{ for all } \nu \in \overline{\mathbb{N}} \ .$$

This implies $\lim\limits_{\nu\to\infty} {}'\varphi_\nu = {}'\varphi_\infty$ uniformly on compacta, hence $\lim\limits_{\nu\to\infty} {}'\pi_\nu = {}'\pi_\infty$. Now assume that this implies $\lim\limits_{\nu\to\infty} {}'l_\nu = {}'l_\infty$ for a suitably selected sequence of $'l_\nu$ in their orbits, $\nu \in \overline{\mathbb{N}}$. As the dT_ν are equivariant, we see that for the correspondingly selected l_ν we also have $'l_\nu('b_j) = l_\nu(b_j)$ for $j < n$, hence $\lim\limits_{\nu\to\infty} l_\nu(b_j) = l_\infty(b_j)$ for $j < n$. From $l_\nu(b_n) = \lambda_\nu \to 0 = l_\infty(b_n)$ it follows that also $\lim\limits_{\nu\to\infty} l_\nu = l_\infty$, proving our claim, that it suffices to treat only the case where $l_\nu(b_n) = 1$ for all $\nu \in \overline{\mathbb{N}}$. In the sequel we make this assumption.

The restrictions of the φ_ν onto ${}_\nu H$, $\nu \in \mathbb{N}$, are integrals over the functions $\tilde{\psi}_{\nu,t} : y \to (\omega_\nu(y^{-1} \cdot e_2^t)|\omega_\nu(e_2^t))_\nu$, which unfortunately are not positive definite. However, for $x = e_2^t \underset{\nu}{\cdot} y \underset{\nu}{\cdot} e_2^{-t} \in e_2^t \cdot {}_\nu H \cdot e_2^{-t}$, i.e. $y \in {}_\nu H$, one has $\tilde{\psi}_{\nu,t}(x) = (\omega_\nu(e_2^t \underset{\nu}{\cdot} y^{-1})|\omega_\nu(e_2^t))_\nu = (\tau_\nu(y)\omega_\nu(e_2^t)|\omega_\nu(e_2^t))_\nu$ thus the $\tilde{\psi}_{\nu,t}$ are positive definite on $e_2^t \cdot_\nu H \cdot e_2^{-t}$. This suggests considering the (variable) subgroup generated by these $e_2^t \cdot {}_\nu H \cdot e_2^{-t}$, but necessarily with a parameter space, which respects also the real parameter t. We start with the following observation: For any $\nu \in \mathbb{N}$ we have in ${}_\nu\mathfrak{g}$

$$\mathrm{Ad}_\nu e_2^t(b_1) \equiv b_1 + \alpha_\nu t b_2 \pmod{\mathfrak{k}} \ .$$

As $\mathfrak{k}$ is $\mathrm{Ad}_\nu G$ invariant this implies

$$\mathrm{Ad}_\nu e_2^t(\mathfrak{h}) = \mathbb{R}(b_1 + \alpha_\nu t b_2) \oplus \mathfrak{k} \tag{17}$$

and consequently

$$e_2^t \cdot {}_\nu H \cdot e_2^{-t} = \exp_\nu \mathbb{R}(b_1 + \alpha_\nu t b_2)_\nu \underset{\nu}{\cdot} K \ .$$

We also note that for $u_\nu = (\alpha_\nu^2 + \beta_\nu^2)^{-\frac{1}{2}}(\alpha_\nu b_{n-1} + \beta_\nu b_n)$ and $\nu \in \mathbb{N}$ we have

$$\mathrm{ad}_\nu b_2(u_\nu) = \eta_\nu b_n \, , \ e_2^t \underset{\nu}{\cdot} \exp_\nu \vartheta u_\nu = e_n^{t\vartheta\eta_\nu} \cdot \exp_\nu \vartheta u_\nu \cdot e_2^t \tag{18}$$

$$\text{with} \quad \eta_\nu := (\alpha_\nu^2 + \beta_\nu^2)^{-\frac{1}{2}} \alpha_\nu \gamma_\nu\,, \quad \text{all} \quad t, \vartheta \in \mathbb{R}\,.$$

From (18), the invariance properties of ω_ν and $\tau_\nu(\exp_\nu \vartheta u_\nu) = \mathbf{1}$ we derive

$$\omega_\nu(x \underset{\nu}{\cdot} e_2^t) e^{is} = \omega_\nu(x \cdot e_n^{t\vartheta\eta_\nu - s} \exp_\nu \vartheta u_\nu e_2^t)$$

for all $x \in {}_\nu G$, $t, s, \vartheta \in \mathbb{R}$. With substitutions $t \to \eta_\nu^{-1} t$, $s \to tw$, $\vartheta \to w$ we get

$$\omega_\nu(x \underset{\nu}{\cdot} e_2^{\eta_\nu^{-1} t}) e^{itw} = \omega_\nu(x \underset{\nu}{\cdot} \exp_\nu w u_\nu \underset{\nu}{\cdot} e_2^{\eta_\nu^{-1} t})\,. \tag{19}$$

The equations (18) suggest considering the subalgebras

$$\operatorname{Ad}_\nu e_2^{\eta_\nu^{-1} t}({}_\nu\mathfrak{h}) := \mathfrak{h}_{\nu,t} \subset {}_\nu\mathfrak{g} \tag{20}$$

which, after (17), are spanned by the basis elements $b_1 + \frac{\alpha_\nu}{\eta_\nu} t b_2, b_3, \ldots, b_n$. The subsequence argument shows that we have to consider two cases:

$$\varepsilon_\infty := \lim_{\nu \to \infty} \frac{\alpha_\nu}{\eta_\nu} \quad \text{exists} \tag{α}$$

$$\lim \frac{\eta_\nu}{\alpha_\nu} = 0\,. \tag{β}$$

Accordingly we define a new $(n-1)$ dimensional VLA $\mathbf{h}^\bullet$: Let $\{b_j^\bullet\}_{j=1,\ldots,n-1}$ be a fixed basis of the vector space $\mathfrak{h}^\bullet$ and put $\varepsilon_\nu = \frac{\alpha_\nu}{\eta_\nu}$ for $\nu \in \mathbb{N}$, hence $\lim \varepsilon_\nu = \varepsilon_\infty$ in case α and $\lim \varepsilon_\nu^{-1} = 0$ in case β. Now define

$$P = \overline{\mathbb{N}} \times \mathbb{R} \quad \text{in case} \quad \alpha\,, \quad P = \overline{\mathbb{N}} \times \mathbb{R} \setminus \{(\infty, 0)\} \quad \text{in case} \quad \beta\,.$$

Based on these parameter spaces we define $\mathbf{h}^\bullet$ as follows:
For $(\nu, t) \in P$ let

$$d\Gamma_{\nu,t}(b_j^\bullet) = \begin{cases} b_1 + \varepsilon_\nu t b_2 & \text{in case } \alpha, \qquad\qquad j = 1 \\ \varepsilon_\nu^{-1} b_1 + t b_2 & \text{in case } \beta, \quad \nu < \infty,\ j = 1 \\ t b_2 & \text{in case } \beta, \quad \nu = \infty,\ j = 1 \\ b_{j+1} & \text{in both cases, all } \nu, \quad 2 \le j \le n-1\,. \end{cases}$$

Then $d\Gamma_{\nu,t}$ extends to a linear isomorphism of $\mathfrak{h}^\bullet$ onto a subalgebra $\mathfrak{h}_{\nu,t}$ of ${}_\nu\mathfrak{g}$ for all $(\nu, t) \in P$. For $\nu < \infty$ this $\mathfrak{h}_{\nu,t}$ is the same as in (20). With

$$[x, y]_{\nu,t} := d\Gamma_{\nu,t}^{-1}([d\Gamma_{\nu,t}(x)\,, d\Gamma_{\nu,t}(y)]_\nu)$$

a Lie product is defined in $\mathfrak{h}^\bullet$, which allows us to define the VLA

$$\mathbf{h}^\bullet = (P \times \mathfrak{h}^\bullet\,, [\,,\,]_{\nu,t})\,.$$

By $d\Gamma((\nu, t), x) = (\nu, d\Gamma_{\nu,t}(x))$ this $\mathbf{h}^\bullet$ is homomorphically mapped into $\mathbf{g}$. If $\{b_j\}_j$ is a Malcev basis in all ${}_\nu\mathfrak{g}$, then the same holds for $\{b_\iota^\bullet\}$ in the ${}_{\nu,t}\mathfrak{h}^\bullet$, hence $\mathbf{h}^\bullet$ is the VLA of a VLG $\mathbf{H}^\bullet = (P \times H^\bullet, \mu^\bullet)$ with $H^\bullet = \mathbb{R}^{n-1}$ and we have a homomorphism $\Gamma : \mathbf{H}^\bullet \to \mathbf{G}$ with a commutative diagram

$$\begin{array}{ccc} \mathbf{h}^\bullet & \xrightarrow{d\Gamma} & \mathbf{g} \\ {\scriptstyle E}\downarrow & & \downarrow{\scriptstyle E} \\ \mathbf{H}^\bullet & \xrightarrow{\Gamma} & \mathbf{G} \end{array}$$

In the usual coordinates in $H^\bullet$, resp. G, defined by the bases $\{b_j^\bullet\}$ and $\{b_j\}$ we have for $y = (y_1, \ldots, y_{n-1}) \in H^\bullet$ and the homomorphism $\Gamma_{\nu,t}$ from ${}_{\nu,t}H^\bullet$ into ${}_\nu G$:

$$\Gamma_{\nu,t}(y) = \begin{cases} \exp_\nu(y_1 b_1 + \varepsilon_\nu y_1 t b_2)\tilde{y} & \text{in case } \alpha \\ \exp_\nu(\varepsilon_\nu^{-1} y_1 b_1 + y_1 t b_2)\tilde{y} & \text{in case } \beta, \quad \nu < \infty \\ e_2^{y_1 t}\tilde{y} & \text{in case } \beta, \quad \nu = \infty \end{cases}$$

with $\tilde{y} = (0, 0, y_2, y_3, \ldots, y_{n-1}) \in {}_\nu G = \mathbb{R}^n$.

The mapping $y \to e_2^{-\eta_\nu^{-1}t} \cdot \Gamma_{\nu,t}(y) \cdot e_2^{\eta_\nu^{-1}t}$ for $\nu \in \mathbb{N}$ is an isomorphism from ${}_{\nu,t}H^\bullet$ onto ${}_\nu H$ in ${}_\nu G$, hence

$$\tau_{\nu,t}(y) = \tau_\nu(e_2^{-\eta_\nu^{-1}t}\Gamma_{\nu,t}(y)e_2^{\eta_\nu^{-1}t})$$

defines a unitary representation $\tau_{\nu,t}$ of ${}_{\nu,t}H^\bullet$, which clearly lies in $\widehat{\mathbf{H}^\bullet}$. A corresponding positive definite function is given by

$$\begin{aligned} \psi_{\nu,t}(y) &= |\eta_\nu|^{-1}(\tau_{\nu,t}(y)\omega_\nu(e_2^{\eta_\nu^{-1}t})|\omega_\nu(e_2^{\eta_\nu^{-1}t}))_\nu = \\ &= |\eta_\nu|^{-1}(\omega_\nu(\Gamma_{\nu,t}(y)^{-1}e_2^{\eta_\nu^{-1}t})|\omega_\nu(e_2^{\eta_\nu^{-1}t}))_\nu \ . \end{aligned}$$

As $\psi_{\nu,t}$ is continuous in t and $\psi_{\nu,t}(e) = |\eta_\nu|^{-1}|\omega_\nu(e_2^{\eta_\nu^{-1}t})|^2$, hence $\int_{\mathbb{R}} \psi_{\nu,t}(e)dt = \int_{\mathbb{R}} |\omega_\nu(e_2^t)|^2 dt = 1$, we can form the positive definite function $\psi_\nu := \int \psi_{\nu,t} dt$ on $\mathbf{H}^\bullet$. For the corresponding positive functional on $\mathbf{L}^\bullet = P \times L^1(H^\bullet)$, the variable group algebra of $\mathbf{H}^\bullet$, we obtain the formula

$$\text{(21)} \qquad \psi_\nu(f) = |\eta_\nu|^{-1} \int_{\mathbb{R}} \int_{H^\bullet} f(\nu, t; y)(\omega_\nu(\Gamma_{\nu,t}(y)^{-1}e_2^{\eta_\nu^{-1}t})|\omega_\nu(e_2^{\eta_\nu^{-1}t}))_\nu \, dydt$$

for $f \in \mathbf{C}_0^\infty(P \times H^\bullet)$. To connect this formula, resp. this ψ_ν, with the positive functional φ_ν of (12) we have to separate the variable t in the exponent of $e_2^{\eta_\nu^{-1}t}$ from the t in f and in $\Gamma_{\nu,t}$. We set

$$\text{(22)} \qquad F_\nu(s, t) = |\eta_\nu|^{-1} \int_{H^\bullet} f(\nu, s; y)(\omega_\nu(\Gamma_{\nu,s}(y)^{-1}e_2^{\eta_\nu^{-1}t})|\omega_\nu(e_2^{\eta_\nu^{-1}t}))_\nu dy \ .$$

The function F_ν is smooth and compactly supported in $\mathbb{R}^2$ and we have $\psi_\nu(f) = \int F_\nu(t,t)dt$. For any $\varphi \in \mathscr{C}_0^\infty(\mathbb{R})$ the properties of the Fourier transform $\varphi \to \widehat{\varphi}$, in particular the inversion formula, yield the formula

$$\varphi(s) = \frac{1}{2\pi}\int_{\mathbb{R}}\int_{\mathbb{R}}(1-\partial^2)\varphi(v)e^{ivw}\frac{e^{isw}}{1+w^2}dv\,dw$$

where ∂ means derivation. So if ∂_1 denotes the derivation of F_ν with respect to the first variable we derive the formula

$$F_\nu(s,t) = \frac{1}{2\pi}\int_{\mathbb{R}^2}[(1-\partial_1^2)F_\nu](v,t)e^{ivw}\frac{e^{isw}}{1+w^2}dv\,dw$$

hence (22) yields

$$(23)\quad \psi_\nu(f) = \frac{1}{2\pi|\eta_\nu|}\int_{\mathbb{R}^3}(1-\partial_1^2)\int_{H^\bullet} f(\nu,v;y)\times$$
$$\times\,(\omega_\nu(\Gamma_{\nu,v}(y)^{-1}e_2^{\eta_\nu^{-1}t})|\omega_\nu(e_2^{\eta_\nu^{-1}t}))_\nu\,dy\;e^{itw}\frac{e^{-ivw}}{1+w^2}dv\,dw\,dt\,.$$

Now we use (19) with $x = \Gamma_{\nu,w}(y)^{-1}$. Since the integrand in (23) as a function in (t,v,w) obviously is absolutely integrable uniformly in y we can first integrate over t. This gives us for all $\nu \in \mathbb{N}$

(24)

$$\psi_\nu(t) = \frac{1}{2\pi}\int_{\mathbb{R}^2}(1-\partial_1^2)\int_{H^\bullet} f(\nu,v;y)\varphi_\nu(\exp_\nu(-wu_\nu)\Gamma_{\nu,v}(y))dy\frac{e^{-ivw}}{1+w^2}dvdw\,.$$

For $\nu = \infty$ we take (24) as definition of a functional ψ_∞ on $\mathbf{C}_0^\infty(P\times H^\bullet)$. Now we can use (16): As the φ_ν, $\nu \in \mathbb{N}$, and all their derivatives converge uniformly on compacta we obtain $\lim\psi_\nu(f) = \psi_\infty(f)$ for all $f \in \mathbf{C}_0^\infty$ and (24) holds also for $\nu = \infty$. To analyze the situation we have to distinguish different cases. The simplest is the case in which

$$\varepsilon_\infty = \lim\frac{\alpha_\nu}{\eta_\nu}\quad\text{exists}\,,\quad \eta_\infty \neq 0\,.$$

Then we can define $\tau_{\nu,t}$, $\psi_{\nu,t}$ and ψ_ν as before for *all* $\nu \in \overline{\mathbb{N}}$ and from (21) we get

$$\lim\psi_\nu(f) = |\eta_\infty|^{-1}\int_{\mathbb{R}}\int_{H^\bullet} f(\infty,t;y)(\omega_\infty(\Gamma_{\infty,t}(y)e_2^{\eta_\infty^{-1}t})|\omega_\infty(e_2^{\eta_\infty^{-1}t}))_\infty dy\,dt =$$
$$= \int_{\mathbb{R}}\psi_{\infty,t}(f)dt = \psi_\infty(f)$$

for $f \in \mathbf{C}_0^\infty$, hence for all $f \in C^*(\mathbf{H}^\bullet)$.

Associated with the ψ_ν are the direct integrals $\overset{\bullet}{\tau}_\nu := \int_{\mathbb{R}}^{\oplus} \tau_{\nu,t} dt$ of the $\tau_{\nu,t} \in C^*(\mathbf{H}^\bullet)^\wedge = \widehat{\mathbf{H}^\bullet}$ and the hull $h(\overset{\bullet}{\tau}_\nu)$ of the kernel $\ker \overset{\bullet}{\tau}_\nu \subset C^*(\mathbf{H}^\bullet)$ is the closure $\{\tau_{\nu,t}\}_t^-$ of the set $\{\tau_{\nu,t}\}_t \subset \widehat{\mathbf{H}^\bullet}$. Then $w^* \lim_{\nu\to\infty} \psi_\nu = \psi_\infty$ implies $\ker \overset{\bullet}{\tau}_\infty \supset \bigcap_{\nu\geq k} \ker \overset{\bullet}{\tau}_\nu$ for every $k \in \mathbb{N}$, thus $h(\overset{\bullet}{\tau}_\infty) \subset (\bigcup_{\nu\geq k} h(\tau_\nu))^-$ and consequently

$$\tau_{\infty,t_0} \in (\bigcup_{\nu\geq k} \{\tau_{\nu,t}\}_t)^- = \{\tau_{\nu,t}; \nu \geq k, t \in \mathbb{R}\}^- .$$

It follows that for every $t_0 \in \mathbb{R}$ there exists a subsequence $\{\nu_j\}_j \subset \mathbb{N}$ and a sequence $\{t_j\}_j \subset \mathbb{R}$, with $\lim(\nu_j, t_j) = (\infty, t_0)$ in P, $\lim \tau_{\nu_j,t_j} = \tau_{\infty,t_0}$ in $\widehat{\mathbf{H}^\bullet}$. Now our induction hypothesis implies that we can choose linear forms $l_j^\bullet \in \mathfrak{h}^{\bullet *}$ with $K(\Omega_{(\nu_j,t_j,l_j^\bullet)}) = \tau_{\nu_j,t_j}$, resp. $K(\Omega_{(\infty,t_0;l_\infty^\bullet)}) = \tau_{\infty,t_0}$, and so that $\lim l_j^\bullet = l_\infty^\bullet$.

Recall the isomorphism $A_{\nu,t}$ from ${}_{\nu,t}H^\bullet$ onto the subgroup H of ${}_\nu G$, defined by

$$A_{\nu,t}(y) = e_2^{-\eta_\nu^{-1}t} \cdot \Gamma_{\nu,t}(y) \cdot e_2^{\eta_\nu^{-1}t}$$

and the corresponding algebra isomorphism $dA_{\nu,t}$ form ${}_{\nu,t}\mathfrak{h}^\bullet$ onto $\mathfrak{h} = \mathbb{R}b_1 \oplus \mathfrak{k} \subset \mathfrak{g}$. Because $\tau_{\nu,t} = \tau_\nu \circ A_{\nu,t}$ it follows that the adjoint isomorphism $(dA_{\nu,t})^*$ maps the orbit of τ_ν in $\mathfrak{h}^*$ onto the orbit of $\tau_{\nu,t}$ in $\mathfrak{h}^{\bullet *}$, in particular we can find l'_{ν_j} in $K^{-1}(\tau_{\nu_j})$ with $(dA_{\nu_j,t_j})^*(l'_{\nu_j}) = l_j^\bullet$, $1 \leq j \leq \infty$. Such a change will not affect (9) because $\mathbb{R}u_k$ is $\mathrm{Ad}_k H$ invariant for all $k \in \mathbb{N}$. Because of the continuity properties of the $A_{\nu,t}$, resp. $dA_{\nu,t}$, this implies $\lim_{j\to\infty} l'_{\nu_j} = l'_\infty$. As we have mentioned before this proves $\lim \Omega_{\nu,j} = \Omega_\infty$ for the orbits in $\Omega(\mathbf{G})$.

In our next case

$$\varepsilon_\infty = \lim_{\nu\to\infty} \frac{\alpha_\nu}{\eta_\nu} \quad \text{exists,} \quad \eta_\infty = 0$$

we see that we also must have $\alpha_\infty = \lim \alpha_\nu = 0$, hence $[b_1, b_2]_\infty \equiv 0 (\mathrm{mod}\ \mathfrak{k})$, i.e. ${}_\infty\mathfrak{g}/\mathfrak{k}$ is commutative and all $\mathfrak{h}_{\infty,t}$, in particular $\mathfrak{h} = \mathfrak{h}_{\infty,0}$, are ideals in ${}_\infty\mathfrak{g}$. Furthermore, it follows from $[b_1, u_\infty]_\infty = \alpha_\infty u_\infty = 0$ and $[b_2, u_\infty]_\infty = \eta_\infty b_n = 0$ that u_∞ is central in ${}_\infty\mathfrak{g}$, hence $\exp_\infty \mathbb{R}u_\infty$ is central in ${}_\infty G$ and consequently

$$\pi_\infty(\exp_\infty wu_\infty) = e^{iwt_0} \cdot \mathbf{1}$$

for some $t_0 \in \mathbb{R}$ and all $w \in \mathbb{R}$. This implies

$$\varphi_\infty(\exp_\infty(-wu_\infty)\Gamma_{\infty,v}(y)) = e^{iwt_0}\varphi_\infty(\Gamma_{\infty,v}(y))$$

and consequently by (24)

$$\psi_\infty(f) = \int_{H^\bullet} f(\infty, t_0; y)\varphi_\infty(\Gamma_{\infty,t_0}(y))dy . \tag{25}$$

Let

$$\mathfrak{m} := d\Gamma_{\infty,t_0}(\mathfrak{h}^\bullet) = \mathbb{R}(b_1 + \varepsilon_\infty t_0 b_2) \oplus \mathfrak{k} \subset \mathfrak{g} .$$

As $\alpha_\infty = 0$, $\mathfrak{m}$ is an ideal in ${}_\infty\mathfrak{g}$ with commutative quotient, hence $M := \exp_\infty \mathfrak{m}$ is a normal subgroup of ${}_\infty G$. Let ρ be the restriction of π_∞ onto M and $l'_\infty = l_\infty|_\mathfrak{m} \in \mathfrak{m}^*$. If the M orbit $\Omega_{l'_\infty}$ of l'_∞ in $\mathfrak{m}^*$ is $\mathrm{Ad}^*{}_\infty G$ invariant, then ρ is irreducible and $\rho = K(\Omega_{l'_\infty})$; otherwise ρ is the direct integral of the $\rho_t = K(\Omega_{\mathrm{Ad}^* e'_2(l'_\infty)})$, $t \in \mathbb{R}$, see ch. I. § 5. It is clear, that the positive definite function $y \to \varphi_\infty(\Gamma_{\infty,t_0}(y))$, appearing at the right hand side of (25), is associated with the representation $y \to \rho(\Gamma_{\infty,t_0}(y))$ of ${}_\infty H^\bullet$, thus the same argument as before shows that

$$\lim \psi_\nu(f) = \int_{H^\bullet} f(\infty, t_0; y)\varphi_\infty(\Gamma_{\infty,t_0}(y))dy$$

implies the existence of a sequence $\{(\nu_{j,t_j})\} \subset P$ with

$$\lim_{j\to\infty} \tau_{\nu_j,t_j} = \check{\tau} \in {}_{\infty,t_0}\widehat{\mathbf{H}^\bullet}$$

where $\check{\tau}$ is $\rho \circ \Gamma_{\infty,t_0}$ if ρ is irreducible or $\check{\tau} = \rho_t \circ \Gamma_{\infty,t_0}$ otherwise, for any fixed $t \in \mathbb{R}$. Then again by induction we know that there exists a sequence $\{l_j^\bullet\}_{j\in\mathbb{N}} \subset \mathfrak{h}^{\bullet *}$ with $\tau_{\nu_j,t_j} = K(\Omega_{(\nu_j,l_j^\bullet)})$ and $\lim l_j^\bullet = l_\infty^\bullet$, where $l_\infty^\bullet = (d\Gamma_{\infty,t_0})^*(l'_\infty)$, the image of l'_∞ under the adjoint isomorphism $(d\Gamma_{\infty,t_0})^* : \mathfrak{m}^* \to \mathfrak{h}^{\bullet *}$. Similarly the $l_j^\bullet$ are related to $l'_j \in \mathfrak{h}^*_{\nu_j,t_j}$, where l'_j is the restriction of a functional $l_j \in \mathfrak{g}^*$ onto $\mathfrak{h}_{\nu_j,t_j}$ for which $\pi_j = K(\Omega_{(\nu_j,l_j)})$. For simplicity let us assume that $\nu_j = j$ for $j \in \mathbb{N}$. Then the relation between the $l_\nu^\bullet$ and the l_ν for all $\nu \in \overline{\mathbb{N}}$ is given by

$$l_\nu^\bullet(b_i^\bullet) = l_\nu(d\Gamma_{\nu,t_i}(b_i^\bullet)) , \quad \text{with} \quad t_\infty = t_0 .$$

For $i \geq 2$ this gives

$$l_j(b_{i+1}) = l_j^\bullet(b_i^\bullet) \to l_\infty^\bullet(b_i^\bullet) = l_\infty(b_{i+1})$$

For $i = 1$;

$$l_j(b_1 + \varepsilon_j t_j b_2) = l_j^\bullet(b_1^\bullet) \to l_\infty(b_1 + \varepsilon_\infty t_0 b_2) .$$

Now observe that because of (11), resp. in our case of $l_j + \mathfrak{h}^\perp_{j,t_j} \subset \Omega_{(j,l_j)}$, we can choose the value $l_j(b_2)$ for all $j \in \mathbb{N}$ arbitrarily. So setting e.g. $l_j(b_2) = (\varepsilon_j t_j)^{-1} l_\infty(\varepsilon_\infty t_0 b_2)$ if all $t_j \neq 0$ or $l_j(b_2) = l_\infty(b_2)$ if infinitely many (hence all, without loss of generality) t_j are 0 we get finally a sequence $\{l_j\}_{j\in\mathbb{N}} \subset \mathfrak{g}^*$ with $\pi_j = K(\Omega_{(j,l_j)})$ and $\lim_{j\to\infty} l_j = l_\infty$.

Now we have to consider case β, i.e.

$$\lim \frac{\eta_\nu}{\alpha_\nu} = 0\,, \quad \text{in particular} \quad \eta_\infty = 0\,.$$

In this case $d\Gamma_{\infty,t}$ maps ${}_{\infty,t}\mathfrak{h}^\bullet$ for all $t \neq 0$ isomorphically onto the ideal

$$\mathfrak{d} := \mathbb{R}b_2 \oplus \mathfrak{k} \subset {}_\infty\mathfrak{g}\,,$$

and $\eta_\infty = 0$ implies that u_∞ is central in $\mathfrak{d}$.

Assume first that $\pi_\infty(\exp_\infty \mathbb{R}u_\infty) \subset \mathbb{C}^\times\mathbf{1}$, i.e. that

$$\pi_\infty(\exp_\infty wu_\infty) = e^{iwt_0}\mathbf{1}\,, \quad t_0 \in \mathbb{R}\,.$$

Then again (25) holds and the same argument as above shows the existence of a convergent sequence in $\mathfrak{g}^*$, *provided t_0 is not zero*, i.e.

(26) $$d\pi_\infty(u_\infty) \neq 0\,.j$$

If π_∞ is not scalar on $\exp_\infty \mathbb{R}u_\infty$, then the restriction $\rho = \pi_\infty|_D$ of π_∞ onto the normal subgroup $D := \exp_\infty \mathfrak{d}$ is reducible, because $\exp_\infty \mathbb{R}u_\infty$ is central in D. Moreover, u_∞ is not central in ${}_\infty\mathfrak{g}$, hence $[b_1, u_\infty]_\infty = \alpha_\infty u_\infty \neq 0$, i.e. $\alpha_\infty \neq 0$. For simplicity of notation let us assume that $\alpha_\infty = 1$. It follows from ch. I, § 5, that π_∞ is induced from the representation $\tau \in \widehat{D}$, corresponding to the restriction $l'_\infty = l_\infty|_\mathfrak{d}$. So there exists an ω in the representation space $L^2(\mathbb{R}, \mathfrak{J})$, $\mathfrak{J}$ the space of τ, with $\varphi_\infty(x) = \int_\mathbb{R}(\omega(x^{-1} \cdot e_1^s)|\omega(e_1^s))ds$.

As $\mathrm{Ad}_\infty e_1^{-s}(u_\infty) = e^{-s}u_\infty$ we see that for any $x \in {}_\infty G$

$$\omega(x \underset{\infty}{\cdot} \exp_\infty wu_\infty \underset{\infty}{\cdot} e_1^s) = \omega(x_1 \underset{\infty}{\cdot} e_1^s \underset{\infty}{\cdot} \exp_\infty(we^{-s}u_\infty)) = \omega(x \underset{\infty}{\cdot} e_1^s)e^{-iwe^{-s}t_0}$$

if $\tau(\exp_\infty wu_\infty) = e^{iwt_0} \cdot \mathbf{1}$. Hence for any $v \in \mathbb{R}^\times$

$$\varphi_\infty(\exp_\infty(-wu_\infty)\Gamma_{\infty,v}(y)) = \int_\mathbb{R} (\omega(\Gamma_{\infty,v}(y)^{-1} \cdot e_1^s)|\omega(e_1^s))e^{-iwe^{-s}t_0}ds.$$

Now again (26) implies $t_0 \neq 0$ and from (24) it follows easily that for $f \in \mathbf{C}_0^\infty(\mathbf{H}^\bullet)$ we have:

$$\psi_\infty(f) = \int_\mathbb{R}\int_\mathbb{R} f(\infty, -e^{-s}t_0; y)(\omega(\Gamma_{\infty,-e^{-s}t_0}(y)^{-1} \cdot e_1^s|\omega(e_1^s))ds\,dy\,.$$

As before we see the existence of a convergent sequence of l_ν's in $\mathfrak{g}^*$.

We have now finished case α and also case β under the additional condition that π_∞ is nontrivial on $\exp_\infty \mathbb{R}u_\infty$, i.e. that (26) holds. If (26) does not hold, i.e. if $d\pi_\infty(u_\infty) = 0$, then u_∞ is linearly independent of b_n, because $d\pi_\nu(b_n) = \mathbf{1}$ for all $\nu \in \overline{\mathbb{N}}$. As $u_\infty = \lim\limits_{\nu\to\infty} u_\nu$ we can assume that $\mathfrak{y} = \mathbb{R}u_\nu \oplus \mathbb{R}b_n$ for all $\nu \in \overline{\mathbb{N}}$. Then it is easy to construct an equivalent VLA $\mathbf{g}^\vee$ with VLG $\mathbf{G}^\vee$ and an equivalence $\theta : \mathbf{G} \to \mathbf{G}^\vee$, $d\theta : \mathbf{g} \to \mathbf{g}^\vee$ with $d\theta_\nu(b_j) = b_j^\vee$, $j \neq n-1$, and $d\theta_\nu(u_\nu) = b_{n-1}^\vee$ where $\{b_j^\vee\}$ is a Malvec basis for $\mathfrak{g}^\vee$. This allows us to assume that $u_\nu = b_{n-1}$ for all $\nu \in \overline{\mathbb{N}}$, which means that finally we have to prove the continuity of K^{-1} in the following situation (observe that $u_\nu = b_{n-1}$ implies $\beta_\nu = 0$, hence $\eta_\nu = \gamma_\nu$ for all $\nu \in \mathbb{N}$):

(27) 1)*The relations (7) (i) – (vi) hold with* $\beta_\nu = 0$, *hence* $u_\nu = b_{n-1}$ *and* $l_\nu(b_{n-1}) = 0$ *for all* $\nu \in \overline{\mathbb{N}}$
2) $\lim\limits_{\nu\to\infty} \frac{\gamma_\nu}{\alpha_\nu} = 0$, *in particular* $\gamma_\infty = \lim\limits_{\nu\to\infty} \gamma_\nu = 0$
3) $d\pi_\infty(b_{n-1}) = 0$, *i.e.* π_∞ *is trivial on* $\exp_\infty \mathbb{R} b_{n-1}$.

Clearly we can also assume that $\gamma_\nu > 0$, $\alpha_\nu > 0$ for all finite ν. First we remark that we can control the supports of the cyclic vectors ω_k in (16): For a Schwartz function φ on $\mathbb{R}$, i.e. $\varphi \in \mathcal{S}(\mathbb{R})$, we define the smooth bounded measure $\varphi^{\bar{}}$ on G by $\varphi^{\bar{}}(f) = \int_{\mathbb{R}} \varphi(s) f(e_{n-1}^s)ds$, thus supp $\varphi^{\bar{}} \subset e_{n-1}^{\mathbb{R}} \subset G$. Then $\pi_k(\varphi^{\bar{}})$ is an operator, bounded by $|\varphi|_1$ and such that

$$\pi_k(\varphi^{\bar{}})\xi(e_2^t) = \int \varphi(s)\xi(e_{n-1}^{-s} \cdot e_2^t)ds = \int \varphi(s)\xi(e_2^t)e^{i\gamma_k st}ds = \widehat{\varphi}(\gamma_k t)\xi(e_2^t)$$

with Fourier transform $\widehat{\varphi}$ of φ. In particular

$$\pi_k(\varphi^{\bar{}})\omega_k(e_2^t) = \widehat{\varphi}(\gamma_k t)\omega_k(e_2^t) \tag{28}$$

As π_∞ is trivial on $e_{n-1}^{\mathbb{R}}$ we have $\pi_\infty(\varphi^{\bar{}}) = \widehat{\varphi}(0)\mathbf{1}$, hence if we take φ with supp $\widehat{\varphi} \subset [-1, 1]$ and $\widehat{\varphi}(0) = 1$ and set $\omega_\nu^{\bar{}} = \pi_\nu(\varphi^{\bar{}})\omega_\nu$, $\nu \in \overline{\mathbb{N}}$, then we have $\omega_\infty^{\bar{}} = \omega_\infty$, supp $\omega_k^{\bar{}} \subset [-\gamma_k^{-1}, \gamma_k^{-1}]$ for $k \in \overline{\mathbb{N}}$ and $\lim\langle\omega_k^{\bar{}}, \omega_k^{\bar{}}\rangle_k = \langle\omega_\infty, \omega_\infty\rangle_\infty = \varphi_\infty$. This shows that we can assume that

$$\text{supp } \omega_k \subset [-\gamma_k^{-1}, \gamma_k^{-1}] \tag{29}$$

holds already for all ω_k, $k \in \mathbb{N}$.

Again we have to distinguish two cases:

A. There exists a constant $T > 0$ such that

$$\inf_k \int_{|t|\le T} |\omega_k(e_2^t)|^2 dt =: C > 0 \,.$$

This is the simpler case whose treatment we will postpone, i.e. we start with the negation of A, which obviously is characterized by the fact that for some subsequence, again denoted by $\{\omega_k\}$, we have

B. There exists an increasing unbounded sequence $\{p_k\}_k \subset \mathbb{R}^+$ with

$$\lim_{k\to\infty} \int_{|t|\leq p_k} |\omega_k(e_2^t)|^2 dt = 0 .$$

Condition B allows us to assume that

$$\text{(30)} \qquad \operatorname{supp} \omega_k \subset \{t \in \mathbb{R};\ p_k \leq |t| \leq \gamma_k^{-1}\} .$$

Now we will show that the condition $d\pi_\infty(b_{n-1}) = 0$ implies that the interval carrying ω_k can be made even smaller. We claim

(31) *There exists a decreasing sequence* $\{c_k\}_k \subset \mathbb{R}$ *with* $1 \geq c_k \searrow 0$ *such that*

$$\lim_{k\to\infty} \int_{c_k \leq \gamma_k |t| \leq 1} |\omega_k(e_2^t)|^2 dt = 0 .$$

Proof. From $\pi_k(e_{n-1}^s)\xi(e_2^t) = \xi(e_2^t)e^{i\gamma_k st}$ for $\xi \in L^2(\mathbb{R}, \mathfrak{J}_k)$ it follows that for smooth ξ

$$d\pi_k(b_{n-1})\xi(e_2^t) = i\gamma_k t\xi(e_2^t)$$

and in particular

$$\text{(32)} \qquad |d\pi_k(b_{n-1})\omega_k|^2 = \gamma_k^2 \int_{-\gamma_k^{-1}}^{\gamma_k^{-1}} t^2 |\omega_k(e_2^t)|^2 dt = \gamma_k^{-1} \int_{-1}^{1} t^2 |\omega_k(e_2^{\gamma_k^{-1} t})|^2 dt .$$

From $\varphi_k(e_{n-1}^s) = (\pi_k(e_{n-1}^s)\omega_k|\omega_k)$ we derive

$$\frac{d^j\varphi_k}{ds^j}(e_{n-1}^s) = (\pi_k(e_{n-1}^s)(d\pi_k(b_{n-1}))^j\omega_k|\omega_k)$$

for $j \in \mathbb{N}$, hence

$$|d\pi_k(b_{n-1})\omega_k|^2 = -\frac{d^2\varphi_k}{ds^2}(e_{n-1}^0) .$$

But we know that $\lim_{k\to\infty} \frac{d^j\varphi_k}{ds^j} = \frac{d^j\varphi_\infty}{ds^j}$ uniformly on compacta for all $j \in \mathbb{N}$. As $d\pi_\infty(b_{n-1}) = 0$, hence $\varphi_\infty(e_{n-1}^s) \equiv 1$ we obtain $\lim |d\pi_k(b_{n-1})\omega_k|^2 = 0$, or after (32)

$$\lim \gamma_k^{-1} \int_{-1}^{1} t^2 |\omega_k(e_2^{\gamma_k^{-1} t})|^2 dt = 0 .$$

This implies the existence of a decreasing sequence $\{c_k\}$ with $\lim c_k = 0$ and such that (possibly for a subsequence) (31) holds, furthermore we can assume that

$$\text{supp } \omega_k \subset \{t; p_k \leq |t| \leq \gamma_k^{-1} c_k\} \tag{33}$$

and also that

$$\lim p_k c_k = 0 \,, \quad p_k c_k < 1 \quad \text{for all} \quad k \tag{34}$$

holds, the last relation, because the p_k can be made smaller (but still tending to ∞), if necessary. Now let

$$q_k = \max\{\frac{2}{\gamma_k p_k}, p_k + 1\}.$$

Then $\gamma_k q_k = \max\{\frac{2}{p_k}, \gamma_k p_k + \gamma_k\}$, hence

$$\lim \gamma_k q_k = 0 \,. \tag{35}$$

Moreover for all $k \in \mathbb{N}$:

$$\gamma_k p_k q_k \geq 2 \quad \text{and} \quad I_k := [p_k, q_k] \neq \emptyset. \tag{36}$$

We have now defined the ingredients for the construction of a new VLG $\widetilde{\mathbf{G}}$: Let

$$\widetilde{P} = (\bigcup_{k \in \mathbb{N}} \{k\} \times I_k) \bigcup \{\infty\}$$

with I_k as in (36). We consider $\widetilde{P}$ as the compact closure of the union of the sets $\{k\} \times [p_k, q_k] \subset \mathbb{R}^2$ in the one point compactification of $\mathbb{R}^2$, i.e. all $\{k\} \times I_k$ are open in $\widetilde{P}$ and $\lim\limits_{j \to \infty} (k_j, t_j) = \infty \in \widetilde{P}$ if and only if $\lim\limits_{j \to \infty} k_j^{-1} = 0$.

First we define the VLA

$$\widetilde{\mathbf{g}} = (\widetilde{P} \times \widetilde{\mathfrak{g}}, [\ ,\ ]_p)$$

by setting

$$\widetilde{\mathfrak{g}} = \mathfrak{g} \oplus \mathbb{R}\widetilde{b}$$

with a new basis element $\widetilde{b}$, added to the "old" Malcev basis $\{b_1, b_2, \ldots, b_n\}$ of $\mathfrak{g}$. For $(k, r) \in \widetilde{P}$, resp. $\infty \in \widetilde{P}$ we define the Lie product $[\ ,\ ]_{k,r}$, resp. $[\ ,\ ]_\infty$ in $\widetilde{\mathfrak{g}}$ by

$$\tag{37} \left\{ \begin{array}{lllllll} [x, y]_{k,r} & = & [x, y]_k \,, & [x, y]_\infty & = & [x, y]_\infty & \text{for} \quad x, y \in \mathfrak{g} \\ [x, \widetilde{b}]_{k,r} & = & [x, \widetilde{b}]_\infty = 0 \,, & & & & \text{for} \quad x \in \mathfrak{k} \\ [b_1, \widetilde{b}]_{k,r} & = & \alpha_k \widetilde{b} \,, & [b_1, \widetilde{b}]_\infty & = & \alpha_\infty \widetilde{b} & \\ [b_2, \widetilde{b}]_{k,r} & = & \gamma_k r b_n \,, & [b_2, \widetilde{b}]_\infty & = & 0 & \end{array} \right.$$

Because of (35) the formulas (37) define a variable Lie algebra structure on $\widetilde{\mathbf{g}}$ with $\{b_1, b_2, \dots, b_{n-2}, \widetilde{b}, b_{n-1}, b_n\}$ as a Malcev basis in all ${}_p\widetilde{\mathfrak{g}}$, $p \in \widetilde{P}$. Furthermore,

$$\widetilde{\mathfrak{y}} := \mathbb{R}\widetilde{b} \oplus \mathbb{R}b_n$$

is an ideal in all ${}_p\widetilde{\mathfrak{g}}$ and minimal noncentral if $p \neq \infty$.

Let $\widetilde{\mathbf{G}} = \widetilde{P} \times \widetilde{G}$ with $\widetilde{G} = G \times \mathbb{R} = \mathbb{R}^{n+1}$ be the corresponding exponential $(n+1)$–dimensional VLG. Clearly we have a natural injection $\widetilde{P} \times G \to \widetilde{\mathbf{G}}$ (with product $\underset{k,r}{\cdot} = \underset{k}{\cdot}$ in $\widetilde{P} \times G$) and a natural projection $\widetilde{P} \times G \to \mathbf{G}$, similarly for the VLA's $\mathbf{g}, \widetilde{\mathbf{g}}$ etc.

In addition to $\mathfrak{k}$, $\mathfrak{h} = \mathbb{R}b_1 \oplus \mathfrak{k}$ and $\mathfrak{d} = \mathbb{R}b_2 \oplus \mathfrak{k}$ we define

$$\widetilde{\mathfrak{k}} = \mathfrak{k} \oplus \mathbb{R}\widetilde{b}\ , \quad \widetilde{\mathfrak{h}} = \mathfrak{h} \oplus \mathbb{R}\widetilde{b}\ , \quad \widetilde{\mathfrak{d}} = \mathfrak{d} \oplus \mathbb{R}\widetilde{b}$$

and denote the corresponding variable subgroups of $\widetilde{G}$ by $\widetilde{K}$, $\widetilde{H}$ and $\widetilde{D}$. So in our coordinates e.g. $\widetilde{D}$ is the subspace of all $\widetilde{x} \in \mathbb{R}^{n+1} = \widetilde{G}$ with first coordinate 0. Further we write

$$e_0^s := \exp_p s\widetilde{b} \in {}_p\widetilde{G}\ ,$$

i.e.

$$S := e_0^{\mathbb{R}} = \exp{}_p\mathbb{R}\widetilde{b}$$

is the one parameter subgroup of ${}_p\widetilde{G}$ defined by $\widetilde{b} \in {}_p\widetilde{\mathfrak{g}}$. Since $[b_1, \widetilde{b}]_{k,r} = \alpha_k\widetilde{b}$ and $[\mathfrak{k}, \widetilde{b}]_{k,r} = 0$ it follows that S is normal in every ${}_p\widetilde{H}$, $p \in \widetilde{P}$ and that

$$(38) \qquad {}_{k,r}\widetilde{H} = {}_kH \ltimes S \subset {}_{k,r}\widetilde{G}\ .$$

Recall that we had $\pi_k = \operatorname{ind}_{{}_kH}^{{}_kG}\tau_k$ with $\tau_k \in {}_k\widehat{H}$. As (38) allows us to consider τ_k also as representation of ${}_{k,r}\widetilde{H}$ we can now define

$$(39) \qquad \widetilde{\pi}_{k,r} := \operatorname{ind}_{{}_{k,r}\widetilde{H}}^{{}_{k,r}\widetilde{G}}\tau_k\ .$$

As ${}_{k,r}\widetilde{G} = e_2^{\mathbb{R}} \cdot {}_{k,r}\widetilde{H}$ we can identify the representation spaces $\mathfrak{H}_{k,r}$ of $\widetilde{\pi}_{k,r}$ and π_k. They consist of all square integrable mappings $\xi : e_2^t \to \xi(e_2^t) \in \mathfrak{J}_k = \mathfrak{H}(\tau_k)$ with $(\widetilde{\pi}_{k,r}(x)\xi)(e_2^t) = \tau_k(y)\xi(e_2^{t'})$ if $x^{-1} \underset{k,r}{\cdot} e_2^t = e_2^{t'} \underset{k,r}{\cdot} y^{-1}$ with $y = y(x,t) \in {}_{k,r}\widetilde{H}$. This implies immediately

$$(40) \qquad \widetilde{\pi}_{k,r}(x) = \pi_k(x) \quad \text{for} \quad x \in G\ , \quad \widetilde{\pi}_{k,r}(e_0^s) = \pi_k(e_{n-1}^{rs})\ ,$$

so in particular all $\widetilde{\pi}_{k,r}$ are irreducible. For the corresponding infinitesimal representations of ${}_{k,r}\widetilde{\mathfrak{g}}$ we obtain

(41) $d\widetilde{\pi}_{k,r}(b_j) = d\pi_k(b_j)$ for $1 \leq j \leq n$, $d\widetilde{\pi}_{k,r}(\widetilde{b}) = rd\pi_{k,r}(\widetilde{b}) = rd\pi_k(b_{n-1})$

and consequently $d\widetilde{\pi}_{k,r}(\widetilde{b} - rb_{n-1}) = 0$, which implies that $\widetilde{\pi}_{k,r}$ is trivial on the normal subgroup

$$Q_{k,r} := \exp_{k,r} \mathbb{R}(\widetilde{b} - rb_{n-1})$$

of ${}_{k,r}\widetilde{G}$.

We will show that the family $\{\widetilde{\pi}_{k,r}\}$ has cluster points in ${}_\infty\widehat{G}$, which are closely related to the representation $\pi_\infty \in {}_\infty\widehat{G}$. For this we construct suitable sequences of positive definite functions on $\widetilde{\mathbf{G}}$.

Let α be a smooth, nonnegative function on $\mathbb{R}$ with

$$\alpha(r) = \alpha(-r), \text{supp } \alpha \in [-2, -1] \cup [1, 2],$$
$$\int_{\mathbb{R}^+} \frac{\alpha(r)^2}{r} dr = 1\,, \; A := \int_{\mathbb{R}^+} \frac{\alpha(r)}{r} dr < 1\,.$$

Then the Fourier transform $\widehat{\alpha} = \beta$ is in $\mathcal{S}(\mathbb{R})$ and $\alpha = \widehat{\beta}$. Let β_0 be the bounded measure on $\widetilde{G}$, defined by $\beta_0(f) = \int_{\mathbb{R}} \beta(s) f(e_0^s) ds$, thus supp $\beta_0 \subset S$. We set

(42) $$\widetilde{\omega}_{k,r} = \widetilde{\pi}_{k,r}(\beta_0)\omega_k$$

hence $|\widetilde{\omega}_{k,r}| \leq |\beta|_1 = \int |b(s)| ds$. Similarly we can define the measure φ_0 for any $\varphi \in \mathcal{S}(\mathbb{R})$ and the same computation leading to (38) shows that

$$\widetilde{\pi}_{k,r}(\varphi_0)\xi(e_2^t) = \widehat{\varphi}(\gamma_k tr)\xi(e_2^t)$$

for any $\xi \in L^2(\mathbb{R}, \mathfrak{J}_k) = \mathfrak{H}_{k,r}$. In particular we find

(43) $$\widetilde{\omega}_{k,r}(e_2^t) = \alpha(\gamma_k tr)\omega_k(e_2^t)$$

and

(44) $$\widetilde{\pi}_{k,r}(\eta_0)\widetilde{\omega}_{k,r} = \widetilde{\omega}_{k,r}$$

for any $\eta \subset \mathcal{S}(\mathbb{R})$ with $\widehat{\eta} \equiv 1$ on $[-2, -1] \cup [1, 2] \subset \mathbb{R}$. Let

(45) $$\widetilde{\varphi}_{k,r}(\widetilde{x}) = (\widetilde{\pi}_{k,r}(\widetilde{x})\widetilde{\omega}_{k,r}|\widetilde{\omega}_{k,r})_k\,,$$

thus $\widetilde{\varphi}_{k,r}$ is a positive definite function on $\widetilde{G}$, associated with $\widetilde{\pi}_{k,r}$ and with norm $|\widetilde{\varphi}_{k,r}| = |\widetilde{\omega}_{k,r}|^2$. It follows that

(46) $$\widetilde{\varphi}_k = \int_{I_k} \widetilde{\varphi}_{k,r}\frac{dr}{r}$$

is a positive definite function, resp. a positive functional on $C^*(\widetilde{\mathbf{G}})$.

To prove that $\widetilde{\varphi}_k$ is bounded by 1 we show the

(47) **Lemma.** *For any* $\xi \in \mathfrak{H}_{k,r}$ *we have*

$$\int_{I_k}(\xi|\widetilde{\omega}_{k,r})\frac{dr}{r} = A(\xi|\omega_k)\,, \quad \int_{I_k}(\widetilde{\pi}_{k,r}(\beta_0)\xi|\widetilde{\omega}_{k,r})\frac{dr}{r} = (\xi|\omega_k)\ .$$

Proof. As α is even, (37) and (43) imply

$$\int_{I_k}(\xi|\widetilde{\omega}_{k,r})\frac{dr}{r} = \int_{p_k}^{q_k}\int_{p_k}^{c_k\gamma_k^{-1}}\frac{\alpha(\gamma_k tr)}{r}\psi(t)dt\,dr$$

with $\psi(t) = (\xi(e_2^t)|\omega_k(e_2^t)) + (\xi(e_2^{-t})|\omega_k(e_2^{-t}))$. If $r \geq q_k$, then (36) implies $\gamma_k tr \geq 2$ for all $t \geq p_k$, hence $\alpha(\gamma_k tr) = 0$. Similarly (34) implies $\gamma_k tr \leq 1$, if $r \leq p_k$ and $t \leq c_k\gamma_k^{-1}$, and again $\alpha(\gamma_k tr) = 0$. It follows that we can write

$$\begin{aligned}\int_{I_k}(\xi|\widetilde{\omega}_{k,r})\frac{dr}{r} &= \int_{p_k}^{c_k\gamma_k^{-1}}\int_{\mathbb{R}^+}\alpha(\gamma_k tr)\frac{dr}{r}\psi(t)dt \\ &= \int_{p_k}^{c_k\gamma_k^{-1}}\int_{\mathbb{R}^+}\alpha(r)\frac{dr}{r}\psi(t)dt = A\int_{\mathbb{R}^+}\psi(t)dt = A(\xi|\omega_k)\ .\end{aligned}$$

The same argument proves the second equation of the lemma. For $\xi = \omega_k$ we get

$$\widetilde{\varphi}_k(e) = \int_{I_k}|\widetilde{\omega}_{k,r}|^2\frac{dr}{r} = |\omega_k|^2 = 1$$

showing that all $\widetilde{\varphi}_k$ have norm $|\widetilde{\varphi}_k| = 1$. It follows that $\{\widetilde{\varphi}_k\}_k$ has convergent subsequences in $\mathcal{P}_1(C^*(\widetilde{\mathbf{G}}))$. For simplicity let us assume that

(48) $$\widetilde{\varphi}_\infty = \lim \widetilde{\varphi}_k \quad \text{in} \quad \mathcal{P}_1(C^*(\widetilde{\mathbf{G}}))$$

exists. Then clearly $\widetilde{\varphi}_\infty$ is a positive definite function on ${}_\infty\widetilde{G}$. Let $D = \exp\mathfrak{d}$ be the (variable) subgroup of all $x \in G$ with $x_1 = 0$, so ${}_kD = \exp_k\mathfrak{d}$ is normal in ${}_kG$ for all k. Recall that $\varphi_\infty = \lim\varphi_k$ was the positive definite function on ${}_\infty G$ with $\varphi_\infty(x) = (\pi_\infty(x)\omega_\infty|\omega_\infty)$. We claim:

(49) *The restrictions* $\widetilde{\varphi}_\infty|_D$ *of* $\widetilde{\varphi}_\infty$ *and* $\varphi_\infty|_D$ *of* φ_∞ *onto* D *coincide, i.e.*

$$\widetilde{\varphi}_\infty(y) = \varphi_\infty(y) \quad \textit{for all} \quad y \in D\ .$$

To prove (49) and for later use we have to study the restrictions of the $\widetilde{\varphi}_\nu$ onto G. As usually we parametrize G by $\mathbb{R}^n$, i.e. write $x = (x_1, x_2, \ldots, x_n)$, hence $K = \{x; x_1 = x_2 = 0\}$ and $x = e_1^{x_1} \underset{k}{\cdot} e_2^{x_2} \underset{k}{\cdot} y$ with $y \in K$; moreover for $x \in G$ we have

$$\widetilde{\varphi}_k(x) = \int_{I_k} (\pi_k(x)\widetilde{\omega}_{k,r}|\widetilde{\omega}_{k,r})\frac{dr}{r} = \int_{I_k} (\pi_k(e_1^{x_1})\pi_k(e_2^{x_2})\pi_k(y)\widetilde{\pi}_{k,r}(\beta_0)\omega_k|\widetilde{\omega}_{k,r})\frac{dr}{r}\ .$$

Using (47) it follows that

$$\widetilde{\varphi}_k(x) = N_k^1(x) + N_k^2(x) + \varphi_k(x) \tag{50}$$

with

$$N_k^1(x) = \int_{I_k} (\pi_k(e_1^{x_1})[\pi_k(e_2^{x_2}), \widetilde{\pi}_{k,r}(\beta_0)]\pi_k(y)\omega_k|\widetilde{\omega}_{k,r})\frac{dr}{r}$$

$$N_k^2(x) = \int_{I_k} ([\pi_k(e_1^{x_1}), \widetilde{\pi}_{k,r}(\beta_0)]\pi_k(e_2^{x_2}y)\omega_k|\widetilde{\omega}_{k,r})\frac{dr}{r}$$

where $[A, B] = AB - BA$ denotes the commutator of A and B.

We claim that $N_k^1(x)$ tends to 0 uniformly on compacta on D. As

$$[\pi_k(e_2^{x_2}),\ \pi_k(e_{n-1}^{rs})] = \pi_k(e_2^{x_2}e_{n-1}^{rs})(1 - e^{i\gamma_k x_2 rs}) \tag{51}$$

and $|1 - e^{it}| \le 2|t|$ for all real t, we get from (51) after multiplication with $\beta(s)$ and integration over s:

$$\|[\pi_k(e_2^{x_2}), \widetilde{\pi}_{k,r}(\beta_0)]\| \le 2\gamma_k|x_2| \cdot |r| \cdot S_\beta$$

for the operator norm $\|\cdot\|$, with $S_\beta = \int_{\mathbb{R}} |s\beta(s)|ds$. Hence

$$|N_k^1(x)| \ \le\ 2\gamma_k|x_2|S_\beta \int_{I_k} |\widetilde{\omega}_{k,r}|dr \ \le\ 2\gamma_k|x_2|S_\beta|\beta|_1(q_k - p_k)$$

and finally $|N_k^1|(x) \le 2\gamma_k q_k|x_2|S_\beta|\beta|_1$, which proves our claim.

As $[\pi_k(e_1^{x_1}), \widetilde{\pi}_{k,r}(\beta_0)]^* = -[\pi_k(e_1^{-x_1}), \widetilde{\pi}_{k,r}(\beta_0)]$ we see that

$$\begin{aligned} N_k^2(x) &= -\int_{I_k} \left(\pi_k(e_2^{x_2}y)\omega_k|[\pi_k(e_1^{-x_1}), \widetilde{\pi}_{k,r}(\beta_0)]\widetilde{\pi}_{k,r}(\beta_0)\omega_k\right)\frac{dr}{r} \\ &= -\left(\pi_k(e_2^{x_2}y)\omega_k\Big|\left\{\int_{I_k} [\pi_k(e_1^{x_1}), \widetilde{\pi}_{k,r}(\beta_0)]\widetilde{\pi}_{k,r}(\beta_0)\frac{dr}{r}\right\}\omega_k\right) \end{aligned}$$

hence with operator norm $\|\cdot\|$:

$$|N_k^2(x)| \leq \| \int_{p_k}^{q_k} [\pi_k(e_1^{x_1}), \widetilde{\pi}_{k,r}(\beta_0)\widetilde{\pi}_{k,r}(\beta_0)\frac{dr}{r}\| \ .$$

Let $\beta_k(s) := \beta(s) - e^{-\alpha_k x_1}\beta(e^{-\alpha_k x_1}s)$. Then

$$[\pi_k(e_1^{x_1}), \widetilde{\pi}_{k,r}(\beta_0)] = \pi_k(e_1^{x_1}) \int_{\mathbb{R}} \pi_k(e_{n-1}^{rs})\beta_k(s)\, ds \ ,$$

hence it suffices to estimate

$$\nu_k := \| \int_{p_k}^{q_k} \int_{\mathbb{R}} \pi_k(e_{n-1}^{rs})\beta_k(s) ds\, \widetilde{\pi}_{k,r}(\beta_0)\frac{dr}{r}\| \ .$$

Let $\{E(\lambda)\}$ be the spectral family of the selfadjoint operator $d\pi_k(b_{n-1})$, thus

$$\pi_k(e_{n-1}^t) = \int_{\mathbb{R}} e^{-it\lambda} dE(\lambda) \ .$$

Then we have

$$\begin{aligned}\widetilde{\pi}_{k,r}(\beta_0) &= \int \pi_k(e_{n-1}^{rt})\beta(t) dt \\ &= \int\int e^{-irt\lambda}\beta(t) dt\, E(\lambda) d\lambda = \int \hat{\beta}(r\lambda) dE(\lambda) = \int \alpha(r\lambda) dE(\lambda)\end{aligned}$$

hence

$$\pi_k(e_{n-1}^{rs})\widetilde{\pi}_{k,r}(\beta_0) = \int e^{-irs\mu} dE(\mu) \int \alpha(r\lambda) dE(\lambda) = \int e^{-irs\lambda}\alpha(r\lambda) dE(\lambda)$$

and consequently

$$\int_{p_k}^{q_k} \int_{\mathbb{R}} \pi_k(e_{n-1}^{rs})\beta_k(s) ds\, \widetilde{\pi}_{k,r}(\beta_0)\frac{dr}{r} = \int_{\mathbb{R}}\int_{\mathbb{R}}\int_{p_k}^{q_k} e^{-irs\lambda}\alpha(r\lambda)\frac{dr}{r}\beta_k(s) ds\, dE(\lambda) \ .$$

As $\| \int F(\lambda) dE(\lambda)\| \leq \sup |F(\lambda)|$ for every continuous function F it follows that

$$\begin{aligned}\nu_k \leq \max_\lambda \left| \int_{\mathbb{R}} \int_{p_k}^{q_k} e^{-irs\lambda}\alpha(r\lambda)\frac{dr}{r}\beta_k(s) ds \right| &\leq \max_\lambda \int_{\mathbb{R}}\int_{\mathbb{R}^+} \alpha(r\lambda)\frac{dr}{r}|\beta_k(s)| ds \\ &= A \int |\beta_k(s)| ds \ .\end{aligned}$$

To estimate $\int |\beta_k(s)| ds$ let C_1 be a constant for which $|\beta(x)|(1+x^2) \leq C_1$ and $|x\beta'(x)|(1+x^2) \leq C_1$ for all $x \in \mathbb{R}$. Then an easy application of the mean value theorem shows that

$$|\beta(s) - e^t\beta(e^t s)| \leq 2C_1 te^t \frac{1}{1+s^2}$$

for all s and all $t \geq 0$. This implies

$$\int |\beta_k(s)|ds \leq \alpha_k |x_1| e^{\alpha_k |x_1|} C$$

with $C = 2C_1 \int_{\mathbb{R}} \frac{ds}{1+s^2}$. We proved

$$|N_k^2(x)| \leq \alpha_k |x_1| e^{\alpha_k |x_1|} AC \tag{52}$$

for all $x = (x_1, x_2, \ldots, x_n) \in G$.

Now we will restrict the $\widetilde{\varphi}_k$ onto D, i.e. we set $x_1 = 0$. Then $N_k^2(x) = 0$ and $\gamma_\infty = 0$ implies $\lim_{k\to\infty} N_k^1(x) = 0$ uniformly on compacta. Because the φ_ν, $\nu \in \overline{\mathbb{N}}$, are pure, i.e. belong to irreducible representations, we know that also $\lim_{k\to\infty} \varphi_k(x) = \varphi_\infty(x)$ uniformly on compacta on G. Hence (50) implies

$$\lim \widetilde{\varphi}_k(x) = \varphi_\infty(x)$$

for all $x \in D$, uniformly on compacta. To prove (49) we have first to show that there exist suitable elements $h \in \mathscr{C}_0(\widetilde{P} \times \widetilde{G}) \subset C^*(\widetilde{\mathbf{G}})$ with $h = h^*$, $|h|_* = 1$ (norm in $C^*(\widetilde{\mathbf{G}})$) and $\lim \widetilde{\varphi}_k(h) = 1$. Let Q be a closed submanifold of $\widetilde{G}$, so that $\widetilde{P} \times \widetilde{Q} = \mathbf{Q}$ is a variable subgroup of $\widetilde{\mathbf{G}}$. Then it is clear that $C^*(\mathbf{Q})$ and in particular $\mathscr{C}_0(\widetilde{P} \times Q)$ is naturally contained in the adjoint algebra $C^*(\widetilde{\mathbf{G}})^\flat$. If $f \in \mathscr{C}_\infty(\mathbf{L}(\mathbf{Q}))$ and $g \in \mathscr{C}_\infty(\mathbf{L}(\widetilde{\mathbf{G}}))$, then $(fg)(p) = f(p) \underset{p}{\star} g(p)$, where we may consider both $f(p)$ and $g(p)$ as elements of the measure algebra $\mathcal{M}({}_p\widetilde{G}) = L^1({}_p\widetilde{G})^\flat$. We apply this to the two cases $Q = G$ and $Q = S = e_0^{\mathbb{R}}$. So let $f \in \mathscr{C}_\infty(\mathbf{L}(\mathbf{G}))$ and $\eta \in \mathscr{C}_\infty(\mathbf{L}(\mathbf{S}))$. Then $(f\eta)(p)$ as measure on $\widetilde{G}$ is given by

$$\begin{aligned}(f\eta)(p)(\psi) &= \int\int \psi(x \underset{p}{\cdot} e_0^{x_0}) f(p)(x)\eta(p)(x_0)dxdx_0 = \\ &= \int \psi(x_1, x_2, \ldots, x_n, x_0) f(p)(x)\eta(p)(x_0)d\widetilde{x}\ ,\end{aligned}$$

which means that $(f\eta)(p)$ is the *function* $F(p)$ on $\widetilde{G}$ defined by

$$F(p)(\widetilde{x}) = f(p)(x)\eta(p)(x_0)\ .$$

This implies $|F|_* \leq |f|_*|\eta|_* \leq |f|_1|\eta|_* = |f|_1|\widehat{\eta}|_\infty$ where $|\widehat{\eta}|_\infty = \sup_p|\widehat{\eta(p)}|_\infty$ is the uniform norm of the "Fourier transform" $\widehat{\eta}$ of η and $|f|_1 = \sup_p|f(p)|_1$ the norm in $\mathscr{C}_\infty(\mathbf{L}(\mathbf{G}))$.

Let $0 < \delta < 1$. As $\lim_{k\to\infty} \varphi_k = \varphi_\infty$ uniformly on compacta, resp. w^*, and $\varphi_\infty(e) = 1$, there exists $f \in \mathscr{C}_0(G)$ with $0 \leq f$, $\int_G f(x)dx = 1$ and $\varphi_k(f) > 1 - \frac{\delta}{2}$ for (almost) all k. Let $\eta \in \mathscr{S}(\mathbb{R})$ as in (44), but additionally with $|\widetilde{\eta}|_\infty = 1$. Then we have for $F = f\eta \in C^*(\widetilde{\mathbf{G}})$:

$$\widetilde{\varphi}_{k,r}(F) = \int_G \int_{\mathbb{R}} f(x)\eta(x_0)(\pi_k(x)\widetilde{\pi}_{k,r}(e_0^{x_0})\widetilde{\omega}_{k,r}|\widetilde{\omega}_{k,r})dx_0 dx =$$
$$= \int_G f(x)(\pi_k(x)\widetilde{\varphi}_{k,r}(\eta_0)\widetilde{\omega}_{k,r}|\widetilde{\omega}_{k,r})dx = (\pi_k(f)\widetilde{\omega}_{k,r}|\widetilde{\omega}_{k,r})$$

hence $\widetilde{\varphi}_k(F) = \widetilde{\varphi}_k(f) = \int_G f(x)\widetilde{\varphi}_k(x)\,dx$.

Now let us further assume that for a given positive $\vartheta \leq 1$ the function f is supported by $\{x \in G; |x_j| \leq \vartheta,\ j = 1, \ldots, n\}$. Then (50) yields

$$|\widetilde{\varphi}_k(f) - \varphi_k(f)| \leq \gamma_k q_k C_1 + \vartheta C_2$$

with constant C_1 and C_2 independent of k and ϑ. For $\vartheta = \min\{1, \frac{\delta}{2C_2}\}$ we get

$$\widetilde{\varphi}_k(F) = \widetilde{\varphi}_k(f) \geq \varphi_k(f) - \frac{\delta}{2} - \gamma_k q_k C_1 \geq 1 - \delta - \gamma_k q_k C_1$$

and for $k \to \infty : \widetilde{\varphi}_\infty(F) \geq 1 - \delta$. As $|F|_* \leq 1$ it follows that $\widetilde{\varphi}_\infty$ has norm 1, i.e. $\widetilde{\varphi}_\infty(e) = 1$.

For $\nu \in \overline{\mathbb{N}}$ the functions $\widetilde{\varphi}_\nu$ can be written

$$\widetilde{\varphi}_\nu(\widetilde{x}) = (\zeta_\nu(\widetilde{x})v_\nu|v_\nu)$$

where ζ_ν is the corresponding cyclic representation of ${}_\nu\widetilde{G}$ and v_ν a normalized cyclic vector, i.e. $|v_\nu| = 1$. Now we fix $x \in D$ and ε with $0 < \varepsilon < 1$. Let δ_x be the Dirac measure in x, hence $f \to f \star \delta_x$ is right translation by x in $L^1({}_\nu\widetilde{G})$. We choose $h \in \mathscr{C}_0(\widetilde{P} \times \widetilde{G}) \subset C^*(\widetilde{\mathbf{G}})$ with $h^* = h$, $|h(p)|_1 = 1$ for all $p \in \widetilde{P}$, and such that

$$(53) \qquad \widetilde{\varphi}_k(h) \geq 1 - \frac{1}{8}\varepsilon^2 \text{ for all finite } k\,, \quad |\widetilde{\varphi}_\infty(h \underset{\infty}{\star} \delta_x) - \widetilde{\varphi}_\infty(x)| < \tfrac{\varepsilon}{2}\,.$$

Then

$$|\widetilde{\varphi}_k(h \star \delta_x) - \widetilde{\varphi}_k(x)|^2 = |(\zeta_k(h)\zeta_k(x)v_k - \zeta_k(x)v_k|v_k)|^2$$
$$\leq |\zeta_k(x)v_k|^2 \cdot |\zeta_k(h)v_k - v_k|^2$$
$$= |\zeta_k(h)v_k|^2 - 2\widetilde{\varphi}_k(h) + 1 \leq 2(1 - \widetilde{\varphi}_k(h)) \leq (\tfrac{\varepsilon}{2})^2\,,$$

i.e. $|\widetilde{\varphi}_k(h\star\delta_x) - \widetilde{\varphi}_k(x)| \leq \frac{\varepsilon}{2}$ for all k. For $k \to \infty$ this gives $|\widetilde{\varphi}_\infty(h\star\delta_x) - \varphi_\infty(x)| \leq \frac{\varepsilon}{2}$ which together with (52) implies $|\widetilde{\varphi}_\infty(x) - \varphi_\infty(x)| < \varepsilon$ hence $\widetilde{\varphi}_\infty(x) = \varphi_\infty(x)$, because ε was arbitrary.

Let X be the support of ζ_∞ in ${}_\infty\widehat{\widehat{G}}$, so that we can write ζ_∞ as a direct integral

(54) $$\zeta_\infty = \int_X^\oplus m(\lambda)\lambda d\mu(\lambda)$$

with multiplicity function m and a certain Borel measure μ, supported on the Borel set $X \subset {}_\infty\widehat{\widehat{G}}$. The subset X_0 of all $\lambda \in X$ which are trivial on S certainly also is Borel. We claim that X_0 has μ measure zero.
For $\varphi \in \mathcal{S}(\mathbb{R})$ and $\lambda \in X_0$ we have

$$\lambda(\varphi_0) = \int \varphi(s)\lambda(e_0^s)ds = \int \varphi(s)ds \cdot \mathbf{1} = \widehat{\varphi}(0) \cdot \mathbf{1} .$$

The vectors ξ in the space $\mathfrak{H}(\zeta_\infty)$ decompose according to (54): $\xi = \int_X^\oplus \xi_\lambda d\mu(\lambda)$ with $\xi_\lambda \in m(\lambda)\mathfrak{H}(\lambda)$. If ξ is supported on X_0, i.e. if $\xi_\lambda = 0$ for $\lambda \notin X_0$, then

$$\zeta_\infty(\varphi_0)\xi = \int_{X_0} m(\lambda)\lambda(\varphi_0)\xi_\lambda d\mu(\lambda) = \widehat{\varphi}(0)\xi .$$

Now we take again our $\vartheta \in \mathcal{S}(\mathbb{R})$ with $\widehat{\vartheta}(t) = 1$ for $1 \le |t| \le 2$ and so that $\widehat{\vartheta}(0) = 0$. If v is the cyclic vector for ζ_∞ with $\widetilde{\varphi}_\infty(\widetilde{x}) = (\zeta_\infty(\widetilde{x})v|v)$, then $\widetilde{\varphi}_\infty(f \underset{\infty}{\star} \vartheta_0) = \lim \widetilde{\varphi}_k(f \underset{k}{\star} \vartheta_0) = \lim \widetilde{\varphi}_k(f) = \widetilde{\varphi}_\infty(f)$ for all $f \in \mathscr{C}_0(\widetilde{G})$ implies $\zeta_\infty(\vartheta_0)v = v$. If $\xi \in \mathfrak{H}(\zeta_\infty)$ is supported by X_0, then also $\zeta_\infty(f)^*\xi$ for $f \in \mathscr{C}_0(\widetilde{G})$ is supported by X_0, hence $\zeta_\infty(\vartheta_0^* \underset{\infty}{\star} f^*)\xi = 0$ and

$$(\zeta_\infty(f)v|\xi) = (\zeta_\infty(f \underset{\infty}{\star} \vartheta_0)v|\xi) = (v|\zeta_\infty(\vartheta_0^* \underset{\infty}{\star} f^*)\xi) = 0 .$$

We have proved:

(55) *The support supp ζ_∞ of the representation ζ_∞ associated to $\widetilde{\varphi}_\infty$, is contained in the Borel subset*

$${}_\infty\widehat{\widehat{G}}_1 = \{\lambda \in {}_\infty\widehat{\widehat{G}};\ d\lambda(\widetilde{b}) \neq 0\}$$

of $\widehat{\widehat{G}}$.

Now we fix λ in supp ζ_∞. Then as before $\lim \widetilde{\varphi}_k = \widetilde{\varphi}_\infty$ implies that there exists a family $\{\widetilde{\pi}_{k,r_k}\}_{k\in\mathbb{N}}$ of irreducible representations $\widetilde{\pi}_{k,r_k} \in {}_{k,r_k}\widehat{\widehat{G}}$ with $\lim \widetilde{\pi}_{k,r_k} = \lambda$ in $\widehat{\widehat{\mathbf{G}}}$. We use the sequence $\{r_k\}$ to construct a new VLG. First we remark that

$\mathbb{R}(\widetilde{b} - r_k b_{n-1})$ is a minimal ideal in ${}_{k,r_k}\widetilde{\mathfrak{g}}$, annihilated by $d\widetilde{\pi}_{k,r_k}$. We set

$$b_j^0 = b_j \quad \text{for} \quad j \neq n-1\,, \quad b_{n-1}^0 = \widetilde{b}\,, \; \mathfrak{g}^0 = \sum_{j=1}^{n} \mathbb{R} b_j^0\,.$$

Then we have

$$(56) \qquad {}_{(k)}\widetilde{\mathfrak{g}} := {}_{k,r_k}\widetilde{\mathfrak{g}} = \mathfrak{g}^0 \oplus \mathbb{R}(\widetilde{b} - r_k b_{n-1})\,.$$

For $x, y \in \mathfrak{g}^0$ the decomposition (56) yields

$$[x, y]_{k,r_k} = [x, y]_k^0 + c_k(x, y)(\widetilde{b} - r_k b_{n-1})$$

with $[x, y]_k^0 \in \mathfrak{g}^0$ and $c_k(x, y) \in \mathbb{R}$. For $k \to \infty$ the products $[x, y]_k^0$ converge, hence we define

$$[x, y]_\infty^0 = \lim_{k \to \infty} [x, y]_k^0 \in \mathfrak{g}^0\,.$$

It is easy to see that in particular

$$[b_1^0, b_{n-1}^0]_\infty^0 = \alpha_\infty b_{n-1}^0\,, \quad [b_2^0, b_{n-1}^0]_\infty^0 = 0\,.$$

Let

$$\mathbf{g}^0 = (\overline{\mathbb{N}} \times \mathfrak{g}^0\,, [\ ,\ ]_\nu^0)$$

be the VLA, defined by the family $\{\mathfrak{g}_\nu^0, [\ ,\ ]_\nu^0\}$, $\nu \in \overline{\mathbb{N}}$. Then $\{b_j^0\}_{j=1,\dots,n}$ is a Malcev basis in $\mathbf{g}^0$. As usually $\mathbf{G}^0$ denotes the corresponding n dimensional VLG : $\mathbf{G}^0 = (\overline{\mathbb{N}} \times G^0\,, \mu^0)$ with $G^0 = \mathbb{R}^n$. For every $k \in \mathbb{N}$ the subgroup $W_k := Q_{k,r_k} = \exp_{k,r_k} \mathbb{R}(\widetilde{b} - r_k b_{n-1})$ is normal in

$${}_{(k)}\widetilde{G} := {}_{k,r_k}\widetilde{G}$$

and there is a commutative diagram with isomorphisms η_k and $d\eta_k$:

$$\begin{array}{ccc}
{}_k\mathfrak{g}^0 & \xrightarrow[d\eta_k]{} & {}_{(k)}\widetilde{\mathfrak{g}}/\mathbb{R}(\widetilde{b} - r_k b_{n-1}) \\
\Big\downarrow \exp & & \Big\downarrow \exp \\
{}_kG^0 & \xrightarrow[\eta_k]{} & {}_{(k)}\widetilde{G}/W_k\,.
\end{array}$$

As $\widetilde{\pi}_{k,r_k}$ is trivial on W_k, hence in $({}_{(k)}\widetilde{G}/W_k)^\wedge$, there is a unique family $\{\pi_k^0\} \subset \widehat{\mathbf{G}}^0$ with $\widetilde{\pi}_{k,r_k} = \pi_k^0 \circ \eta_k$. The $\widetilde{\pi}_{k,r}$ had been induced from the subgroups ${}_{k,r}\widetilde{H}$, see (39). These ${}_{k,r_k}\widetilde{H}$ contain W_k and with $\widetilde{\pi}_{k,r_k}$ also the τ_k are trivial on W_k. It follows easily that we can write

$$\pi_k^0 = \operatorname{ind}_{{}_kH^0}^{{}_kG^0} \tau_k^0$$

with ${}_kH^0 = \exp_k \mathfrak{h}^0$, $\mathfrak{h}^0 = \mathbb{R}b_1^0 \oplus \mathfrak{k}^0$, $\mathfrak{k}^0 = \sum_{j=3}^{n} \mathbb{R}b_j^0$. Furthermore the representation $\lambda \in {}_\infty\widehat{\widetilde{G}}$ is trivial on $W_\infty = \exp_\infty \mathbb{R}b_{n-1}$. This is equivalent to $d\lambda(b_{n-1}) = 0$ and follows from

(57) **Lemma.** *Let* $\mathbf{G}$ *be an arbitrary variable Lie group and* $\mathfrak{g}$ *its VLA. Let* $\{a_k\}_{k\in\mathbb{N}}$ *be a sequence in* $\mathfrak{g}$ *with* $\lim a_k = a_\infty \in \mathfrak{g}$ *and let* $\{\pi_k\}_{k\in\mathbb{N}}$ *be a sequence in* $\widehat{\mathbf{G}}$*, converging to* $\pi_\infty \in \widehat{\mathbf{G}}$*. If* $d\pi_k(a_k) = 0$ *for all* $k \in \mathbb{N}$*, then also* $d\pi_\infty(a_\infty) = 0$.

Proof. The relation $d\pi_k(a_k) = 0$ for $a_k \in {}_k\mathfrak{g}$ is equivalent to the fact that all coefficients $x \to (\pi_k(x)\xi_k|\vartheta_k)_k$ of π_k are invariant under right translation with elements in $\exp_k \mathbb{R}a_k$, i.e. that

$$(58) \qquad (\pi_k(x \underset{k}{\cdot} e^{ta_k})\xi_k|\vartheta_k)_k = (\eta_k(x)\xi_k|\vartheta_k)$$

holds for all $t \in \mathbb{R}$. The same for $k = \infty$. However, any $(\pi_\infty(x)\xi_\infty|\vartheta_\infty)_\infty$ is a limit uniformly on compacta of a sequence $\{(\pi_k(x)\xi_k|\vartheta_k)\}_k$. As $\lim\limits_{k\to\infty} x \underset{k}{\cdot} e^{ta_k} = x \underset{\infty}{\cdot} e^{ta_\infty}$ in G, the relations (58) for all $k \in \mathbb{N}$ imply, that (58) holds also for $k = \infty$, i.e. that $d\pi_\infty(a_\infty) = 0$.

As ${}_\infty\widetilde{G}/W_\infty \cong {}_\infty G^0$, we see that λ defines $\lambda^0 \in {}_\infty\widehat{G}^0 \subset \widehat{\mathbf{G}}^0$. We claim that also $\lambda^0 = \lim\limits_{k\to\infty} \pi_k^0$ in $\widehat{\mathbf{G}}^0$. This is a consequence of the following facts: For every $\nu \in \overline{\mathbb{N}}$ we have epimorphisms

$$dT_\nu = \begin{cases} {}_{(k)}\widetilde{\mathfrak{g}} \;\to\; {}_k\mathfrak{g}^0, & \text{for} \quad \nu = k \in \mathbb{N} \\ {}_\infty\widetilde{\mathfrak{g}} \;\to\; {}_\infty\mathfrak{g}^0, & \text{for} \quad \nu = \infty \end{cases}$$

defined by $dT_\nu(x) = x^0$ if $x = x^0 + \xi_k(\widetilde{b} - r_k b_{n-1})$ for $\nu \in N$, resp. $x = x^0 + \xi_\infty b_{n-1}$ for $\nu = \infty$. Now we identify $\overline{\mathbb{N}}$ with the closed subset $\{(k, r_k)\}_{k\in\mathbb{N}} \bigcup \{\infty\}$ of $\widetilde{P}$, hence $\widetilde{\mathbf{G}}_0 := \overline{\mathbb{N}} \times \widetilde{\mathfrak{g}}$ is a variable subalgebra of $\widetilde{P} \times \widetilde{\mathfrak{g}} = \widetilde{\mathfrak{g}}$ and $\widetilde{\mathbf{G}}_0 := \overline{\mathbb{N}} \times \widetilde{G}$ the corresponding variable subgroup of $\widetilde{\mathbf{G}}$. Gluing together the dT_ν we obtain a commutative diagram

$$(59) \qquad \begin{array}{ccc} \widetilde{\mathfrak{g}}_0 & \xrightarrow{dT} & \mathfrak{g}^0 \\ {\scriptstyle E}\big\downarrow & & {\scriptstyle E}\big\downarrow \\ \widetilde{\mathbf{G}}_0 & \xrightarrow{T} & \mathbf{G}^0 . \end{array}$$

It is obvious that λ and the sequence $\{\widetilde{\pi}_{k,r_k}\}$ can be considered as representations of $\widetilde{\mathbf{G}}_0$, that $\lambda = \lim \widetilde{\pi}_{k,r_k}$ also holds in $\widehat{\widetilde{\mathbf{G}}}_0$ and that $\widetilde{\pi}_{k,r_k} = \pi_k^0 \circ T$, $\lambda = \lambda^0 \circ T$.

We claim that also $\lambda_0 = \lim \pi_k^0$ in $\widehat{\mathbf{G}^0}$. To this end we introduce a new VLG $\mathbf{F}$ equivalent to $\widetilde{\mathbf{G}}_0$: Let $\mathfrak{f}$ be a vector space with basis $c_1, c_2, \ldots, c_n, \widetilde{c}$. For all $\nu \in \overline{\mathbb{N}}$ define

$$d\Phi_\nu(\widetilde{c}) = \widetilde{b}\,, \quad d\Phi_\nu(c_j) = b_j \quad \text{if} \quad j \neq n-1$$

$$d\Phi_\nu(c_{n-1}) = \begin{cases} b_{n-1} - \frac{1}{r_\nu}\widetilde{b} & \text{if} \quad \nu < \infty \\ b_{n-1} & \text{if} \quad \nu = \infty\,. \end{cases}$$

The family $\{\Phi_\nu\}$ defines a bijective mapping

$$d\Phi : \overline{\mathbb{N}} \times \mathfrak{f} \to \overline{\mathbb{N}} \times \widetilde{\mathfrak{g}}$$

and it is obvious that $\overline{\mathbb{N}} \times \mathfrak{f}$ has exactly one variable Lie algebra structure such that $d\Phi$ is an equivalence between $\mathbf{f} = \overline{\mathbb{N}} \times \mathfrak{f}$ and $\widetilde{\mathbf{g}}_0 = \overline{\mathbb{N}} \times \widetilde{\mathfrak{g}}$ and the $c_j, \widetilde{c}$ form a Malcev basis of $\mathbf{f}$. Let $\mathbf{F} = \overline{\mathbb{N}} \times F$ be the associated VLG and Φ the corresponding equivalence from $\mathbf{F}$ onto $\widetilde{\mathbf{G}}_0$.

We set

$$\mathfrak{f}^0 = \sum_{j \neq n-1} \mathbb{R}e_j \oplus \mathbb{R}\widetilde{c}\,.$$

Then the analogue of (56) is

$$\mathfrak{f} = \mathfrak{f}^0 \oplus \mathbb{R}c_{n-1}$$

and $d\Phi_k(\mathfrak{f}^0) = \mathfrak{g}^0$ for all $k \in \mathbb{N}$ and $\mathbb{R}c_{n-1}$ is now an ideal in all ${}_\nu\mathfrak{f}$. This implies that we can take $\widetilde{c}, c_n, c_{n-1}$ in this order as the last three elements of the Malcev basis $\{c_j, \widetilde{c}\}$. For F we get then $\mathbb{R}^{n+1}$ as coordinate system with

$$V := \{x \in \mathbb{R}^{n+1}\,;\ x_1 = \cdots = x_n = 0\} = E(\nu, \mathbb{R}c_{n-1})$$

for all $\nu \in \overline{\mathbb{N}}$, in particular V is a normal one parameter subgroup in every ${}_\nu F$. If $\overline{\pi}_k = \widetilde{\pi}_{k,r_k} \circ \Phi$ and $\overline{\lambda} = \lambda \circ \Phi$, then all $\overline{\pi}_k$ and $\overline{\lambda}$ are trivial on V and obviously

$$\lim \overline{\pi}_k = \overline{\lambda} \quad \text{in} \quad \widehat{\mathbf{F}}$$

because Φ is an equivalence.

The composition $\Psi = T \circ \Phi$ of Φ and T is a homomorphism from $\mathbf{F}$ onto $\mathbf{G}^0$ with $\overline{\pi}_k = \pi_k^0 \circ \Psi$, $\overline{\lambda} = \lambda^0 \circ \Psi$. By construction $d\Psi$ is the projection of F onto $\mathfrak{g}^0$, i.e. $d\Psi_\nu$ is a linear isomorphism from $\mathfrak{f}^0$ onto $\mathfrak{g}^0$, independent of ν, and vanishes on c_{n-1}. It follows that Ψ is the projection of $\mathfrak{f} = \mathbb{R}^{n+1}$ onto $G^0 = \mathbb{R}^n$, obtained by deleting the last coordinate. Hence we can write $x = (x^0, x_{n-1})$ for $x \in F$, with $x^0 \in \mathbb{R}^n = G^0$, $x_{n-1} \in \mathbb{R}$. For the representation $\overline{\pi}_k = \pi_k^0 \circ \Psi$ and $\overline{\lambda} = \lambda \circ \Psi$ this implies $\overline{\pi}_k(x) = \pi_k^0(x^0)$, $\overline{\lambda}(x) = \lambda^0(x^0)$ and similarly for the coefficients, in particular for the associated positive definite functions. As $\lim \overline{\pi}_k = \overline{\lambda}$, we conclude that also $\lim \pi_k^0 = \lambda^0$, as we had claimed. The semidirect decomposition ${}_{(k)}\widetilde{\mathfrak{g}} = {}_k\mathfrak{g} \oplus \mathbb{R}(\widetilde{b} - r_k b_{n-1})$ shows that the extension $\widetilde{l}_k$ of the functional $l_k \in {}_k\mathfrak{g}^*$, defined by $\widetilde{l}_k|_{\mathfrak{g}} = l_k$, $\widetilde{l}_k(\widetilde{b} - r_k b_{n-1}) = 0$ represents the orbit $K^{-1}(\widetilde{\pi}_{k,r_k}) \subset \widetilde{\mathfrak{g}}^*$. As $l_k(b_{n-1}) = 0$ it follows that $\widetilde{l}_k(b_j) = l_k(b_j)$ for $j = 1, 2 \dots, n$ and $\widetilde{l}_k(\widetilde{b}) = 0$. Furthermore this implies that a functional $l_k^0 \in \mathfrak{g}^{0*}$, defining π_k^0, is given by $l_k^0(b_j^0) = l_k(b_j)$, $j = 1, \dots, n$, in particular $l_k^0(b_{n-1}^0) = 0$ for all $k \in \mathbb{N}$. So we are now in the same position as before, i.e. after (7): The sequence of representations $\pi_k^0 = K(\Omega_{(k,l_k^0)})$ converges in $\widehat{\mathbf{G}^0}$ to λ^0, all β_k vanish, hence $u_k = b_{n-1}^0$ for all $k \in \mathbb{N}$ and $l_k^0(b_{n-1}^0) = 0$. Finally we know that $d\lambda^0(b_{n-1}^0) \neq 0$, that means that we have either case α or case β with the additional condition (26). But then we can conclude that a subsequence of the $\mathbf{G}^0$ orbits of the l_k^0 converges to the orbit of λ^0. Let $\{l_k^0\}$ be the subsequence. The diagram (59) allows to shift the situation from $\mathbf{G}^0$ to $\widetilde{\mathbf{G}}_0$ and further to $\widetilde{\mathbf{G}}$: The homomorphisms dT_ν map the ${}_{(\nu)}\mathfrak{g}$ onto ${}_\nu\mathfrak{g}^0$, hence the adjoints $(dT_\nu)^*$ map $\mathfrak{g}^{0*}$ into $\widetilde{\mathfrak{g}}^*$ and define the converging sequence of the $\widetilde{l}_\nu = (dT_\nu)^*(l_\nu^0)$. For $k \in \mathbb{N}$ we have

$$\widetilde{l}_k(b_j) = l_k^0(b_j^0) \text{ if } \quad j \neq n-1 , \qquad \widetilde{l}_k(b_{n-1}) = \frac{1}{r_k} l_k^0(b_{n-1}^0) , \qquad \widetilde{l}_k(\widetilde{b}) = l_k^0(b_{n-1}^0) ,$$

and $\widetilde{l}_k(\widetilde{b} - r_k b_{n-1}) = 0$. Further

$$\widetilde{l}_\infty(b_j) = l_\infty^0(b_j^0) , \; j \neq n-1 , \qquad \widetilde{l}_\infty(\widetilde{b}) = l_\infty^0(b_{n-1}^0) , \qquad \widetilde{l}_\infty(b_{n-1}) = 0 ,$$

in particular $\widetilde{l}_\infty = \lim\limits_{k\to\infty} \widetilde{l}_k$ in $\widetilde{\mathfrak{g}}^*$. The orbit $\widetilde{\Omega}_k := \Omega_{(k,\widetilde{l}_k)} \in \Omega(\widetilde{\mathbf{G}}_0)$ clearly belongs to $\widetilde{\pi}_{k,r_k}$ and $\widetilde{\Omega}_\infty := \Omega_{(\infty,\widetilde{l}_\infty)}$ to λ. So the convergence of $\{\widetilde{l}_k\}$ implies

$$K^{-1}(\lambda) = \lim_{k\to\infty} K^{-1}(\widetilde{\pi}_{k,r_k}) . \tag{60}$$

As the orbits $\widetilde{\Omega}_k$, $\widetilde{\Omega}_\infty$ as subsets of $\widetilde{\mathfrak{g}}^*$ are the same, whether we consider them as elements of $\Omega(\widetilde{\mathbf{G}}_0)$ or of $\Omega(\widetilde{\mathbf{G}})$, we see that we can think of (60) also as a relation in $\Omega(\widetilde{\mathbf{G}})$, hence we have proved

(61) *For every λ occurring in the decomposition (54) of ζ_∞, i.e. $\lambda \in X \subset {}_\infty\widehat{\widehat{\mathbf{G}}}$, there exists a sequence $\{(k, r_k)\}$ in $\widetilde{P}$ such that*

$$\lim K^{-1}(\widetilde{\pi}_{k,r_k}) = K^{-1}(\lambda) .$$

To finish the proof in the present case we somehow have to restrict the different representations onto G, resp. onto D. For this we use first (49). Let

$$\psi = \widetilde{\varphi}_\infty|_D = \varphi_\infty|_D$$

be the restriction of $\widetilde{\varphi}_\infty$, resp. φ_∞ onto ${}_\infty D$ and let σ be a cyclic representation of ${}_\infty D$ with cyclic vector ξ, associated with ψ, hence

$$\widetilde{\varphi}_\infty(y) = \varphi_\infty(y) = (\sigma(y)\xi|\xi) \quad \text{for} \quad y \in {}_\infty D\ .$$

We claim that for the kernels in $C^*({}_\infty D)$ we have

$$\ker \zeta_\infty|_D \bigcup \ker \pi_\infty|_D \subset \ker \sigma. \tag{62}$$

This follows immediately from

(63) **Lemma.** *Let G be an amenable locally compact group, H a closed subgroup, π a cyclic representation of G with cyclic vector ω, σ a cyclic representation of H with cyclic vector ξ. If*

$$(\pi(y)\omega|\omega) = (\sigma(y)\xi|\xi)$$

for all $y \in H$, then

$$\ker \pi|_H \subset \ker \sigma\ .$$

Proof. The amenability of G implies that the C^* hull $C^*(H)$ is a subalgebra of the adjoint algebra $C^*(G)^\flat$, moreover

$$\ker \pi|_H = C^*(H) \bigcap \ker \pi^\flat$$

where $\pi^\flat$ is the extension of π onto $C^*(G)^\flat$, see [18]. Now let a be a positive element in $C^*(H)$ with $\sigma(a) \neq 0$. Then, as σ is cyclic, there exists $u \in \mathscr{C}_0(H)$ with $(\sigma(u^*au)\xi|\xi) > 0$. Let $\{a_j\}$ be a sequence of positive elements in $\mathscr{C}_0(H)$ with $\lim a_j = a$. Then

$$\begin{aligned}
&(\pi^\flat(u^* \star a_j \star u)\omega|\omega) = \\
&= \int_H u^* \star a_j \star u(y)(\pi(y)\omega|\omega)dy = \int_H u^* \star a_j \star u(y)(\sigma(y)\xi|\xi)dy = \\
&= (\sigma(u^* \star a_j \star u)\xi|\xi).
\end{aligned}$$

Taking the limit $j \to \infty$ we get $(\pi^\flat(u^*au)\omega|\omega) = (\sigma(u^*au)\xi|\xi) > 0$, hence $\pi^\flat(u^*au) = \pi^\flat(u)^*\pi^\flat(a)\pi^\flat(u) \neq 0$ and consequently $a \notin \ker \pi^\flat$. For $G = {}_\infty G$, $H = {}_\infty D$, $\pi = \pi_\infty$, resp. $G = {}_\infty\widetilde{G}$, $H = {}_\infty D$, $\pi = \zeta_\infty$ we obtain (62). As ${}_\infty D$

is normal in ${}_\infty G$ its dual ${}_\infty\widehat{D}$ is a ${}_\infty G$–space with locally closed orbits. As π_∞ is irreducible, the hull $h(\pi_\infty|_D)$ of its restriction is a minimal closed ${}_\infty G$ invariant subset of ${}_\infty\widehat{D}$. But $h(\zeta_\infty|_D)$ obviously also is a closed invariant subset. (62) implies that $h(\pi_\infty|_D)\cap h(\zeta_\infty|_D)\neq 0$, thus

$$h(\pi_\infty|_D)\subset h(\zeta_\infty|_D)\,.$$

By (54) we have $h(\zeta_\infty|_D)=(\bigcup_X h(\lambda|_D))^-$. Now because of $[b_2,\widetilde{b}]_\infty=0$, ${}_\infty D$ is a normal subgroup of ${}_\infty\widetilde{G}$ of codimension two, thus every $\lambda|_D$ decomposes into irreducible representations $\kappa\in{}_\infty\widehat{D}$, such that the orbit of κ lies in the restriction of the orbit of λ in $\widetilde{\mathfrak{g}}^*$ onto $\mathfrak{d}$. It follows that $h(\pi_\infty|_D)$ is contained in the closure in ${}_\infty\widehat{D}$ of the set of all these κ, hence if $\rho\in h(\pi_\infty|_D)$, then there exists a sequence $\{\kappa_j\}$ of irreducible representations, occuring in the decomposition of some $\lambda_j|_D$, $\lambda_j\in X$, with $\rho=\lim\kappa_j$. But $\dim D=n-1$, thus by our induction hypothesis we have $K^{-1}(\rho)=\lim_{j\to\infty}K^{-1}(\kappa_j)$ for the corresponding orbits in the orbit space $\Omega({}_\infty D)$. This implies that for $l_\infty\in K^{-1}(\pi_\infty)$ with $\rho=K(\Omega_{(\infty,l_\infty|_\mathfrak{d})})$ we can find $f_j\in K^{-1}(\lambda_j)$ with $l_\infty|_\mathfrak{d}=\lim f_j|_\mathfrak{d}$. Because of $d\lambda_j(\widetilde{b})\neq 0$ we have $f_j(\widetilde{b})\neq 0$. Now let $\alpha_\infty\neq 0$. We claim that

$$f_j+\widetilde{\mathfrak{d}}^\perp\subset\widetilde{\Omega}(f_j) \tag{64}$$

where $\widetilde{\Omega}(f_j)$ is the orbit of f_j in ${}_\infty\widetilde{\mathfrak{g}}^*$. Again it suffices to show $\widetilde{\mathfrak{g}}^{f_j}\subset\widetilde{\mathfrak{d}}$. So assume that $x=\xi b_1+c$ is contained in $\widetilde{\mathfrak{g}}^{f_j}$ with $c\in\widetilde{\mathfrak{d}}$. Because of $\gamma_\infty=0$ we have $[b_2,\widetilde{b}]_\infty=0$, hence $[c,\widetilde{b}]_\infty=0$ and consequently $0=f_j([x,\widetilde{b}]_\infty)=\xi\alpha_\infty f_j(\widetilde{b})$, which implies $\xi=0$, $x\in\widetilde{\mathfrak{d}}$.

We choose a basis $\{b_j^*,b^*\}$ of $\widetilde{\mathfrak{g}}^*$, which is dual to $\{b_j,\widetilde{b}\}$. Then $\mathfrak{g}^*=\sum_1^n\mathbb{R}b_j^*$, $\mathfrak{d}^*=\sum_2^n\mathbb{R}b_j^*$ and $\widetilde{\mathfrak{d}}^\perp=\mathbb{R}b_1^*$. We can write

$$f_j=\sum_{i=1}^n\xi_j^ib_i^*+\xi_jb^*\,,\quad l_\infty=\sum_{i=1}^n\xi_\infty^ib_i^*$$

and $\lim f_j|_\mathfrak{d}=l_\infty|_\mathfrak{d}$ implies $\lim_{j\to\infty}\xi_j^i=\xi_\infty^i$ for $i=2,3,\ldots,n$. Now (64) means, that we can choose ξ_j^1 arbitrarily, without leaving the orbit $\widetilde{\Omega}(f_j)$. So it we set $\xi_j^1=\xi_\infty^1$ for all $j\in\mathbb{N}$ we see that $\lim f_j|_\mathfrak{g}=l_\infty$ in $\mathfrak{g}^*$. As we know already, there exists an increasing sequence $\{k_j\}_j$ in $\mathbb{N}$ and $l_{(j)}$ in the orbit Ω_{k_j}, with

$$\left|f_j|_\mathfrak{g}-l_{(j)}\right|<\frac{1}{j}$$

for all $j\geq 1$, hence $l_\infty=\lim_{j\to\infty}l_{(j)}$ and so finally $\Omega_\infty=\lim_{k\to\infty}\Omega_k$.

Now assume that $\alpha_\infty = 0$. In this case we have even

$$(65) \qquad \widetilde{\varphi}_\infty|_G = \varphi_\infty \, .$$

Proof. (52) shows that there is a constant C_T, depending only on T, such that

$$|N_k^2(x)| < \alpha_k C_T$$

uniformly for all $x \in G$ with $|x_1| \leq T$. Hence it follows from (51) that $\lim\limits_{k\to\infty} \widetilde{\varphi}_k(x) = \varphi_\infty(x)$ uniformly on compacta in G. The rest of the proof goes exactly as in the general case treated earlier. Again with the same arguments as before, formula (64) and Lemma (63) show: π_∞ is contained in the support of the restriction $\zeta_\infty|_G$ of ζ_∞ onto G, i.e. in $(\bigcup\limits_{\lambda\in X} h(\lambda|_G))^-$, hence there exists a sequence $\{\mu_j\} \subset {}_\infty\widehat{G}$ with $\mu_j \in h(\lambda_j|_G)$, $\lambda_j \in X$, such that $\pi_\infty = \lim\limits_{j\to\infty} \mu_j$. We claim that actually $\mu_j = \lambda_j|_G$, i.e. the restrictions of the $\lambda \in X$ onto ${}_\infty G$ in ${}_\infty\widetilde{G}$ are irreducible: From $\alpha_\infty = \gamma_\infty = 0$ it follows that $[v_1, \widetilde{b}]_\infty = [b_2, \widetilde{b}]_\infty = 0$, so that $\mathbb{R}\widetilde{b}$ is central in ${}_\infty\widetilde{\mathfrak{g}}$ and consequently $S = e_0^{\mathbb{R}}$ is central in ${}_\infty\widetilde{G}$. But then ${}_\infty\widetilde{G}$ is the group theoretical direct product of ${}_\infty G$ and S, which implies our claim. Furthermore, also $\mathbb{R}b_{n-1}$ is central in ${}_\infty\mathfrak{g}$, hence ${}_\infty\mathfrak{g}$ and ${}_\infty G$ have centers of dimension at least 2, which means that we are in the case which we had treated before, so we conclude that $\lim \mu_j = \pi_\infty$ implies $\lim K^{-1}(\mu_j) = K^{-1}(\mu_\infty)$, i.e. we have a sequence of forms f_j in the orbits of the λ_j, hence $f_j|_\mathfrak{g} \in K^{-1}(\mu_j)$, with $\lim f_j|_\mathfrak{g} = l_\infty$. As before this implies that we can again find a sequence $\{l_{(j)}\}$, $l_{(j)} \in \Omega_{k_j}$, with $l_\infty = \lim l_{(j)}$, hence $\Omega_\infty = \lim\limits_{k\to\infty} \Omega_k$.

It remains to treat case A, so let us assume that there exist $T > 0$ and $c > 0$ such that

$$\int_{-T}^{T} |\omega_k(e_2^t)|^2 dt \geq c$$

for all $k \in \mathbb{N}$. Again we construct first an $(n+1)$ dimensional VLA $\widetilde{\mathbf{g}} = (\overline{\mathbb{N}} \times \widetilde{\mathfrak{g}}, [\ ,\ ]_\nu)$. Let $\widetilde{\mathfrak{g}} = \sum_{i=1}^n \mathbb{R}b_i \oplus \mathbb{R}\widetilde{b}$ with $\mathfrak{g} = \sum_{i=1}^n \mathbb{R}b_i = \mathbb{R}b_1 \oplus \mathbb{R}b_2 \oplus \mathfrak{k}$ as before and set

$$\begin{array}{lll} [x, y]_\nu = [x, y]_\nu & \text{for } x, y \in \mathfrak{g}\,, & \text{product in } {}_\nu\mathfrak{g}\,, \\ [x, \widetilde{b}]_\nu = 0 & \text{for } x \in \mathfrak{k} & \\ [b_1, \widetilde{b}]_\nu = \alpha_\nu \widetilde{b}\,, & [b_2, \widetilde{b}]_\nu = b_n\,, & \text{all } \nu \in \overline{\mathbb{N}}\,. \end{array}$$

Then $\{b_1, b_2, \ldots, b_{n-1}, \widetilde{b}, b_n\}$ is a Malcev basis which we use to define the corresponding VLG $\widetilde{\mathbf{G}} = \overline{\mathbb{N}} \times \widetilde{G}$ with $\widetilde{G} = \mathbb{R}^{n+1}$. For every $\nu \in \overline{\mathbb{N}}$ the element

$$c_\nu = b_{n-1} - \gamma_\nu \widetilde{b}$$

spans a minimal ideal $\mathbb{R}c_\nu$ in ${}_\nu\widetilde{\mathfrak{g}}$.

Next we define $\mathbf{g}^\bullet = (\overline{\mathbb{N}} \times \mathfrak{g}^\bullet, [\ ,\ ]^\bullet_\nu)$: Let $\mathfrak{g}^\bullet = \sum_{i \neq n-1} \mathbb{R}b_i \oplus \mathbb{R}\widetilde{b}$, hence, similar to (56):

$$ {}_k\widetilde{\mathfrak{g}} = \mathfrak{g}^\bullet \oplus \mathbb{R}c_k \tag{66} $$

for all $k \in \mathbb{N}$. Let P_k be the projection of $\widetilde{\mathfrak{g}}$ onto $\mathfrak{g}^\bullet$, according to the decomposition (66). We set

$$ [x, y]^\bullet_k = P_k([x, y]_k) . $$

Again as before

$$ [x, y]^\bullet_\infty := \lim_{k \to \infty} [x, y]^\bullet_k $$

exists for all $x, y \in \mathfrak{g}^\bullet$ and defines the product $[\ ,\ ]^\bullet_\infty$ on $\mathfrak{g}^\bullet$. We note that

$$ [b_1, \widetilde{b}]^\bullet_\nu = \alpha_\nu \widetilde{b} \ , \quad [b_2, \widetilde{b}]^\bullet_\nu = b_n $$

in *all* ${}_\nu\mathfrak{g}^\bullet$, hence we can define a VLA $\mathbf{g}^\bullet = (\overline{\mathbb{N}} \times \mathfrak{g}^\bullet, [\ ,\ ]^\bullet_\nu)$, in which $\widetilde{b}$ replaces the basis element b_{n-1} in our former $\mathbf{g}$ and all $\gamma^\bullet_\nu$ are 1. So in particular the corresponding $\varepsilon^\bullet_k = \frac{\alpha^\bullet_k}{\gamma^\bullet_k} = \alpha_k$ converge to $\varepsilon^\bullet_\infty = \alpha_\infty$. Obviously the three VLA's are linked by homomorphisms dA, dB, which lift to homomorphisms A und B of the corresponding variable groups:

$$ \begin{array}{ccccc} \mathbf{g} & \xrightarrow{dA} & \widetilde{\mathbf{g}} & \xrightarrow{dB} & \mathbf{g}^\bullet \\ {\scriptstyle E}\downarrow & & \downarrow & & \downarrow \\ \mathbf{G} & \xrightarrow{A} & \widetilde{\mathbf{G}} & \xrightarrow{B} & \mathbf{G}^\bullet \end{array} $$

As $\mathbb{R}c_k$ is an ideal in ${}_k\widetilde{\mathfrak{g}}$, the one parameter subgroup $C_k := E(k, \mathbb{R}c_k)$ is normal in ${}_k\widetilde{G}$ and because ${}_k\widetilde{G} = {}_kG \ltimes C_k$, hence ${}_kG \cong {}_k\widetilde{G}/C_k$, the representations $\pi_k \in {}_k\widehat{G}$ define $\widetilde{\pi}_k \in {}_k\widehat{\widetilde{G}}$, which are trivial on C_k. Hence

$$ \widetilde{\pi}_k|_G = \pi_k \ , \quad \widetilde{\pi}_k(e_0^{\gamma_k s}) = \pi_k(e^s_{n-1}) $$

where again we have set $e^s_0 = \exp_k s\widetilde{b}$ in ${}_k\widetilde{G}$. Furthermore, as C_k is the kernel of the epimorphism B_k from ${}_k\widetilde{G}$ onto ${}_kG^\bullet$, there are unique $\pi^\bullet_k \in {}_k\widehat{G}^\bullet$ with $\widetilde{\pi}_k = \pi^\bullet_k \circ B_k$ for $k \in \mathbb{N}$.

Now we take the cyclic vectors ω_k of π_k to define positive definite functions $\widetilde{\varphi}_k$ for $\widetilde{\pi}_k$:

$$ \widetilde{\varphi}_k(\widetilde{x}) = (\widetilde{\pi}_k(\widetilde{x})\omega_k | \omega_k) \ . $$

As $\widetilde{\varphi}_k(e) = \varphi_k(e) = 1$, all $\widetilde{\varphi}_k$ are in $\mathcal{P}_1(\widetilde{\mathbf{G}})$, hence we can assume that the w^* limit

$$ \widetilde{\varphi}_\infty = \lim_{k \to \infty} \widetilde{\varphi}_k $$

exists. We claim that $\widetilde{\varphi}_\infty \neq 0$.
This is certainly a consequence of the following stronger result:

(67) *There exists a cyclic representation* $(\zeta_\infty, \xi_\infty)$ *of* ${}_\infty\widetilde{G}$ *with* $\widetilde{\varphi}_\infty(\widetilde{x}) = (\zeta_\infty(\widetilde{x})\xi_\infty|\xi_\infty)$ *and*

$$\ker \zeta_\infty|_{C^*(\mathbf{G})} \subset \ker \pi_\infty$$

Proof. The vector $\widetilde{x} \in \mathbb{R}^{n+1}$ was identified with the element $e_1^{x_1} \underset{\nu}{\cdot} e_2^{x_2} \underset{\nu}{\cdot} \ldots \cdot e_0^{x_0} \underset{\nu}{\cdot} e_n^{x_n}$ in ${}_\nu\widetilde{G}$, hence because $e_0^{x_0} \underset{\nu}{\cdot} e_n^{x_n} = e_n^{x_n} \underset{\nu}{\cdot} e_0^{x_0}$ for all $\nu \in \overline{\mathbb{N}}$, we can write $\widetilde{x} = x \underset{\nu}{\cdot} e_0^{x_0}$ with $x = (x_1, x_2, \ldots, x_n) \in \mathbb{R}^n = {}_\nu G = G$ for all ν. So if $f \in \mathscr{C}_\infty(\mathbf{L}(\mathbf{G}))$ and $h \in \mathscr{C}_\infty(\mathbf{L}(\mathbf{S}))$ (where clearly $\mathbf{S}$ is the trivial VLG $\overline{\mathbb{N}} \times \mathbb{R}$, $\mathbb{R} = S \stackrel{\sim}{=} e^{\mathbb{R}\widetilde{b}} = e_0^{\mathbb{R}}$), then

$$f \odot h(\nu, \widetilde{x}) = f(\nu, x)h(\nu, x_0)$$

defines an element $f \odot h$ in $\mathscr{C}_\infty(\mathbf{L}(\widetilde{\mathbf{G}}))$. Observe that f and h both lie in $C^*(\widetilde{\mathbf{G}})^\flat$, hence for any unitary representation π of $C^*({}_\nu\widetilde{G})$ we have

$$\begin{aligned}\pi(f \odot h) &= \int f(\nu, x)h(\nu, x_0)\pi(\widetilde{x})d\widetilde{x} = \\ &= \int f(\nu, x)\pi(x)dx \int h(\nu, x_0)\pi(e_0^{x_0})dx_0 = \\ &= \pi^\flat(f)\pi^\flat(h) \ .\end{aligned}$$

This implies $|f \odot h|_* \leq |f|_*|h|_*$ for the C^* norms and shows that $(f, h) \to f \odot h$ extends to a bounded bilinear $C^*(\mathbf{G})$–left $C^*(\mathbf{S})$–right mapping of $C^*(\mathbf{G}) \times C^*(\mathbf{S})$ into $C^*(\widetilde{\mathbf{G}})$.

Now let us assume that we have an element $a \in C^*(\mathbf{G})$ with $a^* = a$ and

$$\pi_\infty(a)\omega_\infty = \omega_\infty\,, \quad |a|_* = 1 \ .$$

Further, let h be in $\mathscr{S}(\mathbb{R}) \subset C^*(S)$ with $0 \leq \hat{h} \leq 1$, $\hat{h}(s) = 1$ for $|s| \leq T$. Let $h(\nu, x_0) = h(x_0)$ for all $\nu \in \overline{\mathbb{N}}$. Then

$$\widetilde{\pi}_k(h)\xi(e_2^t) = \hat{h}(t)\xi(e_2^t)$$

for all $k \in \mathbb{N}$ and $\xi \in \mathfrak{H}(\widetilde{\pi}_k) = \mathfrak{H}(\pi_k)$, hence

$$\widetilde{\varphi}_k(a \odot h) = (\widetilde{\pi}_k(a \odot h)\omega_k|\omega_k) = (\hat{h}\omega_k|\pi_k(a)\omega_k)\,.$$

From $\lim \widetilde{\varphi}_k(a) = \lim(\pi_k(a)\omega_k|\omega_k) = (\pi_\infty(a)\omega_\infty|\omega_\infty) = 1$ and $|a|_* = 1$ we get as usually $\lim\limits_{k\to\infty} |\pi_k(a)\omega_k - \omega_k| = 0$. As $|\hat{h}\omega_k| \leq 1$ for all k it follows that

$\lim |(\hat{h}\omega_k|\pi_k(a)\omega_k - \omega_k)| = 0$ and consequently

$$|\widetilde{\varphi}_\infty(a \odot h)| = \lim_{k\to\infty} |\widetilde{\varphi}_k(a \odot h)| = \lim_{k\to\infty} |(\hat{h}\omega_k|\omega_k) + (\hat{h}\omega_k|\pi_k(a)\omega_k - \omega_k)|$$
$$\geq \liminf(\hat{h}\omega_k|\omega_k) \geq c \ ,$$

in particular $\zeta_\infty(a \odot h) = \zeta_\infty(a)\zeta_\infty(h) \neq 0$, $\zeta_\infty(a) \neq 0$. Let $\mathcal{K}$ be the kernel of $\zeta_\infty|_{C^*(\mathbf{G})}$ in $C^*(\mathbf{G})$. If $\pi_\infty(\mathcal{K}) \neq 0$, then the restriction $\pi_\infty|_{\mathcal{K}}$ is irreducible, hence the transitivity theorem (see [7]) guarantees the existence of an hermitian $a \in \mathcal{K}$ with $\pi_\infty(a)\omega_\infty = \omega_\infty$. It is easy to see that one also can assume that $|a|_* = 1$. But, as we have seen, this implies $\zeta_\infty(a) \neq 0$, i.e. $a \notin \mathcal{K}$. Thus (67) follows.

Let us decompose ζ_∞ into irreducible representations

$$\zeta_\infty = \int_X^\oplus m(\lambda)\lambda d\mu(\lambda) \tag{68}$$

with multiplicity function $m(\lambda)$ and the support X of the measure μ in ${}_\infty\widehat{\widetilde{G}}$. From lemma (57) we see that the λ's in X are trivial on $\ker B_\infty = e_{n-1}^{\mathbb{R}} \subset {}_\infty\widetilde{G}$, hence for fixed $\lambda \in X$ there exists $\lambda^\bullet \in {}_\infty\widehat{G}^\bullet$ with $\lambda = \lambda^\bullet \circ B_\infty$. The same reasoning as before shows that also in the present case for $\lambda \in X$ there exists a subsequence $\{\pi_{k_j}^\bullet\}_j$ of the $\pi_k^\bullet$ with $\lambda^\bullet = \lim_{j\to\infty} \pi_{k_j}^\bullet$. In the definition of $d\varphi_\nu(c_{n-1})$ we only have to take $b_{n-1} - \gamma_\nu \widetilde{b}$ instead of $b_{n-1} - \frac{1}{r_\nu}\widetilde{b}$. But, as we had mentioned already, now $\lim_{k\to\infty} \varepsilon_k^\bullet = \varepsilon_\infty^\bullet$ exists, i.e. we are in case α and so we conclude that $\lim_{j\to\infty} K^{-1}(\pi_{k_j}^\bullet) = K^{-1}(\lambda^\bullet)$ holds for the orbits, which implies

$$K^{-1}(\lambda) = \lim_{j\to\infty} K^{-1}(\widetilde{\pi}_{k_j}) \tag{69}$$

for the orbits in $\widetilde{\mathfrak{g}}^*$.

Now we can proceed as before: From (67) and (68) we see that there exist sequences $\{\mu_j\}_j \subset {}_\infty\widehat{G}$ and $\{\lambda_j\}_j \subset \operatorname{supp} \zeta_\infty \subset {}_\infty\widehat{\widetilde{G}}$ with $\mu_j \in \operatorname{supp} \lambda_j|_{{}_\infty G}$ and $\pi_\infty = \lim_{j\to\infty} \mu_j$. But π_∞ and ζ_∞, hence all μ_j are trivial on the normal subgroup $e_{n-1}^{\mathbb{R}}$ in ${}_\infty G$, consequently $K^{-1}(\pi_\infty) = \lim K^{-1}(\mu_j)$, i.e. $l_\infty = \lim_{j\to\infty} f'_j$ for suitably choosen f'_j in the orbits $K^{-1}(\mu_j) = \Omega(\mu_j) \subset {}_\infty\mathfrak{g}^*$. It follows from theorem 9, ch. I, that $\Omega(\mu_j)$ is contained in the restriction of the ${}_\infty\widetilde{G}$–orbit $\widetilde{\Omega}(\lambda_j)$ of λ_j in ${}_\infty\widetilde{\mathfrak{g}}^*$, thus $f'_j = f_j|_{\mathfrak{g}^*}$ for some $f_j \in \widetilde{\Omega}(\lambda_j)$. From (69) we derive with the same arguments as before the existence of a sequence $\{l_{(j)}\}_j$, $l_{(j)} \in \Omega_{k_j}$, with $l_\infty = \lim_{j\to\infty} l_{(j)}$. So we have proved $\Omega_\infty = \lim_{j\to\infty} \Omega_{k_j}$, i.e. $\Omega_\infty = \lim_{k\to\infty} \Omega_k$, in all possible subcases of case 1.

§ 3 The continuity of the inverse of the Kirillov map, part II

We start with case 2. Again we have the VLA $\mathbf{g} = \overline{\mathbb{N}} \times \mathfrak{g}$ and an orthonormal basis $\{b_j\}_{1\leq j\leq n}$ of $\mathfrak{g}$, such that $\mathbb{R}b_n$ is the center of ${}_k\mathfrak{g}$ for all finite k and is contained in every nonvanishing ideal of ${}_k\mathfrak{g}$. Further,

$$\mathfrak{y} = \mathbb{R}b_{n-2} \oplus \mathbb{R}b_{n-1} \oplus \mathbb{R}b_n$$

is a minimal noncentral ideal in every ${}_k\mathfrak{g}$. Let us fix $k \in \mathbb{N}$, and let us write $\mathfrak{g}$, $[\ ,\ ]$ etc. for ${}_k\mathfrak{g}$, $[\ ,\ ]_k$ etc. Again let $\mathfrak{k}$ be the centralizer of $\mathfrak{y}$ and $\mathfrak{d}$ the annihilator of $\mathfrak{y}/\mathbb{R}b_n =: \mathfrak{w}$. This $\mathfrak{w}$ is an irreducible two dimensional faithful $\mathfrak{g}/\mathfrak{d}$ module and no ad x, $x \in \mathfrak{g}$, has purely imaginary eigenvalues, hence it follows that $\dim \mathfrak{g}/\mathfrak{d} = 1$ and we can write

$$\mathfrak{g} = \mathbb{R}b_1 \oplus \mathfrak{d}\ .$$

We claim that $\dim \mathfrak{d}/\mathfrak{k} = 2$. As $\mathfrak{w}$ is irreducible we can find w_1 and w_2 in $\mathfrak{y}$, λ_1, $\lambda_2 \in \mathbb{R}$ with $\lambda_1\lambda_2 \neq 0$ and

$$[b_1, w_1] = \lambda_1 w_1 - \lambda_2 w_2$$
$$[b_1, w_2] = \lambda_2 w_1 + \lambda_1 w_2\ .$$

Then there exists γ_1, $\gamma_2 \in \mathfrak{g}^*$ with

$$[x, w_i] = \gamma_i(x) b_n\,, \quad i = 1, 2\,, \quad x \in \mathfrak{d}\ .$$

We have to show that γ_1 and γ_2 are independent. If this were not the case, then, if necessary after rotating w_1, w_2, we may assume that $\gamma_2 = 0$, i.e. $[x, w_2] = 0$ for all $x \in \mathfrak{d}$. Now take $c \in \mathfrak{d}$ with $\gamma_1(c) = 1$, i.e. $[c, w_1] = b_n$. Then

$$[c, [b_1, w_2]] = \lambda_2 b_n = [[c, b_1], w_2] + [b_1, [c, w_2]] = 0\ ,$$

because $[c, b_1] \in \mathfrak{d}$. This contradiction proves our claim.

It follows that $\dim \mathfrak{d}/\mathfrak{k} = 2$ and that $(x, y) \to [x, y] = \gamma(x, y) b_n$ for $x \in \mathfrak{d}$, $y \in \mathfrak{y}$ defines a nondegenerate bilinear form γ on $(\mathfrak{d}/\mathfrak{k}) \times \mathfrak{w}$, resp. on $(\mathbb{R}b_2 \oplus \mathbb{R}b_3) \times (\mathbb{R}w_1 \oplus \mathbb{R}w_2)$. Let $\{v_1, v_2\}$ be a dual basis of $\mathfrak{u} = \mathbb{R}w_1 \oplus \mathbb{R}w_2$ with respect to b_2, b_3 and γ, i.e. let $\gamma(b_{1+i}, v_j) = \delta_{ij}$. We can assume that $\gamma_1 = |v_1|^{-1} \leq \gamma_2 = |v_2|^{-1}$, hence for $v_j^0 = \gamma_j v_j$ we have $|v_j^0| = 1$ and $\gamma(b_{1+i}, v_j^0) = \gamma_j \delta_{ji}$, $\frac{\gamma_1}{\gamma_2} \leq 1$.

Now we set

$$b_2' = b_2 - b_3\,, \quad u_1 = v_1^0$$
$$b_3' = b_3\,, \qquad u_2 = v_1^0 + \frac{\gamma_1}{\gamma_2} v_2^0$$

Then obviously $\gamma(b'_{1+i}, u_j) = \gamma_1\delta_{ij}$, $|u_1| = 1$, $|u_2| \leq 2$. As also $\{b_1, b'_2, b'_3, \ldots, b_n\}$ is a Malcev basis, generating the same subalgebras as the basis $\{b_j\}$, we see that we can assume that the subspace $\mathfrak{u}$ is spanned by elements u_1 and u_2, which can be renormalized so that

$$(70) \qquad [b_{1+i}, u_j] = \delta_{ij}\gamma b_n\,, \quad |u_1| + |u_2| = 1\,.$$

Let c_1 and c_2 be the elements in $\mathbb{R}b_2 \oplus \mathbb{R}b_3$ with

$$(71) \qquad [b_1, b_{1+i}] \equiv -\alpha_i c_i (\text{mod } \mathfrak{k}) \quad |c_i| = 1\,,\ \alpha_i > 0$$

for $i = 1, 2$. We can assume that $\alpha_2 \geq \alpha_1$, hence $|\beta| \leq 1$ for $\beta = \frac{\alpha_1}{\alpha_2}(c_1|c_2)$. If $|(c_1|c_2)| \geq \frac{1}{2}$ we exchange the basis element b_2 by $b_2 - \beta b_3 =: b'_2$. This gives new c'_1, c'_2 in (70), for which $(c'_1|c'_2) = 0$. If we change the metric of $\mathfrak{g}$, so that $\{b_1, b'_2, b_3, \ldots, b_n\}$ becomes orthonormal, then it follows easily that $|(c'_1|c'_2)| < \frac{1}{2}$ in this new metric. In this new basis we have to substitute $\beta u_1 + u_2$ for u_2 and renormalize, so that (70) still holds.

Moreover, as $\mathfrak{u}$ is ad b_1 invariant, we see that

$$[b_1, u_1] = \alpha_{11}u_1 + \alpha_{12}u_2$$
$$[b_1, u_2] = \alpha_{21}u_1 + \alpha_{22}u_2\,.$$

Let $[b_1, b_2] \equiv \xi_1 b_2 + \xi_2 b_3 (\text{mod } \mathfrak{k})$. Applying the Jacobi identity to $[[b_1, b_2], u_j]$ we get $\xi_1\sigma_{1j} + \xi_2\sigma_{2j} = -\alpha_{j1}$, i.e. $\xi_1 = -\alpha_{11}$, $\xi_2 = -\alpha_{21}$; similarly for $[b_1, b_3]$. It follows that

$$[b_1, b_2] \equiv -\alpha_{11}b_2 - \alpha_{21}b_3 (\text{mod } \mathfrak{k})$$
$$[b_1, b_3] \equiv -\alpha_{12}b_2 - \alpha_{22}b_3 (\text{mod } \mathfrak{k})\,.$$

Now we recall that we have indeed *sequences* of elements $u_1^k, u_2^k \in \mathfrak{y}$, $\gamma_k > 0$ and matrices $A_k = (\alpha_{ij}^k)$ and such that $|u_1^k| + |u_2^k| = 1$ for all k. Hence we can assume that the following limits exist:

$$u_j^\infty = \lim_{k\to\infty} u_j^k \in \mathfrak{y}\,, \quad \gamma_\infty = \lim_{k\to\infty}\gamma_k \geq 0\,, \quad A_\infty = (\alpha_{ij}^\infty) = \lim_{k\to\infty} A_k\,.$$

For $\mathbf{g} = \overline{\mathbb{N}} \times \mathfrak{g}$ this means that $\mathfrak{g}$ has an orthonomal Malcev basis $\{b_1, b_2, \ldots, b_n\}$ such that in all $_\nu\mathfrak{g}$, $\nu \in \overline{\mathbb{N}}$, we have the chain of ideals

$$\mathfrak{d} := \sum_2^n \mathbb{R}b_j \supset \mathfrak{k} := \sum_4^n \mathbb{R}b_j \supset \mathfrak{y} := \sum_{n-2}^n \mathbb{R}b_j \supset \mathfrak{z} = \mathbb{R}b_n\,,$$

where for all finite k the subspace $\mathfrak{z}$ is the center of ${}_k\mathfrak{g}$, $\mathfrak{y}$ is a noncentral minimal ideal, $\mathfrak{k}$ the centralizer of $\mathfrak{y}$ and $\mathfrak{d}$ the annihilator of $\mathfrak{y}/\mathfrak{z}$. Moreover,

$$\mathfrak{u}^k = \mathbb{R}u_1^k \oplus \mathbb{R}u_2^k$$

is the irreducible eigenspace for $\mathrm{ad}_k b_1$ of $\mathfrak{y}$ for the complex eigenvalues $\lambda_\pm^k = \lambda_1^k \pm i\lambda_2^k$ with $\lambda_1^k\lambda_2^k \neq 0$ and for all $\nu \in \overline{\mathbb{N}}$ the following relations hold

(72)
$$[b_1, u_1^\nu]_\nu = \alpha_{11}^\nu u_1^\nu + \alpha_{12}^\nu u_2^\nu$$
$$[b_1, u_2^\nu]_\nu = \alpha_{21}^\nu u_1^\nu + \alpha_{22}^\nu u_2^\nu$$

(73)
$$[b_1, b_2]_\nu \equiv -\alpha_{11}^\nu b_2 - \alpha_{21}^\nu b_3 (\mathrm{mod}\ \mathfrak{k})$$
$$[b_1, b_3]_\nu \equiv -\alpha_{12}^\nu b_2 - \alpha_{22}^\nu b_3 (\mathrm{mod}\ \mathfrak{k})$$

(74)
$$|(c_1^\nu | c_2^\nu)| < \frac{1}{2} \text{ for the } c_i^\nu \text{ defined by (71)}$$

(75)
$$[b_2, u_1^\nu]_\nu = [b_3, u_2^\nu]_\nu = \gamma_\nu b_n\,,\ [b_2, u_2^\nu]_\nu = [b_3, u_1^\nu]_\nu = 0\,.$$

The possible need to change the basis $\{b_i\}$ infinitely often, i.e. to substitute $b_2 \to b_2' = b_2 - \beta^\nu b_3$ to obtain (74) causes no difficulty, because we may assume the sequence $\{\beta^k\}$ to be convergent.

For the ℓ_ν in the orbits $\Omega_\nu = K^{-1}(\pi_\nu)$ we had $\ell_\nu(b_n) = 1$ for all $\nu \in \overline{\mathbb{N}}$. For $\xi_1, \xi_2 \in \mathbb{R}$ let $x(\xi) = \xi_1 b_2 + \xi_2 b_3$. Then

$$(\mathrm{Ad}_\nu^* \exp(-x(\xi))\ell_\nu)(u_j^\nu) = \ell_\nu(u_j^\nu + [x(\xi), u_j^\nu]_\nu) = \ell_\nu(u_j^\nu) + \xi_j\gamma_\nu\,,$$

hence if $\gamma_\nu \neq 0$, in particular for all finite ν, we can find an $x_\nu \in \mathfrak{g}$ with $(\mathrm{Ad}_\nu^* \exp x_\nu \ell_\nu)(u_j) = 0$, which proves

(76) *For all* $\nu \in \overline{\mathbb{N}}$ *with* $\gamma_\nu \neq 0$ *there exist* $\ell_\nu \in \Omega_\nu$ *with* $\ell_\nu(\mathfrak{u}^\nu) = 0$.

In the following we will always make this assumption on the ℓ_ν. Furthermore we remark that $\gamma_\infty \neq 0$ and (75) implies that also for $\nu = \infty$ we have $\mathfrak{y} = \mathfrak{u}^\infty \oplus \mathfrak{z}$.

Next we show:

(77) *Let* $\gamma_\nu \neq 0$ *and* ℓ_ν *as in (76). There exists a Pukanszky polarization* $\mathfrak{p}$ *for* ℓ_ν *with*

$$\mathfrak{p} \subset \mathfrak{h} := \mathbb{R}b_1 \oplus \mathfrak{k}\,,$$

in particular π_ν *is induced from a representation* τ_ν *of the subgroup*

$${}_\nu H = H := E(\nu, \mathfrak{h}) \subset {}_\nu G\,,$$

i.e.

$$\pi_\nu = ind^{\nu G}_{\nu H} \tau_\nu \; .$$

Proof. As $\mathfrak{y}$ is an abelian ideal in ${}_\nu\mathfrak{g}$ we can find a polarization $\mathfrak{p}$ for ℓ_ν with $\mathfrak{y} \subset \mathfrak{p}$. Let $x = \xi_1 b_1 + \xi_2 b_2 + \xi_3 b_3 + c \in \mathfrak{p}$ with $c \in \mathfrak{k}$. Then $[x, u_1^\nu]_\nu = y + \xi_2 \gamma_\nu b_n$ with $y \in \mathfrak{u}^\nu$, hence $0 = \ell_\nu([x, u_1^\nu]) = 0 + \xi_2 \gamma_\nu$, i.e. $\xi_2 = 0$, similarly $\xi_3 = 0$ and consequently $x \in \mathfrak{h}$.

Let $\mathfrak{J}_\nu$ be the representation space of τ_ν. Then similar to case 1 the space $\mathfrak{H}_\nu$ of π_ν consists of the measurable functions ξ from G into $\mathfrak{J}$ with $\xi(x \underset{\nu}{\cdot} y) = c_\nu(y)^{1/2} \tau_\nu(y)^* \xi(x)$ for all $x \in G$, $y \in H$ and $|\xi|_2^2 = \int_G^\nu |\xi(x)| B_\nu(x) dx < \infty$ for a suitable Bruhat function B_ν. Here one computes easily that

$$c_\nu(y) = e^{-2\lambda_1^\nu y_1}, \quad y = (y_1, 0, 0, y_4, \dots, y_n) \in H$$

where $\lambda_1^\nu = \mathrm{Re}\ \lambda^\nu$ with λ^ν the eigenvalue of $\mathrm{ad}_\nu b_1$ in $\mathfrak{w}$. For $s = (s_1, s_2) \in \mathbb{R}^2$ we define

$$e_\nu(s) := E(\nu, s_1 b_2 + s_2 b_3) = (0, s_1, s_2, \dots) \in {}_\nu G \; ,$$

hence $e_\nu(s) = e_2^{s_1} \underset{\nu}{\cdot} e_3^{s_2} \underset{\nu}{\cdot} z$ with an element $z \in K$, uniquely defined by s. Again as in case 1 one proves the formula

$$(\xi|\vartheta) = \int_{\mathbb{R}^2} (\xi(e_\nu(s)) | \vartheta(e_\nu(s))) ds$$

for the inner product in $\mathfrak{H}_\nu$. We remark that we also have

$$(\xi|\vartheta) = \int_{\mathbb{R}^2} (\xi(e_2^{s_1} \cdot e_3^{s_2}) | \vartheta(e_2^{s_1} \cdot e_3^{s_2})) ds \; .$$

The formula for the positive definite function φ_ν associated to π_ν becomes now

$$\varphi_\nu(x) = \int_{\mathbb{R}^2} (\omega_\nu(x^{-1} \underset{\nu}{\cdot} e_\nu(s)) | \omega(e_\nu(s))) ds$$

where the ω_ν are unit vectors in $\mathfrak{H}_\nu$. The proof of proposition (16) is independent of $\dim G/H$, thus (16) holds also in case 2, in particular we can assume that the $\omega_\nu(e_\nu(s))$ are compactly supported in $\mathbb{R}^2$.

We derive from (71) and (74) that we can assume that the limits $\alpha_i^\infty = \lim_{k\to\infty} \alpha_i^k$ and $c_i^\infty = \lim_{k\to\infty} c_i^k$ exist and that

(78) $$[b_1, b_{1+i}]_\infty \equiv -\alpha_i^\infty c_i^\infty (\mathrm{mod}\ \mathfrak{k}) \, , \quad |c_i^\infty| = 1 \, , \quad |(c_1^\infty | c_2^\infty)| < \frac{1}{2} \, ,$$

which implies that $\mathbb{R}c_1^\infty + \mathbb{R}c_2^\infty = \mathbb{R}b_2 \oplus \mathbb{R}b_3$. Furthermore

$$(79) \qquad 0 < \eta_k' := \frac{\alpha_1^k}{\alpha_2^k} \le 1\,, \quad \eta_\infty' = \lim_{k\to\infty} \eta_k' \ge 0\,.$$

We also introduce

$$(80) \qquad \varepsilon_k' := \frac{\alpha_2^k}{\gamma_k}\,, \quad k \in \mathbb{N}\,.$$

As before we will have to distinguish the two cases α and β. For this purpose we set for $\nu \in \overline{\mathbb{N}}$, $k \in \mathbb{N}$:

$$\mu_\nu = \eta_\nu' |u_1^\nu| + |u_2^\nu|\,, \quad \varepsilon_k = \mu_k \varepsilon_k'$$

and define:

$$(\alpha) \qquad \text{The limit} \quad \varepsilon_\infty := \lim_{k\to\infty} \varepsilon_k \quad \text{exists}\,,$$

$$(\beta) \qquad \lim \varepsilon_k^{-1} = 0\,.$$

Furthermore we set

$$\eta_\nu = 1 \text{ in case } \alpha\,, \qquad \eta_\nu = \eta_\nu' \text{ in case } \beta$$

and

$$(81) \qquad v_1^\nu = \mu_\nu^{-1} \eta_\nu u_1^\nu\,, \quad v_2^\nu = \mu_\nu^{-1} u_2^\nu \quad \text{if} \quad \mu_\nu > 0\,,$$

in particular for all finite ν. Observe that $|v_1^k| + |v_2^k| = 1$ for all $k \in \mathbb{N}$, hence we can assume that $v_i^\infty = \lim_{k\to\infty} v_i^k$ if $\mu_\infty > 0$ and take it as definition of the v_i^∞, $i = 1, 2$, if $\mu_\infty = 0$. Again (81) and Jacobi imply

$$(82) \qquad \begin{aligned} [b_1, v_1^\nu]_\nu &= \alpha_{11}^\nu v_1^\nu + \eta_\nu \alpha_{12}^\nu v_2^\nu \\ [b_1, v_2^\nu]_\nu &= \eta_\nu^{-1} \alpha_{21} v_1^\nu + \alpha_{22}^\nu v_2^\nu\,, \end{aligned}$$

and with

$$c(t) = t_1 b_2 + t_2 b_3\,,\ v^\nu(s) = s_1 v_1^\nu + s_2 v_2^\nu\,,\ (t|s)_\nu = \eta_\nu t_1 s_1 + t_2 s_2$$

we get

$$(83) \qquad [c(t), v^\nu(s)]_\nu = \gamma_\nu^\vee (t|s)_\nu b_n \quad \text{with} \quad \gamma_\nu^\vee = \mu_\nu^{-1} \gamma_\nu$$

Observe that in case α clearly all $\mu_\nu = 1$, $v_i^\nu = u_i^\nu$, $v^\nu(s) = u^\nu(s)$ etc. We also remark that the sequence $\{\gamma_\nu^\vee\}$ is bounded, because $[b_3, v_2^\nu]_\nu = \gamma_\nu^\vee b_n$, hence we suppose that $\gamma_\infty^\vee = \lim \gamma_k^\vee$ exists. For $k \in \mathbb{N}$ it is convenient to introduce the following notation:

$$t^{(k)} = \mu_k \gamma_k^{-1}(\eta_k^{-1} t_1, t_2) \in \mathbb{R}^2$$

Now, as $e_\nu(t) = \exp_\nu c(t)$, (83) implies

$$e_\nu(-t) \underset{\nu}{\cdot} \exp_\nu v^\nu(s) \underset{\nu}{\cdot} e_\nu(t) = \exp_\nu v^\nu(s) \cdot e_n^{\gamma_\nu^\vee (t|s)_\nu} \tag{84}$$

thus, with $ts = t_1 s_1 + t_2 s_2$, for all finite k:

$$\omega_k(x \underset{k}{\cdot} e_k(t^{(k)}))e^{its} = \omega_k(x \underset{k}{\cdot} \exp_k v^k(s) \underset{k}{\cdot} e_k(t^{(k)})) \tag{85}$$

If $\gamma_\infty > 0$, hence $\gamma_\infty^\vee > 0$, then formula (85) holds also for $\nu = \infty \in \overline{\mathbb{N}}$.

Now we start with the definition of the $(n-2)$ dimensional VLA $\mathbf{h}^\bullet$. We define for $t = (t_1, t_2) \in \mathbb{R}^2$, $\nu \in \overline{\mathbb{N}}$:

$$b_1^\nu(t) = b_1 + \varepsilon_\nu(t_1 c_1^\nu + t_2 c_2^\nu) \qquad \text{in case} \quad \alpha$$
$$b_1^\nu(t) = \varepsilon_\nu^{-1} b_1 + (t_1 c_1^\nu + t_2 c_2^\nu) \qquad \text{in case} \quad \beta$$

where of course $\varepsilon_\infty^{-1} b_1$ means 0 in case β. Furthermore, if

$$P = \begin{cases} \overline{\mathbb{N}} \times \mathbb{R}^2 & \text{in case } \alpha\ , \\ \overline{\mathbb{N}} \times \mathbb{R}^2 \setminus \{(\infty, 0)\} & \text{in case } \beta \end{cases} \tag{86}$$

then we see that $b_1^\nu(t) \neq 0$ for all $(\nu, t) \in P$ in both cases. For $(k, t) \in P$ let

$$\mathfrak{h}_{k,t} = \mathbb{R} b_1^k(t) \oplus \mathfrak{k}\ .$$

Evidently

$$\mathfrak{h}_{k,t} = \mathrm{Ad}_k e_k(t^{(k)})(\mathfrak{h}) = \mathbb{R} e^{\mathrm{ad}_k c(t^{(k)})}(b_1) \oplus \mathfrak{k} \subset {}_k\mathfrak{g}$$

Now we construct the VLA $\mathbf{h}^\bullet = P \times \mathfrak{h}^\bullet$ with P as in (86) and $\mathfrak{h}^\bullet = \sum_{i=1}^{n-2}{}^{\oplus} \mathbb{R} b_i^\bullet$. As in case 1 we define injective linear mappings $d\Gamma_{\nu,t} : \mathfrak{h}^\bullet \to \mathfrak{g}$ for every $(\nu, t) \in P$ by

$$d\Gamma_{\nu,t}(b_j^\bullet) = \begin{cases} b_1^\nu(t) & \text{for} \quad j = 1 \\ b_{j+2} & \text{for} \quad j \geq 2\,. \end{cases}$$

As before the family $\{d\Gamma_{\nu,t}; (\nu, t) \in P\}$ defines a VLA $\mathbf{h}^\bullet = (P \times \mathfrak{h}^\bullet, [\ ,\ ]_{\nu,t})$ and a homomorphism $d\Gamma$ of $\mathbf{h}^\bullet$ into $\mathbf{g}$. Furthermore, $\{b_j^\bullet\}_j$ is a Malcev basis in $\mathbf{h}^\bullet$ and again there exists a VLG $\mathbf{H}^\bullet$ with Lie algebra $\mathbf{h}^\bullet$ and a commutative diagram, connecting $\mathbf{h}^\bullet$, $\mathbf{g}$, $\mathbf{H}^\bullet$ and $\mathbf{G}$. If $\Gamma_{\nu,t}$ is the homomorphism of ${}_{\nu,t}H^\bullet$ into ${}_\nu G$, then the formula for the representation $\tau_{\nu,t} \in {}_{\nu,t}\widehat{H^\bullet} \subset \widehat{\mathbf{G}^\bullet}$ for all ν with $\gamma^\nu > 0$ is

$$\tau_{\nu,t}(y) = \tau_\nu(e_\nu(t^{(\nu)})^{-1} \underset{\nu}{\cdot} \Gamma_{\nu,t}(y) \underset{\nu}{\cdot} e_\nu(t^{(\nu)}))$$

and the corresponding positive definite function is (with $\Delta_\nu = \eta_\nu^{-1}\mu_\nu^2\gamma_\nu^{-2}$):

$$\psi_{\nu,t}(y) = \Delta_\nu(\omega_\nu(\Gamma_{\nu,t}(y)^{-1}e_\nu(t^{(\nu)}))|\omega_\nu(e_\nu(t^{(\nu)}))) \ .$$

Let $\psi_\nu = \int_{\mathbb{R}^2}\psi_{\nu,t}dt$, hence $\psi_\nu(e) = \int_{\mathbb{R}^2}|\omega_\nu(e_\nu(t))|^2dt = 1$. For $f \in \mathbf{L}^0 = P \times L^1(H^\bullet)$ we have

$$\psi_\nu(f) = \Delta_\nu \int_{\mathbb{R}^2}\int_{H^\bullet} f(\nu, t; y)(\omega_\nu(\Gamma_{\nu,t}(y)^{-1}e_\nu(e^{(\nu)}))|\omega_\nu(e^{(\nu)})))dy\,dt \tag{87}$$

which is the exact analogue of (21).

For $f \in \mathscr{C}_0^\infty$ one can now introduce the function F_ν as in (22). Then (85) and the same arguments leading to (24) show that

$$\begin{aligned}&\psi_\nu(f) = \\ &(2\pi)^{-2}\int_{\mathbb{R}^4}(1-\partial_1^4)\int_{H^\bullet} f(\nu, s; y)\varphi_\nu(\exp_\nu(-v^\nu(t)\Gamma_{\nu,s}(y))dy\frac{e^{-ist}}{1+t^4}ds\,dt\end{aligned} \tag{88}$$

where now of course $\partial_1^4 = (\frac{\partial^2}{\partial s_1^2} + \frac{\partial^2}{\partial s_2^2})^2$, applied to functions $\phi(s_1, s_2, t_1, t_2)$, $t^4 = (t_1^2+t_2^2)^2$. As $\lim \varphi_k = \varphi_\infty$ uniformly on compacta with all derivates we obtain

$$\lim \psi_k(f) = \psi_\infty(f)$$

where ψ_∞ is defined as the functional given by the right hand side in (88) for $\nu = \infty$.

If $\gamma_\infty > 0$, then, as we have already mentioned, we are in case α and also ψ_∞ is given by (87). This allows us to proceed exactly as in the first case in § 2, in particular (76) and (77) show that also formula (21) holds. So in this case $\lim \Omega_k = \Omega_\infty$ follows.

The subcase α, $\gamma_\infty = 0$, is equally simple: In this case also $\alpha_i^\infty = \lim \alpha_i^k = 0$, i.e. $\alpha_{ij}^\infty = 0$, which means that $\mathfrak{u}^\infty$ is central in ${}_\infty\mathfrak{g}$. This implies

$$\varphi_\infty(\exp_\infty(-u^\infty(s))\Gamma_{\infty,r}(y)) = \varphi_\infty(\Gamma_{\infty,r}(y))e^{it_0s} \tag{89}$$

for some $t_0 \in \mathbb{R}^2$ and allows us to proceed as in case 1.

We will now discuss the subcase β, $\eta_\infty > 0$. Then it is allowed to assume $\eta_\nu = \mu_\nu = 1$ for all $\nu \in \overline{\mathbb{N}}$, hence $v_i^\nu = u_i^\nu$ for all $\nu \in \overline{\mathbb{N}}$. If $\alpha_2^\infty = 0$, then $[b_1, \mathfrak{u}^\infty] = 0$ and $\mathfrak{u}^\infty = \mathfrak{v}^\infty$ is central in ${}_\infty\mathfrak{g}$. This implies $\pi_\infty(u^\infty(s)) = e^{it_0s}$ with $t_0 \in \mathbb{R}^2$. If $t_0 \neq 0$, i.e. if

$$d\pi_\infty(\mathfrak{v}^\infty) \neq 0 \ , \tag{90}$$

then $(\infty, t_0) \in P$ and our theorem follows as before. If $\alpha_2^\infty > 0$ then there is no loss of generality in assuming $\alpha_2^\nu = 1$, for all $\nu \in \overline{\mathbb{N}}$. Then $\eta_\infty = 1$ and $|(c_1^\infty|c_2^\infty)| < \frac{1}{2}$ implies that the restriction A_∞ of $\mathrm{ad}_\infty b_1$ acts regularly on $\mathfrak{u}^\infty$. Again let us assume that (90) holds. As $\mathfrak{u}^\infty$ is central in $\mathfrak{d} = \sum_2^n \mathbb{R}b_j$ it follows that π_∞ is induced from $D = \exp_\infty \mathfrak{d}$, say $\pi_\infty = \mathrm{ind}_D^G \sigma$. We proceed as in case 1 in the paragraph following (26): We have $\mathfrak{H}(\pi_\infty) = L^2(\mathbb{R}, \mathfrak{J})$, with representation space $\mathfrak{J}$ of σ, and for some $\omega \in \mathfrak{J}$ we have $\varphi_\infty(x) = \int_{\mathbb{R}} (\omega(s^{-1} \underset{\infty}{\cdot} e_1^s)|\omega(e_1^s))ds$. Next,

$$\exp_\infty v^\infty(t) \cdot e_1^s = e_1^s \cdot \exp_\infty \mathrm{Ad}\, e_1^s(v^\infty(t)) = e_1^s \cdot \exp_\infty v^\infty(tA(s))$$

where $A(s) = e^{sA_\infty} \in GL(\mathbb{R}^2)$. Because $\exp_\infty \mathfrak{v}^\infty$ is central in D there exists $t_0 \in \mathbb{R}^2$ with $\sigma(\exp_\infty v^\infty(t)) = e^{it_0t}$, hence

$$\omega(x \exp v^\infty(t) e_1^s) = \omega(x e_1^s) e^{it_0(tA(s))}$$

and

$$\varphi_\infty(\exp_\infty(-v^\infty(t))\Gamma_{\infty,r}(y)) = \int (\omega(\Gamma_{\infty,r}(y)^{-1} e_1^s)|\omega(e_1^s)) e^{it_0A(s)^*t} ds$$

which implies

$$\psi_\infty(f) = \int_{\mathbb{R}} \int_{H^\bullet} f(\infty, t_0A(s)^*; y)(\omega(\Gamma_{\infty,t_0A(s)^*}(y)^{-1} e_1^s)|\omega(e_1^s)) dy\, ds\,.$$

Because (90) implies that $t_0 \neq 0$, the rest follows exactly as in case 1.

Now we treat subcase β, $\eta_\infty = 0$ under the assumption that (90) holds. Then if $\mathfrak{v}^\infty$ is central we just repeat the arguments from subcase α, $\gamma_\infty = 0$, we only have to observe that now (90) implies that the t_0 in (89) is nonzero, hence $(\infty, t_0) \in P$.

More difficult is subcase β, $\eta_\infty = 0$ and $\mathfrak{v}^\infty$ noncentral, i.e. $[b_1, \mathfrak{v}^\infty]_\infty \neq 0$. From $\eta_\infty = 0$ we derive $\alpha_1^\infty = 0$, hence $\alpha_{11}^\infty = 0$ and so (82) implies $[b_1, v_1^\infty]_\infty = 0$. For $\nu \in \overline{\mathbb{N}}$ let $V_\nu = ('\alpha_{ij}^\nu)$ be the matrix appearing in (82), in particular $'\alpha_{ij}^k = \eta_k^{j-i}\alpha_{ij}^k$ for finite k.

As clearly $|\eta_k^{-1}\alpha_{21}^k| \le \alpha_2^k$ we can assume that $\lim {}'\alpha_{21}^k = {}'\alpha_{21}^\infty$ exists, hence that $V_\infty = \lim V_k$. All V_k have conjugate complex eigenvalues with nonvanishing real and imaginary parts, hence $4 \det V_k > (\mathrm{trace}\, V_k)^2$ for all $k \in \mathbb{N}$, which implies $4 \det V_\infty \ge (\mathrm{trace}\, V_\infty)^2 \ge 0$. But $'\alpha_{1i}^\infty = 0$, thus $\det V_\infty = 0$ and consequently also $\mathrm{trace}\, V_\infty = 0$ which implies $'\alpha_{22}^\infty = 0$, $[b_1, v_2^\infty]_\infty = {}'\alpha_{21}^\infty v_1^\infty \neq 0$. Let $d\pi_\infty(v_1^\infty) \neq 0$.

As $\mathfrak{v}_\infty$ is central in $\mathfrak{d}$ it follows that π_∞ is induced from some $\sigma \in \widehat{D}$, so as in previous cases we are done. If $d\pi_\infty(v_1^\infty) = 0$ then (90) implies $d\pi_\infty(v_2^\infty) \neq 0$ and π_∞ is trivial on $\exp_\infty \mathbb{R}v_1^\infty =: Y \subset {}_\infty G$. Hence, as Y is normal in ${}_\infty G$, π_∞ is a representation of the quotient $\overline{G} = {}_\infty G/Y$, nontrivial on the image X of $\exp_\infty \mathbb{R}v_2^\infty$

in $\overline{G}$. But X is central in $\overline{G}$, thus (89) holds with $t_0 \neq 0$. So in either case the theorem is proved.

Finally we have to treat case β without the assumption (90), i.e. we assume that $d\pi_\infty(\mathfrak{v}^\infty) = 0$. As in case 1 we will distinguish two subcases:

A. There exists $T > 0$ with

$$C := \inf_k \int\limits_{\substack{|t_1| \leq \eta_k^{-1} T \\ |t_2| \leq T}} |\omega_k(e_k(t))|^2 dt > 0$$

B. There exists a sequence $\{p_k\}$ with $0 < p_k < p_{k+1}$, $\lim p_k = \infty$ and

$$\lim \iint\limits_{\substack{|t_1| \leq \eta_k^{-1} p_k \\ |t_2| \leq p_k}} |\omega_k(e_k(t))|^2 dt = 0$$

Again we start with B. Let φ be a Schwartz function on $\mathbb{R}^2$. As before we define the corresponding "variable measure" $\varphi^\natural$ on $\overline{\mathbb{N}} \times G$ by setting

$$\varphi^\natural(\nu)(f) = \int_{\mathbb{R}^2} f(\exp_\nu v^\nu(s))\varphi(s)\, ds \;,$$

thus $\varphi^\natural(\nu) \in \mathcal{M}(G)$, the finite Radon measures on G, and $\varphi^\natural \in \mathscr{C}(\overline{\mathbb{N}}, \mathcal{M}(G))$. Obviously $\varphi^\natural$ is a multiplier of $\mathscr{C}_\infty(\mathbf{L}(\mathbf{G}))$ and $C^*(\mathbf{G})$, i.e. $\varphi^\natural \in C^*(\mathbf{G})^\flat$ hence using former arguments, in particular (85), we get the analogue of (28):

$$(\pi_k(\varphi^\natural)(\xi)(e_k(t^{(k)})) = \widehat{\varphi}(t)\xi(e_k(t^{(k)}))$$

for $\xi \in \mathfrak{H}(\pi_k)$. For $\nu = \infty$ (90) implies $\pi_\infty(\exp_\infty v^\infty(s)) = \mathbf{1}$, hence $\pi_\infty(\varphi^\natural) = \widehat{\varphi}(0) \cdot \mathbf{1}$. Now we take φ so that $\widehat{\varphi}(0) = 1$, supp $\widehat{\varphi} \subset [-1, 1]^2$ and set $\omega'_\nu = \kappa_\nu \pi_\nu(\varphi^\natural)\omega_\nu$ with a positive factor κ_ν for which $|\omega'_\nu| = 1$. Then $\omega'_\infty = \omega_\infty$, the corresponding positive functionals still converge to φ_∞ and the ω'_k now are supported in $[-\mu_k(\eta_k\gamma_k)^{-1}, \mu_k(\eta_k\gamma_k)^{-1}] \times [-\mu_k\gamma_k^{-1}]$. This justifies supposing

$$\operatorname{supp} \omega_k \subset \{t \in \mathbb{R}^2 ;\, |t_1| \leq \frac{\mu_k}{\eta_k\gamma_k} \,,\, |t_2| \leq \frac{\mu_k}{\gamma_k}\}$$

already for the vectors ω_k. We claim that condition (90) implies that these supports can be made even smaller.

(91) *There exist a decreasing sequence* $\{c_k\}$ *with* $\lim c_k = 0$ *and so that*

$$\lim_{k \to \infty} \iint\limits_{\substack{c_k\mu_k \leq \eta_k\gamma_k|t_1| \\ c_k\mu_k \leq \gamma_k|t_2|}} |\omega_k(e_k(t)))|^2 dt = 0 \,.$$

Proof. With $\Delta_{1k} = \eta_k \frac{\gamma_k}{\mu_k}$, $\Delta_{2k} = \frac{\gamma_k}{\mu_k}$, $\Delta_k = \Delta_{1k}\Delta_{2k}$ the integral in (91) can be written as

$$i_k := \Delta_k^{-1} \int\int_{c_k \le |t|_1,\, |t_2| \le 1} |\omega_k(e_k(t^{(k)}))|^2 dt\,.$$

As before we see that (91) will follow when we can show that the sequences of the integrals

$$I_k^i := \Delta_k^{-1} \int\int_{|t_1|,|t_2| \le 1} t_i^2 |\omega_k(e_k(t^{(k)}))|^2\, dt\,, \quad i = 1, 2$$

converge to 0. Now from (84) we obtain

$$(d\pi_k(v_i^k)\xi)(e_k(t)) = \Delta_{ik} t_i \xi(e_k(t))\,, \quad i = 1, 2$$

for $\xi \in \mathfrak{H}(\pi_k)$. Moreover, condition B implies that we can assume that

$$\operatorname{supp} \omega_k \subset W_k := \{t \in \mathbb{R}^2; \eta_k^{-1} p_k \le |t_1| \le \Delta_{1k}^{-1}\,, p_k \le |t_2| \le \Delta_{2k}^{-1}\}$$

hence

$$|d\pi_k(v_i^k)\omega_k|^2 = \Delta_{ik}^2 \int_{W_k} t_i^2 |\omega_k(e_k(t))|^2 dt = \Delta_k^{-1} \int_{|t_1|,|t_2| \le 1} t_i^2 |\omega_k(e_k(t^{(k)}))|^2 dt = I_k^i$$

Now again the same arguments give what we want: $\lim_{k\to\infty} I_k^i = \lim_{k\to\infty} |d\pi_k(v_i^k)\omega_k|^2 = 0$ for $i = 1, 2$. It follows from (91) that we can assume

(92) $$\operatorname{supp} \omega_k \subset \{t;\ p_k \le |\eta_k t_1| \le c_k \frac{\mu_k}{\gamma_k}\,,\ p_k \le |t_2| \le c_k \frac{\mu_k}{\gamma_k}\}$$

At this place it is convenient to modify the notation: Since μ_k and γ_k in case B never appear separated, but only connected as $\frac{\gamma_k}{\mu_k}$, and also $\lim \frac{\gamma_k}{\mu_k} = 0$, from now on we suppress the μ_k and write simply γ_k, so in particular ω_k is supported in $\{p_k \le |\eta_k t_1|\,, |t_2| \le c_k \gamma_k^{-1}\}$. Having this in mind we can presume that in the present case also the relations (34), (35) and (36) hold, i.e.:

(93) $$\lim p_k c_k = 0\,, \quad p_k c_k < 1 \quad \text{for} \quad k \in \mathbb{N}$$

(94) $$\lim \gamma_k q_k = 0 \quad \text{with} \quad q_k = \max\,\{\frac{2}{\gamma_k p_k}\,,\ p_k + 1\}$$

(95) $$\gamma_k p_k q_k \ge 2\,,\ I_k := [p_k, q_k] \ne \emptyset\,.$$

We proceed as before: Let $\widetilde{P} = (\bigcup_{k \in \mathbb{N}} \{k\} \times I_k) \cup \{\infty\}$ the compact parameter space,

$$\widetilde{\mathfrak{v}} = \mathbb{R}\widetilde{v}_1 \oplus \mathbb{R}\widetilde{v}_2 \,,\ \widetilde{\mathfrak{g}} = \mathfrak{g} \oplus \widetilde{\mathfrak{v}} \,,\ \widetilde{\mathbf{g}} = (\widetilde{P} \times \widetilde{\mathfrak{g}} \,, [\ ,\ ]_p)$$

with Lie products in $\widetilde{\mathfrak{g}}$ defined by

$$(96)\qquad \begin{cases} [x, y]_{k,r} &= [x, y]_k \,,\ [x, y]_\infty = [x, y]_\infty \quad \text{for} \quad x, y \in \mathfrak{g} \\ [\mathfrak{k}, \widetilde{\mathfrak{v}}]_p &= 0 \\ [b_1, \widetilde{v}_1]_{k,r} &= \alpha_{11}^k \widetilde{v}_1 + \eta_k \alpha_{21}^k \widetilde{v}_2 \\ [b_1, \widetilde{v}_2]_{k,r} &= \eta_k^{-1} \alpha_{21}^k \widetilde{v}_1 + \alpha_{22}^k \widetilde{v}_2 \\ [b_1, \widetilde{v}_i]_\infty &= \lim_{k \to \infty} [b_1, \widetilde{v}_i]_{k,r} \,,\ i = 1, 2 \\ [b_2, \widetilde{v}_1]_{k,r} &= \eta_k \gamma_k r b_n \,,\ [b_3, \widetilde{v}_2]_{k,r} = \gamma_k r b_n \\ [b_2, \widetilde{v}_2]_{k,r} &= [b_3, \widetilde{v}_1]_{k,r} = 0 \\ [b_2, \widetilde{\mathfrak{v}}_\infty]_\infty &= [b_3, \widetilde{\mathfrak{v}}]_\infty = 0 \end{cases}$$

Observe that the limits $[b_1, \widetilde{v}_i]_\infty$ exist and are 0 if $\alpha_2^\infty = 0$. If $\eta_\infty > 0$, i.e. without restriction $\eta_\nu = 1$ for all ν, then $[b_1, \widetilde{v}_i]_\infty = \alpha_{i1}^\infty \widetilde{u}_1 + \alpha_{i2}^\infty \widetilde{u}_2$. If $\eta_\infty = 0$, then $[b_1, \widetilde{v}_2]_\infty = \alpha_{21}^\infty \widetilde{v}_1$, $[b_1, \widetilde{v}_1]_\infty = 0$.

It is clear that $\widetilde{\mathbf{g}}$ is a VLA with Malcev basis $b_1, b_2, \dots, b_{n-1}, \widetilde{v}_1, \widetilde{v}_2, b_n$. Let $\widetilde{\mathbf{G}} = \widetilde{P} \times G$, with $\widetilde{G} = G \times \mathbb{R}^2$, be its corresponding VLG. There are natural injections $\widetilde{P} \times G \to \widetilde{\mathbf{G}}$ and projections $\widetilde{P} \times G \to \mathbf{G}$, similarly for the algebras.

Also as before we define

$$\widetilde{\mathfrak{h}} = \mathfrak{h} \oplus \widetilde{\mathfrak{v}}, \quad \widetilde{\mathfrak{d}} = \mathfrak{d} \oplus \widetilde{\mathfrak{v}}, \quad \widetilde{\mathfrak{k}} = \mathfrak{k} \oplus \widetilde{\mathfrak{v}} = \widetilde{\mathfrak{h}} \cap \widetilde{\mathfrak{d}} \,,$$

with variable subgroups $\widetilde{\mathbf{H}}$, $\widetilde{\mathbf{D}}$ and $\widetilde{\mathbf{R}}$ of $\widetilde{\mathbf{G}}$. Then $\widetilde{\mathbf{S}} = \widetilde{P} \times S$ with $S = \{(0, 0, \dots, 0, s_1, s_2, 0)\,; s_1, s_2 \in \mathbb{R}\} \simeq \mathbb{R}^2$ is a variable subgroup with ${}_pS = \exp_p \widetilde{\mathfrak{v}}$ for all $p \in \widetilde{P}$, and ${}_pS$ is normal in ${}_p\widetilde{H}$, with

$${}_{k,r}\widetilde{H} = {}_kH \ltimes S \,, \qquad (k, r) \in \widetilde{P} \,.$$

The representation $\tau_k \in {}_k\widehat{H}$ (see (77)) defines naturally a representation $\tau_{k,r} \in {}_{k,r}\widehat{\widetilde{H}}$, trivial on S. Let

$$\widetilde{\pi}_{k,r} = \operatorname{ind}_{{}_{k,r}\widetilde{H}}^{{}_{k,r}\widetilde{G}} \tau_{k,r} \in {}_{k,r}\widehat{\widetilde{G}} \,.$$

Since ${}_kG \subset {}_{k,r}\widetilde{G}$ and $\widetilde{\pi}_{k,r}|_{{}_kG} = \pi_k$ one computes easily for $\widetilde{v}(s) = s_1\widetilde{v}_1 + s_2\widetilde{v}_2$ and $v^k(s)$:

$$d\widetilde{\pi}_{k,r}(\widetilde{v}(s))\xi(e_k(t)) = ir\gamma_k(s|t)_k\xi(e_k(t))$$
$$d\widetilde{\pi}_{k,r}(v^k(s))\xi(e_k(t)) = i\gamma_k(s|t)_k\xi(e_k(t)) \,,$$

hence $d\pi_{k,r}(\widetilde{v}(s) - rv^k(s)) = 0$ and

$$(97) \qquad Q_{k,r} := \exp_{k,r}(\mathbb{R}(\widetilde{v}_1 - rv_1^k) \oplus \mathbb{R}(\widetilde{v}_2 - rv_2^k))$$

is a closed normal subgroup in ${}_{k,r}\widetilde{G}$ with

$$\widetilde{\pi}_{k,r|Q_{k,r}} = \mathbf{1} \ .$$

Any $\varphi \in \mathcal{S}(\mathbb{R}^2)$ defines a bounded measure φ_0 on $\widetilde{G}$, supported by $\mathcal{S}$, by setting

$$\varphi_0(f) = \int\int f(\exp \widetilde{v}(s))\varphi(s)\, ds \ .$$

As to be expected we have

$$(\widetilde{\pi}_{k,r}(\varphi_0)\xi)(e_k(t)) = \widehat{\varphi}(\gamma_k r(\eta_k t_1, t_2))\xi(e_k(t))$$

Our next steps are the same as before lemma (47): We take a smooth nonnegative $\alpha \in \mathcal{C}_0^\infty(\mathbb{R})$ with $\alpha(r) = \alpha(-r)$, $\alpha(r) = 0$ if $|r| \notin [\sqrt{2}, 2\sqrt{2}]$ and such that

$$\int_{\mathbb{R}^+} \frac{\alpha(r)^2}{r}\, dr = 1 \,, \quad A := \int \frac{\alpha(r)}{r}\, dr < 1 \,.$$

Then there exists a unique radial function $\beta \in \mathcal{S}(\mathbb{R}^2)$ with $\widehat{\beta}(x) = \alpha(|x|)$ (with $|x|^2 = x_1^2 + x_2^2$) for $x \in \mathbb{R}^2$. Let

$$\widetilde{\omega}_{k,r} = \widetilde{\pi}_{k,r}(\beta_0)\omega_k \ ,$$

which means

$$\widetilde{\omega}_{k,r}(e_k(t)) = \alpha(\gamma_k r|(\eta_k t_1, t_2)|)\omega_k(e_k(t)) \ .$$

Let $t = (t_1, t_2) \in \operatorname{supp} \omega_k$, i.e.

$$\gamma_k p_k \leq \gamma_k \eta_k |t_1| \leq c_k \,, \quad \gamma_k p_k \leq \gamma_k |t_2| \leq c_k \,.$$

Then $r > q_k$ implies

$$\gamma_k r|(\eta_k t_1, t_2)| > q_k \sqrt{2} \gamma_k p_k \geq 2\sqrt{2}$$

after (95), and $r < p_k$ implies similarly

$$\gamma_k r|(\eta_k t_1, t_2)| < p_k c_k \sqrt{2} < \sqrt{2}$$

after (93).

Hence the same proof shows that also in the present case lemma (47) holds.

We define the positive definite functions $\widetilde{\varphi}_{k,r}$ and $\widetilde{\varphi}_k$ on $\widetilde{G}$ as in (45) and (46), i.e. $\widetilde{\varphi}_{k,r}(\widetilde{x}) = (\widetilde{\pi}_{k,r}(\widetilde{x})\widetilde{\omega}_{k,r}|\widetilde{\omega}_{k,r})$, $\widetilde{\varphi}_k = \int_{p_k}^{q_k} \widetilde{\varphi}_{k,r} \frac{dr}{r}$ etc. Then again $\widetilde{\varphi}_k(e) = 1$ and we can suppose that a limit $\widetilde{\varphi}_\infty = \lim \widetilde{\varphi}_k$ exists in $\mathcal{P}_1(C^*(\widetilde{\mathbf{G}}))$. We will show (with the same, appropriately adjusted proof), that also now (49) holds. For this we have to modify the formulas for $N_k^i(x)$, $i = 1, 2$ in (50) by substituting $e_k(x_2, x_3)$ for $e_2^{x_2}$ in the integrands. Instead of (51) we have (with $e_k(x) = e_k(x_2, x_3)$)

$$[\pi_k(e_k(x)), \widetilde{\pi}_{k,r}(\exp_k \widetilde{v}(s))] = \pi_k(e_k(x)\exp(rv^k(s)))(1 - e^{i\gamma_k r(x|s)_k})$$

which, because of $|1 - e^{it}| \leq 2|t|$ as before implies

$$\|[\pi_k(e_k(x))\,,\ \widetilde{\pi}_{k,r}(\beta_0)]\| \leq 2\gamma_k r|(\eta_k x_2, x_3)| \int_{\mathbb{R}^2} |s||\beta(s)|\, ds$$

with 2 norm $|(y_1, y_2)| = (y_1^2 + y_2^2)^{\frac{1}{2}}$ on $\mathbb{R}^2$. With constant $C = |\beta|_1 \int |s||\beta(s)|\, ds$ this gives

$$|N_k^1(x)| \leq 2\gamma_k|(x_2, x_3)| \cdot C$$

and proves $\lim_{k\to\infty} N_k^1(x) = 0$ uniformly on compacta.

To estimate $N_k^2(x)$ we have to compute

$$[\pi_k(e_1^{x_1})\,,\ \widetilde{\pi}_{k,r}(\beta_0)] = \pi_k(e_1^{x_1})(J_1 - J_2)$$

with

$$J_1 = \int \pi_k(\exp rv^k(s))\beta(s)\, ds\,, \quad J_2 = \int (\exp r \ \mathrm{Ad}_k e_1^{-x_1}(v^k(s)))\beta(s)\, ds$$

Now $\mathrm{Ad}_k e^{-x_1}(v^k(s)) = e^{-x_1 \ \mathrm{ad}\ b_1}(v^k(s)) = v^k(se^{-x_1 B})$, where B is the matrix of the restriction ad $b_1|_{\mathfrak{v}^k}$, appearing in (82). The linear transformation $s \to se^{x_1 B}$ of $\mathbb{R}^2$ yields

$$J_2 = \int \pi_k(\exp rv^k(se^{-x_1 B}))\beta(s)\, ds = e^{x_1(\alpha_{11}^k + \alpha_{22}^k)} \int \pi_k(\exp rv^k(s))\beta(se^{x_1 B})\, ds$$

because $\det e^{x_1 B} = e^{x_1 \ \mathrm{trace}\ B} = e^{x_1(\alpha_{11}^k + \alpha_{22}^k)}$. With

$$\beta_k(x_1, s) = \beta(s) - e^{x_1(\alpha_{11}^k + \alpha_{22}^k)}\beta(se^{x_1 B})$$

it follows that

$$[\pi_k(e_1^{x_1}),\ \widetilde{\pi}_{k,r}(\beta_0)] = \pi_k(e_1^{x_1}) \int_{\mathbb{R}^2} \pi_k(\exp_k rv^k(s))\beta_k(x_1, s)\, ds$$

By Bochner's theorem there is a spectral family $\{E(\lambda)\}_{\lambda\in\mathbb{R}^2}$ of projections $E(\lambda)$ of $\mathfrak{H}(\pi_k)$ with

$$\pi_k(\exp_k v^k(s)) = \int e^{-is\lambda} dE(\lambda)$$

hence $\widetilde{\pi}_{k,r}(\beta_0) = \int \alpha(r\lambda)dE(\lambda)$. The same arguments as in the proof of (52) show that

$$|N_k^2(x)| \le A \int_{\mathbb{R}^2} |\beta_k(x_1, s)|\, ds\ .$$

To estimate β_k let $f(t) = \det(e^{tx_1B})\beta(se^{tx_1B})$, thus $\beta_k(x_1, s) = f(0) - f(1)$ and $|\beta_k(x_1, s) \le \max\limits_{0\le t\le 1} |f'(t)|$. With $\sigma = \text{trace } B = \alpha_{11}^k + \alpha_{22}^k$ we have

$$f'(t) = x_1\sigma e^{tx_1\sigma}\beta(se^{tx_1B}) + e^{tx_1\sigma}\beta'(se^{tx_1B})(sx_1Be^{tx_1B})$$

hence

$$|f'(t)| \le |x_1|e^{t|x_1||\sigma|}\{|\sigma||\beta(se^{tx_1B})| + |\beta'(se^{tx_1B})||se^{tx_1B}|\}$$

with matrix norm $|B| = B^{|k|} := |\alpha_{11}^k| + |\eta_k\alpha_{12}^k| + |\eta_k^{-1}\alpha_{21}^k| + |\alpha_{22}^k|$.

Furthermore, since $\beta' = (\frac{\partial\beta}{\partial s_1}, \frac{\partial\beta}{\partial s_2})$ and β decrease rapidly, there exists $C_1 > 0$ with

$$|\beta(s)| + |\beta'(s)||s| \le \frac{C_1}{1+|s|^4}, \qquad s \in \mathbb{R}^2\ .$$

It follows that with another constant C_2

$$|f'(t)| \le C_2|x_1|B^{|k|}e^{|x_1|B^{|k|}} \cdot \frac{1}{1+|se^{tx_1B}|^4}\ .$$

Now $|se^{tx_1B}| \ge |s||e^{-tx_1B}|^{-1} \ge |s|e^{-|x_1|B^{|k|}}$ implies

$$|\beta_k(x_1, s)| \le C_3|x_1|B^{|k|}e^{5|x_1|B^{|k|}} \cdot \frac{1}{1+|s|^4}$$

with constant C_3 independent of k, x_1 and s. So finally with $C = AC_3 \int \frac{ds}{1+s^4}$ we obtain

$$|N_k^2(x)| \le C|x_1|B^{|k|}e^{5|x_1|B^{|k|}}\ ,$$

i.e. the estimate (52) holds also in the present case. The rest of the proof of (49) is a verbatim repetition of the arguments used in case 1. The same is true for most of the arguments leading to (55). We start again with the decomposition (54) of the cyclic representation ζ_∞, corresponding to $\widetilde{\varphi}_\infty$, i.e. we set $\zeta_\infty = \int_X^\oplus m(\lambda)\lambda d\mu(\lambda)$, $\widetilde{\varphi}(\widetilde{x}) = (\zeta_\infty(\widetilde{x})\widetilde{\omega}|\widetilde{\omega})$ with a cyclic vector $\widetilde{\omega} \in \mathfrak{H}(\zeta_\infty)$. Then we claim:

(98) *The support of* ζ_∞ *in* ${}_\infty\widehat{\widetilde{G}}$ *is contained in*

$${}_\infty\widehat{\widetilde{G}}_1 := \{\lambda \in {}_\infty\widehat{\widetilde{G}};\ d\lambda(\widetilde{\mathfrak{v}}) \neq 0\} ,$$

in other words, the intersection X_0 of the set X with the complement of ${}_\infty\widehat{\widetilde{G}}$ has measure 0. To prove this we take a radial Schwartz function $\vartheta \in \mathcal{S}(\mathbb{R}^2)$, i.e. $\vartheta(s) = \Theta(|s|)$ for some $\Theta \in \mathcal{S}(\mathbb{R})$, with $\widehat{\vartheta}(0) = 0$ and $\widehat{\vartheta}(s) = 1$ if $\sqrt{2} \le |s| \le 2\sqrt{2}$. Then, with this new ϑ, the proof of (55) in § 2 works also for (98).

Now let us fix λ in supp ζ_∞, hence $d\lambda(\widetilde{\mathfrak{v}}) \neq 0$. Then again we see that there exists a sequence $\{\widetilde{\pi}_{j_k,r_k}\}$ with limit λ in $\widehat{\widetilde{\mathbf{G}}}$. For simplicity we assume that $j_k = k$ hence $\lambda = \lim\limits_{k\to\infty} \widetilde{\pi}_{k,r_k}$. Then – see (97) –

$$\mathfrak{q}_k := \mathbb{R}(\widetilde{v}_1 - r_k v_1^k) \oplus \mathbb{R}(\widetilde{v}_2 - r_k v_2^k)$$

is an ideal in ${}_{k,r}\widetilde{\mathfrak{g}}$, such that $\widetilde{\pi}_{k,r_k}$ is trivial on $Q_k := \exp_k \mathfrak{q}_k \subset {}_{k,r_k}\widetilde{G}$. Furthermore

$$\widetilde{\mathfrak{g}} = \mathfrak{g} + \mathfrak{q}_k , \qquad {}_{k,r_k}\widetilde{\mathfrak{g}}/\mathfrak{q}_k \simeq {}_k\mathfrak{g}$$

and also ${}_{k,r_k}\widetilde{G}/Q_k \simeq {}_kG$.

We are now in a situation strictly analogous to the one with which we started in this paragraph: Let

$$\overset{\bullet}{\mathbf{g}} = (\overline{\mathbb{N}} \times \widetilde{\mathfrak{g}}, [\ ,\ ]_\nu)$$

where now $[x, y]_k := [x, y]_{k,r_k}$ for $k \in \mathbb{N}$, $x, y \in \widetilde{\mathbf{g}}$, i.e. ${}_k\mathfrak{g}^\bullet$ is identified with ${}_{k,r_k}\widetilde{\mathfrak{g}}$. Also for $\nu = \infty$ we set ${}_\infty\mathfrak{g}^\bullet = {}_\infty\widetilde{\mathfrak{g}}$. So $\mathbf{g}^\bullet$ is a variable subalgebra of $\widetilde{\mathbf{g}}$. Then $\mathbb{R}b_n$ is the center of ${}_k\mathfrak{g}^\bullet$ for all finite k and $\mathfrak{y}^\bullet := \widetilde{\mathfrak{v}} \oplus \mathbb{R}b_n$ is a minimal noncentral ideal. Equations (96) holds with indices k instead of k, r_k and $\gamma_k r_k$ instead of $\gamma_k r$. Observe that (94) implies $\lim \gamma_k r_k = 0$.

Here the basis elements $\widetilde{v}_i$ play the same role as previously the b_{n-2}, b_{n-1} in $\mathfrak{g}$. The equations (96) for the $\widetilde{v}_i$ show that the new elements $\overset{\bullet}{u}{}_i^k$, which satisfy equations (72) and $\overset{\bullet}{(}75)$ with the old α_{ij}^k and new γ's and with normalization $|\overset{\bullet}{u}{}_1^k| + |\overset{\bullet}{u}{}_2^k| = 1$, are given by

$$\overset{\bullet}{u}{}_1^k = \frac{1}{1+\eta_k}\widetilde{v}_1 , \quad \overset{\bullet}{u}{}_2^k = \frac{\eta_k}{1+\eta_k}\widetilde{v}_2$$

consequently $\overset{\bullet}{v}{}_1^k = \frac{1}{2}\widetilde{v}_i$ are the new $\overset{\bullet}{v}{}_i^k$ and $\widetilde{\mathfrak{v}} = \overset{\bullet}{\mathfrak{v}}{}^\infty = \mathbb{R}\overset{\bullet}{v}{}_1^\infty \oplus \mathbb{R}\overset{\bullet}{v}{}_2^\infty$. For the $\overset{\bullet}{\gamma}_k$ we get

$$\overset{\bullet}{\gamma}_k = \frac{\eta_k}{1+\eta_k}\gamma_k r_k ,$$

hence $\lim \overset{\bullet}{\gamma}_k = \overset{\bullet}{\gamma}_\infty = 0$.

With $\overset{\bullet}{\pi}_k = \widetilde{\pi}_{k,r_k}$ and $\overset{\bullet}{\pi}_\infty = \lambda$ we obtain a sequence $\{\overset{\bullet}{\pi}_k\}$ in $\widehat{\overset{\bullet}{\mathbf{G}}}$, the dual of the VLG $\overset{\bullet}{\mathbf{G}} = \exp \overset{\bullet}{\mathbf{g}}$, with $\lim \overset{\bullet}{\pi}_k = \overset{\bullet}{\pi}_\infty$. All $\overset{\bullet}{\pi}$, $k \in \mathbb{N}$, are induced: $\overset{\bullet}{\pi}_k = \operatorname{ind}_{{}_k\overset{\bullet}{H}}^{{}_k\overset{\bullet}{G}} \overset{\bullet}{\tau}_k$, where the $\overset{\bullet}{\tau}_k$ are our old $\tau_{k,r_k} \in {}_{k,r_k}\widehat{\widetilde{H}} = {}_k\widehat{\overset{\bullet}{H}}$. Now with these data the construction of the VLA

$$\mathbf{H}^\dagger = P \times \mathfrak{h}^\dagger$$

with P as in (86) and $\mathfrak{h}^\dagger = \sum_{i=1}^n \mathbb{R} b_i^\bullet$ goes exactly along the same lines as in the case of $\mathfrak{h}^\bullet$ after (86). Also the definition of the positive functionals $\overset{\bullet}{\psi}_\nu$ is the same as before, hence we obtain the analogue of (87) and (88) and moreover $\lim \overset{\bullet}{\psi}_k(f) = \overset{\bullet}{\psi}_\infty(f)$ holds. Based on this relation and assuming that (90) holds, in the previous cases we could conclude that there are sequences $\{k_j, t_j\}_j$ and $t_0 \in \mathbb{R}^2$ with $\lim \overset{\bullet}{\tau}_{k_j,t_j} = \overset{\bullet}{\tau}_{\infty,t_0}$ in $\widehat{\mathbf{H}}^\bullet$ for some $\overset{\bullet}{\tau}_{\infty,t_0} \in {}_\infty\widehat{H}^\bullet$. These arguments were independent of $\dim \mathfrak{h}^\bullet$. However, now we have $\mathfrak{v}^\infty = \widetilde{\mathfrak{v}}$ and $d\lambda(\widetilde{\mathfrak{v}}) \neq 0$, so (90) is satisfied, and consequently there exists again a sequence $\{k_j, t_j\}$ such that $\lim \overset{\bullet}{\tau}_{k_j,t_j} = \overset{\bullet}{\tau}_{\infty,t_0}$, resp. $= \overset{\vee}{\tau} \in {}_\infty\widehat{H}^\dagger$ exists. Since $\dim \mathfrak{h}^\dagger = n$, we cannot use induction to prove $\lim K^{-1}(\overset{\bullet}{\tau}_{k_j,t_j}) = K^{-1}(\overset{\bullet}{\tau}_{\infty,t_0})$, but now all $\overset{\bullet}{\tau}_{k,t}$ are trivial on the normal subgroups Q_k. This means that we are in the situation treated in § 2, so we see that we have indeed convergence also for the orbits.

Finally we have to step down from $\widetilde{\mathbf{G}}$ onto $\mathbf{G}$. If $B^{|\infty|} = 0$, i.e. $B := \operatorname{ad}_\infty b_1|_{\tilde{\mathfrak{v}}} = 0$, then our estimates show that $\widetilde{\varphi}_\infty|_G = \varphi_\infty$ and we can use the arguments after (65) to finish the proof.

If B is regular, then (64) is true also in our present situation. Indeed $d\lambda(\widetilde{\mathfrak{v}}) \neq 0$ implies the existence of some $w \in \widetilde{\mathfrak{v}}$ and $f \in K^{-1}(\lambda)$ with $f(w) = 1$. For $v = B^{-1}(w) \in \widetilde{\mathfrak{v}}$ this gives $[b_1, v]_\infty = w$ and the former proof – with v instead of $\widetilde{b}$ – works also here.

In any case, if $B^{|\infty|} > 0$, then $\alpha_2^\infty > 0$ and since $\overset{\bullet}{\varepsilon}_k = 2\frac{\alpha_2^k}{\gamma_k r_k}$ tends to ∞, we are in case β. So in this case the $t_k \in \mathbb{R}^2$, appearing in $\overset{\bullet}{\tau}_{k,t_k}$ (assume that $k = k_j$), converge to some $t_\infty \neq 0$. This implies that

$$\lim b_1^k(t_k) = b_1^\infty \neq 0\,, \quad b_1^\infty \in \mathfrak{d}\,.$$

Now let $f \in \widetilde{\Omega}_\infty = K^{-1}(\lambda)$ and $h_k \in \widetilde{\Omega}_k = K^{-1}(\overset{\bullet}{\pi}_k)$ with $f = \lim h_k$. With

$$\widetilde{\mathfrak{h}}_{k,t} = \operatorname{Ad}_k e_k(t^{(k)})(\widetilde{\mathfrak{h}})\,, \quad \widetilde{H}_{k,t} = \exp_{k,r_k} \widetilde{\mathfrak{h}}_{k,t} \text{ etc.}$$

our construction shows that the h_k are extensions of forms h_k' in the orbits of the $\overset{\bullet}{\pi}_{k,t_k} \in \widehat{H}_{k,t_k}$, so they vanish on $\operatorname{Ad}_k e_k(t_k^{(k)})(\widetilde{\mathfrak{v}})$ and consequently Ω_k contains

$h_k + \widetilde{\mathfrak{h}}^{\perp}_{k,t_k}$. Let

$$b^k = |b_1^k(t_k)|^{-1} b_1^k(t_k)\,, \quad b^\infty = |b_1^\infty|^{-1} b_1^\infty\,,$$

hence $\widetilde{\mathfrak{h}}_{(k)} := \widetilde{\mathfrak{h}}_{k,t_k} = \mathbb{R}b^k \oplus \widetilde{\mathfrak{k}}$, $\widetilde{\mathfrak{h}}_\infty := \mathbb{R}b^\infty \oplus \widetilde{\mathfrak{k}} \subset \widetilde{\mathfrak{d}}$ and $\lim_{k\to\infty} b^k = b^\infty$. Clearly we can find a sequence of unitaries U_k of $\widetilde{\mathfrak{g}}$ with $\lim_{k\to\infty} U_k = \mathbf{1}$, $U_k b^\infty = b^k$ and $U_k(\widetilde{\mathfrak{h}}_\infty) = \widetilde{\mathfrak{h}}_{(k)}$. With these U_k it follows that $h_k + \widetilde{\mathfrak{h}}^{\perp}_{(k)} = h_k + U_k(\widetilde{\mathfrak{h}}^{\perp}_\infty) \subset \widetilde{\Omega}_k$, hence

$$\widetilde{\mathfrak{h}}^{\perp}_\infty \subset \bigcap_{k\in\mathbb{N}} U_k^*(\widetilde{\Omega}_k - h_k)$$

This means that, given $z \in \widetilde{\mathfrak{h}}^{\perp}_\infty$, there exists a sequence $\{z_k\}$ of elements $z_k \in \widetilde{\Omega}_k$ with $U_k z = z_k - h_k$. Since $\lim U_k z = z$ and $\lim h_k = f$ it follows that for every $k_0 \in \mathbb{N}$:

$$f + z = \lim(h_k + U_k z) = \lim z_k \in \overline{\bigcup_{k\geq k_0} \widetilde{\Omega}_k}\,,$$

so finally, because $\widetilde{\mathfrak{h}}^{\perp}_\infty \supset \widetilde{\mathfrak{d}}^{\perp}$:

$$f + \widetilde{\mathfrak{d}}^{\perp} \subset \overline{\bigcup_{k\geq k_0} \widetilde{\Omega}_k}\,. \tag{99}$$

The inclusion (99) suffices to proceed as in case 1, § 2: For $\ell_\infty \in K^{-1}(\pi_\infty)$ there exists a sequence $\{f_j\}$ with $f_j \in K^{-1}(\lambda_j)$, $\lambda_j \in \operatorname{supp} \zeta_\infty$, such that $\ell_\infty|_{\mathfrak{d}} = \lim f_j|_{\mathfrak{d}}$. As $\mathfrak{g}^* = \widetilde{\mathfrak{d}}^{\perp} + \mathfrak{d}^*$, where $\mathfrak{d}^*$ is identified with $(\mathbb{R}b_1 \oplus \widetilde{\mathfrak{v}})^{\perp}$ in $\widetilde{\mathfrak{g}}^*$, there exists a sequence $\{g_j\}$, $g_j \in \widetilde{\mathfrak{d}}^{\perp}$, with $\ell_\infty = \lim_{j\to\infty} (f_j + g_j)|_{\mathfrak{g}}$. Finally (99) shows that we can choose $\widetilde{\ell}_j \in \overline{\bigcup_{k\geq k_0} \widetilde{\Omega}_k}$ with $|f_j + g_j - \widetilde{\ell}_j| < \frac{1}{j}$, hence $\ell_\infty = \lim_{j\to\infty} \ell_j$ for $\ell_j = \widetilde{\ell}_j|_{\mathfrak{g}} \in \overline{\bigcup_{k\geq k_0} \widetilde{\Omega}_k}$, which means $\Omega_\infty = \lim \Omega_k$ and proves the theorem in case 2.B.β.

It remains to treat case A. In this case we take $P = \overline{\mathbb{N}}$ and again $\widetilde{\mathfrak{g}} = \mathfrak{g} \oplus \widetilde{\mathfrak{v}}$ with $\widetilde{\mathfrak{v}} = \mathbb{R}\widetilde{v}_1 \oplus \mathbb{R}\widetilde{v}_2$. For $\nu \in \overline{\mathbb{N}}$ we set

$$\begin{cases} [x,y]_\nu &= [x,y]_\nu \text{ for } x,y \in \mathfrak{g} \\ [\widetilde{\mathfrak{v}}, \mathfrak{k}]_\nu &= 0 \\ [b_1, \widetilde{v}_1]_\nu &= \alpha^\nu_{11}\widetilde{v}_1 + \eta_\nu \alpha^\nu_{12}\widetilde{v}_2 \\ [b_1, \widetilde{v}_2]_\nu &= \eta_\nu^{-1}\alpha^\nu_{21}\widetilde{v}_1 + \alpha^\nu_{22}\widetilde{v}_2 \\ [b_2, \widetilde{v}_1]_\nu &= \eta_\nu b_n\,,\ [b_3, \widetilde{v}_2]_\nu = b_n \\ [b_2, \widetilde{v}_2]_\nu &= [b_3, \widetilde{v}_1]_\nu = 0 \end{cases} \tag{100}$$

where as before $\eta_\infty^{-1}\alpha^\infty_{21}$ means $\lim_{k\to\infty} \eta_k^{-1}\alpha^k_{21}$ if $\eta_\infty = 0$

We see that the elements $v_i^\nu - \gamma_\nu \widetilde{v}_i$ span an ideal $\mathfrak{q}_\nu$ in ${}_\nu\widetilde{\mathfrak{g}}$. For finite k the algebra ${}_k\widetilde{\mathfrak{g}}$ is the semidirect product ${}_k\mathfrak{g} \oplus \mathfrak{q}_k$, similarly for the groups: ${}_k\widetilde{G} = {}_kG \ltimes Q_k$ with $Q_k = \exp \mathfrak{q}_k$.

Let $\widetilde{\pi}_k \in {}_k\widehat{\widetilde{G}}$ be defined by $\widetilde{\pi}_k|_{{}_kG} = \pi_k$, $\widetilde{\pi}_k|_{Q_k} = \mathbf{1}$ and $\widetilde{\varphi}_k(\widetilde{x}) = (\widetilde{\pi}_k(\widetilde{x})\omega_k|\omega_k)$, hence $\widetilde{\varphi}_k|_{{}_kG} = \varphi_k$. Then again we can assume that

$$\widetilde{\varphi}_\infty = w^* \lim_{k\to\infty} \widetilde{\varphi}_k$$

exists.

The basis $\{b_1, b_2, \ldots, b_{n-1}, \widetilde{v}_1, \widetilde{v}_2, b_n\}$ defines in the VLG $\widetilde{\mathbf{G}}$ coordinates $\widetilde{x} = (x_1, x_2, \ldots, x_{n-1}, s_1, s_2, x_n)$ such that the manifold S of all elements $s = (0, 0, \ldots, 0, s_1, s_2, 0)$ is the subgroup $\exp_\nu \widetilde{\mathfrak{v}} = {}_\nu S$ in every ${}_\nu\widetilde{G}$ and so that ${}_\nu\widetilde{G} = {}_\nu G \underset{\nu}{\cdot} S$ for all $\nu \in \overline{\mathbb{N}}$. Now we take $h \in \mathcal{S}(\mathbb{R}^2)$ with $0 \leq \widehat{h} \leq 1$ and $\widehat{h}(s) = 1$ for $|s_1| \leq T$, $|s_2| \leq T$.

This function h defines an element in the C^* hull of the variable subgroup $P \times S$ of $\widetilde{\mathbf{G}}$, again denoted by h, for which

$$\widetilde{\pi}_k(h)\omega_k(e_k(t)) = \widehat{h}(t)\omega_k(e_k(t)) ,$$

so in particular $(\widetilde{\pi}_k(h)\omega_k|\omega_k) \geq C$ for all $k \in \mathbb{N}$. With this h, a suitable $a = a^* \in C^*(\mathbf{G})$ and the composed element $a \odot h \in C^*(\widetilde{\mathbf{G}})$ we repeat the proof of (67) and see that also in this case (67) is true.

With our old notation let $\widetilde{\varphi}_\infty(\widetilde{x}) = (\zeta_\infty(\widetilde{x})\widetilde{\omega}|\widetilde{\omega})$ and $\zeta_\infty = \int_X m(\lambda)\lambda d\mu(\lambda)$ with $X = \operatorname{supp} \zeta_\infty \subset {}_\infty\widehat{\widetilde{\mathbf{G}}}$. Because $d\widetilde{\pi}_k(b_n) = i\mathbf{1}$ for all k it follows that $d\zeta_\infty(b_n) = i\mathbf{1}$ and consequently $d\lambda(b_n) = i\mathbf{1}$ for all $\lambda \in X$. Moreover, $[b_3, \widetilde{v}_2]_\infty = b_n$ implies $[d\lambda(b_3), d\lambda(\widetilde{v}_2)] = d\lambda(b_n) \neq 0$, thus $d\lambda(\widetilde{v}_2) \neq 0$ for all $\lambda \in X$.

For a fixed $\lambda \in X$ there exists a subsequence of $\{\widetilde{\pi}_k\}$ converging to λ. So let us assume that $\lambda = \lim \widetilde{\pi}_k$ in $\widehat{\widetilde{\mathbf{G}}}$. Let $\mathfrak{q}$ be the limit ideal in ${}_\infty\widetilde{\mathfrak{g}}$ of the $\mathfrak{q}_k \subset {}_k\widetilde{\mathfrak{g}}$, i.e. the set of all limits $x = \lim x_k$, $x_k \in \mathfrak{q}_k$. Clearly $\mathfrak{q}_\infty \subset \mathfrak{q}$ and $d\lambda(x) = 0$ for all $x \in \mathfrak{q}$, see lemma (57), hence $b_n \notin \mathfrak{q}$. Now observe that $\mathfrak{y} \oplus \widetilde{\mathfrak{v}} = \widetilde{\mathfrak{v}} \oplus \mathfrak{q}_k \oplus \mathbb{R}b_n$ for all $k \in \mathbb{N}$. We use proposition 1 in § 1 and the arguments in its proof to twist the coordinates of $\mathfrak{y} \oplus \widetilde{\mathfrak{v}}$ (i.e. to define an equivalent VLA) so that

$$\mathfrak{y} \oplus \widetilde{\mathfrak{v}} = \mathbb{R}w_1 \oplus \mathbb{R}w_2 \oplus \mathfrak{q} \oplus \mathbb{R}b_n$$

and all $\mathfrak{q}_k$ are transformed into $\mathfrak{q}$. We set $\mathfrak{w} = \mathbb{R}w_1 \oplus \mathbb{R}w_2$ and

$$\mathfrak{g}^\bullet = \sum_{i=1}^{n-3}{}^{\oplus} \mathbb{R}b_i \oplus \mathfrak{w} \oplus \mathbb{R}b_n ,$$

hence $\widetilde{\mathfrak{g}} = \mathfrak{g}^\bullet \oplus \mathfrak{q}$, and define P to be the orthogonal projection from $\widetilde{\mathfrak{g}}$ onto $\mathfrak{g}^\bullet$. As after (66) we define the product $[\ ,\ ]^\bullet_\nu$ in $\mathfrak{g}^\bullet$ by

$$[x, y]^\bullet_\nu = P([x, y]_\nu)$$

for all $\nu \in \overline{\mathbb{N}}$. Then $(\overline{\mathbb{N}} \times \mathfrak{g}^\bullet, [\ ,\ ]^\bullet_\nu) = \mathfrak{g}^\bullet$ is a n dimensional VLA and P defines an epimorphism $dB : \widetilde{\mathbf{g}} \to \mathbf{g}^\bullet$, leading to the same diagram as in § 2.

The representations λ and $\widetilde{\pi}_k$ all vanish on $Q = \exp \mathfrak{q}$, hence they define $\lambda^\bullet$ and $\pi^\bullet_k$ with $\widetilde{\pi}_k = \pi^\bullet_k \circ A$, $\lambda = \lambda^\bullet \circ A$. We quote again diagram (59) and the paragraph thereafter and obtain $\lambda^\bullet = \lim \pi^\bullet_k$ in $\widehat{\mathbf{G}}^\bullet$. Now put $\rho_k = \sqrt{1+\eta_k^2}$. From (100) one derives readily that in $\widetilde{\mathbf{g}}$ the elements u^k_i, $i = 1, 2$, defined by (75), are given by $u^k_1 = \rho_k^{-1}\widetilde{v}_1$, $u^k_2 = \eta_k\rho_k^{-1}\widetilde{v}_2$, hence, see (81), we have again $v^k_i = \frac{1}{2}\widetilde{v}_i$ and $\mathfrak{v}^\nu = \widetilde{\mathfrak{v}}$ for all $\nu \in \overline{\mathbb{N}}$. Passing to our equivalent algebra and projecting with P onto $\mathbf{g}^\bullet$ we obtain the corresponding elements $\overset{\bullet}{v}{}^\nu_i$. From $[v_3, v^\infty_2]_\infty = \frac{1}{2}b_n$ we derive $[\overset{\bullet}{b}_3, \overset{\bullet}{v}{}^\infty_2]^\bullet_\infty = \frac{1}{2}\overset{\bullet}{b}_n$ for the images in ${}_\infty\overset{\bullet}{\mathfrak{g}}$, hence $d\lambda^\bullet(\overset{\bullet}{v}{}^\infty_2) \neq 0$, which implies, that now condition (90) is fulfilled, i.e. $d\lambda^\bullet(\overset{\bullet}{\mathfrak{v}}{}^\infty) \neq 0$. As $\dim \overset{\bullet}{\mathfrak{g}} = n$ this implies $\lim K^{-1}(\overset{\bullet}{\pi}_k) = K^{-1}(\overset{\bullet}{\lambda})$, hence also $\lim K^{-1}(\widetilde{\pi}_k) = K^{-1}(\lambda)$. We mentioned already that (67) holds, so there exist a sequence $\{\lambda_j\}_j \subset X$ and for every j a μ_j in $\operatorname{supp} \lambda_j|_{\infty G} \subset {}_\infty\widehat{G}$ with $\lim \mu_j = \pi_\infty$. All μ_j and π_∞ are trivial on ${}_\infty Q \subset {}_\infty G$, hence $\lim\limits_{j\to\infty} K^{-1}(\mu_j) = \Omega_\infty$. Finally, $K^{-1}(\lambda_j) = \lim\limits_{i\to\infty} K^{-1}(\widetilde{\pi}_{k_i})$ for some subsequence $\{\widetilde{\pi}_{k_i}\}_i$, depending on j, implies as before that also $K^{-1}(\mu_j) = \lim\limits_{i\to\infty} Q_{k_i}$ for a suitable subsequence of $\{Q_k\}$ and so eventually ends our proof of Theorem 1.

Definitions and Symbols

References

[1] Bernat, P., Sur les représentations unitaires des groupes de Lie résolubles. Ann. Ec. Norm. Sup. **82** (1965), 37–99.

[2] Bernat, P., Conze, N. et. al. Représentations des groupes de Lie résolubles. Dunod, Paris 1972.

[3] Blattner, R., On induced representations. Amer. J. Math. **83** (1961), 79–98.

[4] —.— Positive definite measures. Proc. Amer. Math. Soc. **14** (1963), 423–428.

[5] Bourbaki, N., Groupes et algebres de Lie, Chapitres 2 et 3. Act. Sci. et Ind. 1349, Hermann, Paris 1972.

[6] Brown, I., Dual topology of a nilpotent Lie group. Ann. Sci. Ècole Norm. Sup. **6** (1973), 407–411.

[7] Dixmier, J., Les C^*–algèbres et leurs représentations. Gauthier–Villars, Paris 1964

[8] Eymard, P., Moyennes invariantes et représentations unitaires. LNM 300, Springer, Heidelberg 1972.

[9] Fell, J.M.G., Weak containment and induced representations of groups. Canad. J. Math. **14** (1962), 237–268.

[10] Fujiwara, H., Sur le dual d'un groupe de Lie résoluble exponentiel. J. Math. Soc. Japan **36** (1984), 629–636.

[11] —.— Sur les restrictions des représentations unitaires des groupes de Lie resoluble exponentiels. Invent. Math. **104** (1991), 647–654.

[12] Grauert, H. u. Fritzsche, K., Einführung in die Funktionentheorie mehrerer Veränderlicher. Springer, Heidelberg 1974.

[13] Guichardet, A., Caractères des algèbres de Banach involutives. Ann. Inst. Fourier **13** (1963), 1–81.

[14] Howe, R., On a connection between nilpotent groups and oscillatory integrals associated to singularities. Pacific J. Math. **73** (1977), 329–363.

[15] Johnson, B.E., An introduction to the theory of centralizers. Proc. London Math. Soc. **14** (1964), 299–320.

[16] Kehlet, E.T., Cross sections for quotient maps of locally compact groups. Math. Scand. **55** (1984), 152–160.

[17] Kirillov, A.A., Unitary representations of nilpotent Lie groups. Uspeki Mat. Nauk. **17** (1962), 53–104.

[18] Leptin, H., Verallgemeinerte L^1–Algebren und projektive Darstellungen lokal kompakter Gruppen I, II. Invent. Math. **3** (1967), 257–281; **4** (1967), 68–86.

[19] —.— Darstellungen verallgemeinerter L^1-Algebren. Invent. Math. **5** (1968), 192–215.

[20] —.— The structure of primitive quotients of certain group algebras. Symposia Math. **29** (1987), 79–96.

[21] Ludwig, J., Irreducible representations of exponential solvable Lie groups and operators with smooth kernels. J. reine angew. Math. **339** (1983), 1–26.

[22] Mathieu, O., Bicontinuity of the Dixmier map. J. Amer. Math. Soc. **4** (1991), 837–863.

[23] Moore, C.C., Representations of solvable and nilpotent groups and harmonic analysis on nil and solvmanifolds. Proceedings of Symposia in Pure Mathematics **XXVI**, Amer. Math. Soc. (1973), 1–44.

[24] Müller–Römer, P., A note of Mackey's imprimitivity theorem. Bull. Amer. Math. Soc. **77** (1971), 1089–1090.

[25] Poguntke, D., Operators of finite rank in unitary representations of exponential Lie groups. Math. Annalen **259** (1982), 371–383.

[26] —.— An example of a generalized completely continuous representation of a locally compact group. Studia Math. **105** (1993), 189–205.

[27] —.— Unitary representations of Lie groups and operators of finite rank. Ann. of. Math., to appear.

[28] Pukanszky, L., On the theory of exponential groups. Trans. Amer. Math. Soc. **126** (1967), 487–507.

[29] —.— On the unitary representations of exponential groups. J. Funct. Anal. **2** (1968), 73–113.

[30] —.— Unitary representations of solvable Lie groups. Ann. Sci. Ècole Norm. Sup. **4** (1971), 457-608.

[31] —.— The primitive ideal space of solvable Lie groups. Invent. Math. **22** (1973), 75–118.

[32] Reiter, H., Classical harmonic analysis and locally compact groups. Oxford Math. Monographs, Oxford (1968).

[33] Rickart, C.E., General theory of Banach algebras. Van Nostrand, Princeton, N.J. (1960).

[34] Takenouchi, O., Sur la facteur représentation des groupes de Lie de type (E). Math. J. Okayama Univ. **7** (1957), 151–161.

Bei Fragen zur Produktsicherheit wenden Sie sich bitte an:
If you have any questions regarding product safety,
please contact:

Walter de Gruyter GmbH
Genthiner Straße 13
10785 Berlin
productsafety@degruyterbrill.com